Mathematical Models for Estimating Occupational Exposure to Chemicals, 2nd edition

Edited by
Charles B. Keil, PhD, CIH
Catherine E. Simmons, CIH
T. Renée Anthony, PhD, CIH, CSP

Disclaimer

This book was developed by experts with background, training, and experience in mathematical models for estimating occupational exposure to chemicals, working with information and conditions existing at the time of publication. The American Industrial Hygiene Association (AIHA), as publisher, and the authors have been diligent in ensuring that the material and methods addressed in this book reflect prevailing occupational health and safety and industrial hygiene practices. It is possible, however, that certain procedures discussed will require modification due to advances in science and practical knowledge and experience; changing or evolving definitions of acceptable risk; or changes in federal, state, and local regulations.

AIHA and the authors disclaim any liability, loss, or risk resulting directly or indirectly from the use of the practices and/or theories discussed in this book. Moreover, it is the reader's responsibility to stay informed of developments in science, changes in levels of acceptable risk or changes in federal, state, and local regulations that might affect the material contained herein, and the policies adopted specifically in the reader's workplace.

Specific mention of manufacturers and products in this book does not constitute an endorsement by AIHA.

Stock Number: AEAM09-379
ISBN: 978-1-935082-10-1

American Industrial Hygiene Association
2700 Prosperity Avenue, Suite 250
Fairfax, VA 22031
Tel: (703) 849-8888
E-mail: Infonet@aiha.org
www.aiha.org

Printed in the United States of America

Table of Contents

Acknowledgements

A special thanks to Thomas W. Armstrong and Daniel Drolet for completing the task to update the modeling spreadsheet that was developed for and previously accompanied the Mathematical Modeling Professional Development Course. However, their work exceeded all expectations by adding to the list of models contained in the software, including multi-lingual options for international users, and improving the user-friendliness of the spreadsheets. Readers can download this Windows Excel-based software (IHMod), free of charge, at the Exposure Assessment Strategies Committee page of the American Industrial Hygiene Association's website, http://www.aiha.org/Content/InsideAIHA/Volunteer+Groups/EAScomm.htm.

Spreadsheet users are urged to use this software only after understanding the model, its parameters, and limitations as described in this book.

The editors would also like to thank the many individuals who peer-reviewed chapters for this second edition. They not only provided feedback on technical aspects of the chapters, but recommended examples to provide the reader with ideas on how to apply these models to their workplaces.

Reviewers:
T. Renée Anthony, PhD, CIH, CSP
Susan F. Arnold, CIH
Russell Bowen, CIH, CSP
M. Cathy Fehrenbacher, CIH
Michael Jayjock, PhD, CIH
Charles B. Keil, PhD, CIH
Perry W. Logan, CIH
Mark Nicas, PhD, CIH
Ralph Rogers, CIH
Catherine E. Simmons, CIH
Mark Stenzel, CIH
James Stewart, PhD, CIH, CSP

Preface

By
John R. Mulhausen, PhD, CIH

Mathematical Modeling Tools: Critical Elements of an Effective and Efficient Exposure Assessment and Management Program

Exposure models are extremely important to the science and practice of industrial hygiene and critical to effective and efficient exposure risk management programs. While models vary in their scope and complexity, they are generally inexpensive, immediate, and able to estimate exposures prospectively or retrospectively. They are extremely useful as tools for performing quick exposure estimates that can provide early warning about potential exposures that need further investigation or enable exposure situations to be "put to rest" because the exposures they present are very low or trivial. Used appropriately, they inform an industrial hygienist's judgment and allow exposure decisions to be made when monitoring data are not available. Used systematically in a comprehensive exposure assessment and management strategy, they can reduce bias and increase decision accuracy.

In *A Strategy for Assessing and Managing Occupational Exposures*, 3rd edition[1] the AIHA Exposure Assessment Strategies Committee presents a systematic and comprehensive approach to evaluating and controlling occupational health hazards. By necessity, this practical approach is cyclic, tiered and continuously improving (Figure 1).

In this approach, available information summarized in the basic characterization is used to assess exposures and classify them as acceptable, unacceptable or uncertain. Acceptable exposures are documented as such, unacceptable exposures are prioritized for control, and uncertain exposures are prioritized for further information gathering. All new information about existing or newly controlled exposures is added to the basic characterization for use in subsequent cycles through the process to update exposures and determine appropriate follow up actions and priorities.

Exposure estimates are generally based on 1) modeling (physical chemical properties and/or environmental information and/or process usage

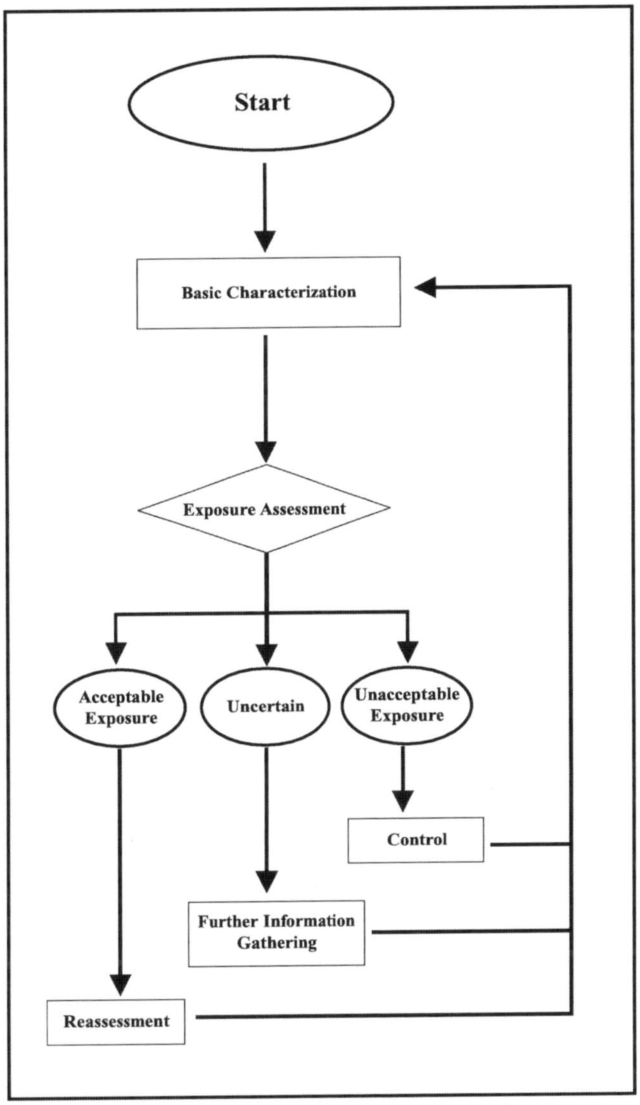

Figure 1 — A strategy for assessing and managing occupational exposures.[1]

information); 2) surrogate data (from another agent, from another operation), and 3) monitoring data (personal, area). In the early cycles through the exposure assessment process, there is typically little monitoring data available. The initial exposure assessments are therefore based primarily on modeling and/or surrogate data. In most exposure assessment programs the majority of assessments can be resolved in this manner — and the majority of those are likely to be based on mechanistic modeling or the application of model-like frameworks based on relational experiences.

For example, when an industrial hygienist reviews a situation and judges the amount of chemical used to be insufficient to result in a significant airborne exposure, he or she is relating the physical/chemical properties of the chemical and his/her past experience with similar processes to estimate the exposure. No monitoring has been performed, yet the industrial hygienist may be very certain of his/her exposure estimate. While the decision framework used by the industrial hygienist did not include application of the more formal, reproducible models for quantitative exposure estimates described in this text, it was nonetheless based on modeling. It is agreed that often these exposure situations are "trivial" in the sense that the amounts of chemical used are so small that the exposure estimate is nearly obvious and can therefore be performed almost subconsciously by the industrial hygienist. Nevertheless, it is this ability to construct relationships between potential exposures and the chemical's properties and usage amounts and conditions that enable the judgment in the absence of monitoring data.

By formalizing the exposure decision framework to systematically incorporate the modeling tools and techniques in this text, the industrial hygienist will strengthen the accuracy and reproducibility of his/her exposure judgments in several ways. First, the systematic exploration and documentation of exposure determinants needed to fuel the model will lead to better understanding of the exposure environment. Second, the structure imposed by the model parameters can counter biases that are inherent in typical decision-making heuristics. Third, the model outcome itself, even if presented in ranges of potential exposures, informs the judgment.

Additional benefit can be gained by integrating a systematic approach to modeling into a comprehensive exposure assessment program through the regular feedback loop of the overall strategy. That feedback loop, whereby initial assessments get modified based on additional information gathering such as monitoring results, allows models to be finely tuned and validated for specific operations. That fine-tuning and validation builds confidence in model outputs for the specific operations and allows confident exposure decisions to be made with less monitoring data.

It is the desire to increase the power and flexibility of exposure judgments made in the absence of monitoring data that pushes the development and understanding of more accurate quantitative modeling techniques like those described in this text. They are much less expensive than monitoring and there is no need to wait for laboratory analysis. The results are immediately available so an exposure estimate can be made and appropriate follow-up defined quickly.

Appropriate follow-up actions and priority will depend on the exposure estimate and on the degree of uncertainty that the industrial hygienist has in that estimate. If the estimate is extremely low, and the model has been properly constructed to be appropriately conservative (over-estimating of exposures), the industrial hygienist may be willing to accept a great deal of uncertainty and judge the exposure acceptable.

If the modeled exposure estimate is near the occupational exposure limit (OEL), and there is significant uncertainty associated with the model, then the industrial hygienist may judge the exposure uncertain. In such cases better exposure information may be needed for the next tier cycle through the exposure assessment process. This additional exposure information may be generated by monitoring or by employing a more sophisticated (and less uncertain) model.

If the modeled exposure estimate is high (e.g. > OEL), then the industrial hygienist may make a variety of judgments, depending on the degree of uncertainty in the model, the expense associated with various control options, the expense associated with monitoring, and the toxicity associated with the particular agent of concern. If monitoring costs are high, and control costs relatively low, then the industrial hygienist may judge the exposure unacceptable in spite of high uncertainty in the model and pursue control strategies. If control costs are high and monitoring costs low, then the industrial hygienist may decide to generate additional exposure information through monitoring to reduce the uncertainty in the exposure estimate. If the hazards of short-term exposure to the particular toxic agent are severe, then the industrial hygienist may choose to require short-term controls, such as respiratory protection, while the monitoring is being performed.

Modeling has a special advantage over other exposure estimating techniques in that it can be used to estimate exposures in the absence of the physical presence of the process or operation. That is one of the reasons that models are being widely applied in product exposure risk assessment – in particular to address the regulatory requirements of REACH in the European Union. Models can by used in a tiered approach to screening and prioritizing product use scenarios based on potential exposure.

Models can be used as a means for prospectively assessing exposures for new operations that are not yet installed. This can be extremely useful for planning needed control measures for new processes or materials. Models can even be useful for estimating the effect that various process or material changes might have on exposures and thereby aid in making the best selection from a variety of alternatives — all without the time and expense of trials involving the actual alternative process changes or materials.

Models are also useful for estimating exposures from operations that were performed far in the past. In fact, they may be the only viable option for estimating exposures when the actual past conditions are extremely expensive or impossible to recreate.

Because of their many advantages, techniques for worker exposure modeling are improving rapidly and are being employed more frequently and creatively. This book captures many of those improved techniques as well as most of the basic "old standby" exposure modeling approaches. Any industrial hygienist will find these techniques to be useful additions to his or her box of exposure assessment and management tools.

John R. Mulhausen, PhD, CIH

References

1. **Ignacio, J. and W.H. Bullock (eds.):** *A Strategy for Assessing and Managing Occupational Exposures,* 3rd edition. Fairfax, VA: AIHA, 2006.

Why Model?

By
Michael Jayjock, PhD, CIH
Mark Stenzel, CIH
T. Renée Anthony, PhD, CIH, CSP
Susan F. Arnold, CIH

1.0 Overview

This book is intended to provide users with additional tools to assess exposures to indoor contaminants, supplementing and enhancing traditional air monitoring techniques. Although sampling and analytical techniques are integral to understanding workplace and community exposures, the practical limitations of sampling exposure to *all* possible contaminants can overwhelm our ability to quantitatively assess exposures in a world where a person is potentially exposed to hundreds or thousands of chemicals.

This chapter provides readers with background motivation on why modeling is useful to the science of exposure assessment. It provides justification to the proactive use of models when anticipating exposures, such as when the reader is asked to approve the use of a new chemical at a worksite or to recommend controls for a new manufacturing process, and provides a framework for using models to enhance and support professional judgment, including providing data to support decisions of low and high risk exposure potentials. The use of models is not intended to replace air monitoring efforts, but rather supports making better decisions on when air sampling is necessary and when analytical costs can be shifted to quantifying higher-risk exposures, lending to better focused exposure monitoring plans. This chapter contains a discussion on the benefits of modeling and attempts to provide a new paradigm to motivate industrial hygienists and exposure assessors to see the value of using modeling tools to enhance professional judgment, qualitative exposure assessment techniques, and sampling strategies.

1.1 Exposure Modeling in the Context of Risk Assessment

Perhaps the primary function of an industrial hygienist is to estimate worker exposure to chemical and physical agents. This is usually performed to evaluate the potential health risk to the workers from this exposure. Thus, this critical aspect of the work of industrial hygienists is a type of human health risk assessment.

The inhalation exposure to workers is often determined by directly measuring the airborne concentration in the breathing zone of the worker and assuming that the worker breathes this air for at least some known portion of the working day. An extrapolation of this data to represent a full- or partial-shift exposure results in the commonly known time-weighted average exposure computation.

The time-weighted exposure is then compared with toxicological dose-response or, more typically, an exposure limit for that particular compound or chemical species. This exposure limit is based on health-effect data that incorporates toxicological or epidemiological studies. Indeed, the measured and computed exposure has no particular meaning or context relative to its potential health effect without this comparison. This combination of data from these two sources (exposure and health effect) is the essence of human health risk assessment as practiced by the industrial hygienist.(1)

Thus, the act of exposure assessment represents half of the process of risk assessment in the following equation:

$$\text{Risk} = f\,(\text{exposure})\ \frac{\text{Health effect}}{\text{exposure}} \qquad (1\text{-}1)$$

Determination of the term health effects given exposure (health effects/exposure) is the stock-in-trade of the toxicologist or the epidemiologist. The work products of this group of scientists are *exposure limits* as exemplified in the ACGIH® Threshold Limit Values (TLVs®) and other occupational exposure limits (OELs). This reckoning is NOT the subject of this chapter.

Determination of the actual *exposure* is the stock-in-trade of industrial hygienists. Mathematical modeling to provide a quantitative estimation of this exposure IS the subject of this chapter.

The comparison of *exposure* to the *exposure limit* is another way of framing the risk assessment process typically used in industrial hygiene in a modified form of the above equation:

$$\text{Risk} = f\left[\frac{\text{exposure}}{\text{exposure limit}}\right] \qquad (1\text{-}2)$$

Whenever the term "exposure assessment" or exposure estimate from modeling is mentioned in this chapter, it should be understood that these evaluations always occur within this exposure — exposure limit — risk framework. Also, it should be noted that these toxicological benchmarks (i.e., exposure limits) are critical to the determination or estimation of risk. Even though they are not the subject of this chapter, they are mentioned briefly within to remind the reader of their critical role in the overall process.

Another way of looking at this subject is to examine the relationship of the risk assessment process with what should be the familiar elements of the practice of industrial hygiene; namely, the well-known steps of:

1. Anticipation
2. Recognition
3. Evaluation
    a. Exposure Assessment
    b. Comparison to OEL
4. Control

Elements 1, 2, and 3 of the industrial hygiene process are also the elements of the *risk assessment process*. Element 4 should be recognized as *risk management*.[2]

One vital function of the industrial hygienist it to determine or estimate the actual exposure to the worker: this represents fully half of the risk assessment picture. We should, however, always be conscious of the fact that the quality of the risk assessment we perform is driven by the quality of our estimate of exposure and the appropriateness of the exposure limit. We should examine the documentation and basis for the limit and discuss the details of the limit with our colleagues in toxicology. If there are questions about whether the limit is appropriate for your evaluation, you should seek consensus within your organization for what might be an acceptable limit. In some cases, you might decide on additional safety factors with a concomitantly lower exposure limit to accommodate the uncertainty. Indeed, the obvious historical trend, at least in the realm of the ACGIH® TLVs®, is a reduction of exposure limits over time.[3]

1.2 Why Has the Industrial Hygiene Community Resisted Exposure Modeling?

Heretofore, only a small (but dedicated) group of mathematically-oriented hygiene professionals have adopted the use of models to estimate exposure. The reason for this may be that many IHs view models as being very time consuming, highly complicated and limited in application or utility. This perception persists despite the fact that models and modeling are used extensively in other fields with a few prominent examples, including business and finance, weather forecasting, insurance, mathematics, natural sciences (chemistry and physics), and engineering.

Modeling already plays a prominent role in fields that are closely allied with Industrial Hygiene, including toxicology, epidemiology, environmental risk assessment, and actuarial risk assessment. In toxicology, stochastic models have been developed to extrapolate from high doses given to animals to estimate what might happen to people at typically much lower doses. Epidemiological data or quantitative information on human disease rates is modeled to provide estimations of risk or a determination as to whether an exposure was associated with a particular disease state.

Exposure modeling should not be seen as mysterious, difficult, or daunting. These models are based on fundamental laws we learned in chemistry and physics and are merely the application of common-sense to scientific principles that anyone can learn. The secret is to learn and apply the models incrementally. It is the purpose and design of this book to do just that: you can learn and think about the simpler models presented herein and then proceed slowly but surely to add to your knowledge and skill base with the more sophisticated models. Each model links to the next in a progression. Even the simpler models will be of value to you and assist in your understanding. So, go ahead and give it a try!

1.3 Why Model When You Can Measure?

Assuming we have a reasonable and acceptable exposure limit, one can focus the exposure half of the risk equation by simply measuring exposures. Indeed, measurements provide a quick answer and the approach is one that most industrial hygienists have relied on for years. We simply determine the worker's exposure potential directly by measuring the concentration of contaminant in his or her breathing zone. For hygienists, the process is certainly a relatively straight forward process requiring instrumentation and an analytical chemist.

As a result, the practice of industrial hygiene has come to mean the direct measurement of the work place concentrations/exposure. It is a methodology that has the advantage of striking right at the heart of the question. It provides direct evidence and an answer to the immediate question at hand. But, is it the most efficient and cost effective method? Can monitoring answer all of the questions regarding exposures? Can we, as a profession, be more proactive in exposure estimation? Should we be?

Consider the typical IH reality: typical large modern manufacturing plant has hundreds of workers performing perhaps thousands of tasks. It is probably safe to say that under the current operating system the exposure associated with a majority of tasks are never monitored because the industrial hygienist judges them to be safe.[4,5] The most common number of samples collected to make this determination is zero.[5] Currently, the hygienist is not required to present a formalized or systematic analysis to support these decisions. That is, he or she has observed the universe of scenarios and has, for the most part and probably correctly, concluded that the exposure limit is not exceeded. When asked how that determination was arrived at, the typical answer is that he or she applied expert judgment. This means that the occupational hygienist used his or her combined experience to make this decision. When pressed further, the hygienist might say that it is because the system or scenario under consideration is relatively closed, that the vapor pressure is low, the exposure limit is relatively high, etc. This combination of factors tells an experienced occupational hygienist that overexposure will not occur. When professional judgment leads the hygienist to believe exposures may approach or exceed the exposure limit, a monitoring plan is put into action. The results of that monitoring determine whether controls are implemented.

Much of exposure assessment in general, and in industrial hygiene in particular, has been practiced using the reactive, reflective, qualitative, and undefined expert judgment as outlined above. Within the IH world, this manner and technique of working has generally protected many workers from overexposure and subsequent adverse health effects. However, it has a number of serious flaws, including:

(1) It is difficult or impossible to explain objectively.
(2) It is typically not supported by explicit quantified facts relating specific cause and effect.
(3) It is not amenable to technology transfer (i.e., those new to the field find it hard to learn).
(4) It is often insufficient to provide convincing evidence to affected workers or to defend against litigation or other legal challenges.

Thus, the standard method of direct measurement is clearly not the best we can do as industrial hygienists. Sometimes, measurements cannot be taken. Consider the following cases:

- You want to monitor exposures, but there is NO method available
- You cannot measure exposures "right now" when they are occurring
- You cannot measure exposures because you cannot be present, such as when they happen at another location, they happened previously (retrospective), or they have not happened yet (prospective)
- A small sample size of exposure monitoring events leads to a heavy bias toward concluding unacceptable exposures are acceptable[3]
- The financial burden associated with technician's time to collect sample and analytical fees are real-world challenges that restrict monitoring efforts

What about the science of human exposure assessment? How does the dominance of direct measure within the field of industrial hygiene affect this critical need? To examine this question, we will get into the exact nature of modeling in order to compare and contrast it with the monitoring paradigm. Later in this chapter we will view the monitoring versus the modeling paradigm within the context of the scientific method.

1.4 Exactly What Is a Model?

We are going to ask that you consider the nature of the physical world and the logical progression of cause and effect as they relate to worker exposure in the following diagram:

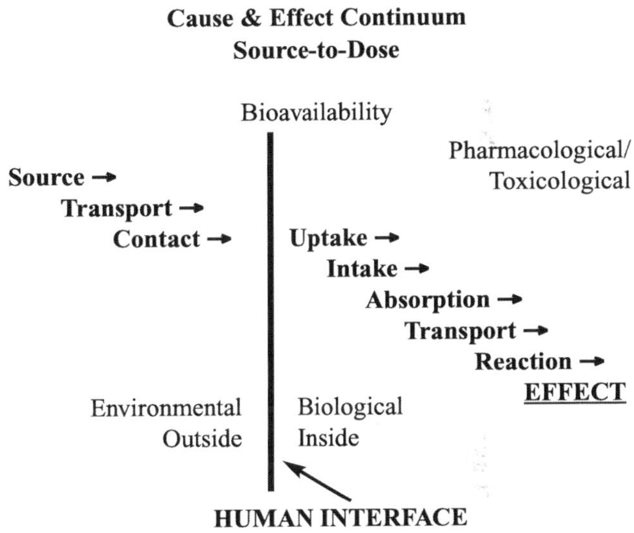

Clearly, our focus as exposure assessors and industrial hygienists is on the left-hand side of this continuum and that the exposure that the workers experience is caused by elements or determinants of source generation, transport and contact within their work environment.

For example, in the case of inhalation exposure from a vaporizing liquid, we understand that this exposure is initially caused by such source generation factors as the vapor pressure of the liquid, the surface area of the room air/liquid interface, and the velocity of air over the air/liquid surface. We also understand that the vapor pressure is dependent on the temperature of the liquid. All of these (and perhaps some additional) factors *caused* the vaporization, which, after transport and contact, ultimately *caused* the exposure to the worker.

In its simplest form inhalation modeling can be described by the 3 conceptual equations below:

First the airborne concentration potentially in the breathing zone of exposed persons is simply a function of the rate at which the material enters the air and how it is transported from the source to the person.

Concentration =
f (Source generation rate, Transport mechanisms) (1-3)

Actual physical models describe the concentration occurring over any period of time with actually measured or estimated determinants (input variables). The relationship below illustrates variables needed in a simple well-mixed room model (to be detailed in Chapter 3):

Concentration= f (input variables) = for example f (G,Q,V,t) (1-4)

To illustrate, this well-mixed room model requires us to know as input:

G = source rate entering the room air (mass/time)
Q = the general mixing ventilation rate of fresh air entering into the room air (volume/time)
V = volume of room air (volume)
t = time from the start of the exposure (time)

Of course, the estimated airborne concentration does not equal exposure unless one is there to experience or actually contact the concentration. Thus, combining the above estimates of exposure with the specifics of contaminant transport and contact (e.g., how long the person was there, how fast was he or she breathing, when the source started) renders the final relationship for determining the human exposure:

Exposure = f (Concentration, Contact) (1-5)

The above is the sum and substance of modeling. To perform this effectively, we need to learn and understand quite a bit about the factors that actually caused the exposure to happen. Please note that the same general principles apply to modeling other routes of exposure, including dermal or incidental hand-to-mouth transfer.

If we simply and directly measure the airborne concentration in a worker's breathing zone and make some assumptions about the time spent in that area at that concentration we have a more or less direct measure of "exposure" as described above. What we do not have is a reasonable quantification or understanding of the variables that *caused* the exposure to happen. Unfortunately, the industrial hygiene community has evolved by focusing primarily on the direct measurement of "exposure" and has not focused on identifying the predictors of an exposure. As such, our exposure databases are typically rich in information about exposure concentration without much knowledge about the factors that caused the exposure, limiting our ability to estimate exposures in similar situations. Thus, the science of exposure assessment has not developed much beyond the technology associated with its direct measure. As a result of this approach, our ability to estimate exposures in unique situations is limited to our more or less unquantitated (or under-qualified) experience and judgment.

1.5 Believe It or Not: You Are Already a Modeler!

The reality is that every industrial hygienist is already a modeler! We have already discussed that a majority of potential exposures in the workplace are never monitored because the occupational hygienist judges them to be safe. How does he or she arrive at this conclusion? We believe that most IHs do it by applying a subliminal "model" while using the basic scientific method. To explain this further, consider the basic elements of the scientific method:

1. State the problem or premise
2. Form a hypothesis
3. Experiment and observe
4. Interpret the data
5. Draw conclusions and make estimations

In the course of evaluating the exposure/risk to workers, the industrial hygienist forwards a hypothesis or "model" of what he or she thinks is happening in the world relative to cause and effect. Their observations and interpretations in the vast majority of cases (or thought "experiments" in this example) support the hypothesis that exposures are below hazardous levels, and conclude that the situation is safe.

When asked how they arrived at this conclusion for a majority of the scenarios under their responsibility, the usual answer is "expert judgment" which is defined elsewhere as:

> *"The application and appropriate use of knowledge gained from formal education, observation, experimentation, inference, and analogy."*

As noted previously in this chapter, if asked to further explain their decision not to monitor, the industrial hygienist would provide qualitative or quantitative information on the ability of the chemical to be released into the work environment, including chemical properties (e.g., "it has a low vapor pressure" or "it is present in low concentrations") or ventilation conditions ("it's a closed process"). Clearly, the hygienist has some model in mind when forming this hypothesis. A subliminal mathematical relationship or an algorithm was used to hypothesize, *"given the characteristics of all the causes of exposure in this situation, the resulting exposure will be less than the exposure limit."* Unfortunately, much more often than not the hygienist is so sure of the hypothesis (the model) that he or she does not do the actual "experiment," namely, there is no physical measure of the exposure.

In some of the scenarios, and using the same subliminal model, the hygienist concludes that overexposure is possible. He or she is then compelled to test that hypothesis of relative safety. Thus, after observation, the industrial hygienist may conduct the experiment: someone measures the concentration and computes exposures. These monitoring data then provide direct and objective experimental evidence to accept or reject the hypothesis and to draw conclusions about the risk of this exposure. Perhaps more importantly, this process feeds the hygienist's internal database (i.e., it increases experience) and allows for improvements in estimations about similar exposures in the future. That is, the use of data to evaluate the exposure hypothesis actually develops expert judgment. When the hypothesis that "exposures are low" is rejected by monitoring data, it is imperative to learn what went wrong in judgment to not only improve expertise but to ensure the health of persons served.

The hygienist should realize that the expert judgment is an application of some "model," even if you have not considered it as such. As we were developing our IH skills, we remember that with more experience, we got better at understanding what we needed to know (the "inputs") to make better qualitative decisions on "safe" or "let's monitor." This is no different from learning the mathematical models to be presented in this book. As we progress through the world of modeling, we will learn what models give us good enough answers and when we need to use more complex inputs to get better answers. This is left to the readers to discover through the remaining chapters.

1.6 Are You Ready to Start?

The specific aim of this book is to provide industrial hygienists with better tools to make decisions regarding exposure when working without the luxury of extensive monitoring data to make decisions. Rather than simply invoking a claim of unsubstantiated professional judgment relative to their decisions, the use of explicit models enables the industrial hygienist to understand and display the scientific rationale behind those decisions.

This book is designed to provide tools needed to perform computations to estimate exposures and, ultimately, reduce risk. As readers progress through the book, the complexity of models increases to accommodate more complex contaminant generation or transport scenarios; as the complexity of the models increase, so too does the amount of input parameters needed to run the model. As you progress from simple to complex, your skills at making assumptions and evaluating the impact of your input will improve, allowing you to make better estimates of exposures and provide better uncertainty bounds on your estimates of exposures.

But we hope readers realize that the book isn't merely a tool box of equations: It provides a different perspective on the field of exposure estimation and provides examples how we, as industrial hygienists, can use the underlying chemistry and physics behind exposure science to advance the profession. This book was written to encourage readers to not only use the models but to adopt a modeling approach as the scientific basis for your industrial hygiene "assumptions" when qualitatively assessing exposures and risk.

So, read on to learn the current state of exposure modeling as applied to industrial hygiene. Identify conditions in your operation that are similar to the examples in the text and start evaluating exposures as you progress through the book.

1.7 References

1. **Jayjock, M.A., J.R. Lynch, and D.I. Nelson:** *Risk Assessment Principles for the Industrial Hygienist.* Fairfax, VA: American Industrial Hygiene Association, 2000.
2. **National Research Council:** Risk Assessment in the Federal Government: Managing the Process. Washington, D.C.: National Academy Press, 1983.
3. **Mulhausen, J.R., and J. Damiano (eds.):** *A Strategy for Assessing and Managing Occupational Exposures*, 2nd edition. Fairfax, VA: American Industrial Hygiene Association, 1998.
4. **Mulhausen, J.R:** "Interpreting Monitoring Data: Are We Making the Right Judgments?" presented at the Professional Conference of Industrial Hygiene. Denver, CO, Oct. 25, 2005.
5. **Mulhausen, J.R.:** "Exposure Judgments: Continuously Improving Accuracy Using the AIHA Strategy with its Exposure Control Banding Approach" in Roundtable RT 233 at AIHce 2006. Chicago, IL. May 18, 2006.

General Principles of Physical — Chemical Models

2

By
Charles B. Keil, PhD, CIH

2.0 Overview

Physical-chemical models relate the properties about the release of contaminants and the characteristics of the air movement in a room to the resulting concentrations. In these models, an airborne contaminant enters an air space by any of a number of mechanisms. The contaminant mass is moved within the room by air currents, creating concentrations that may vary within the room. Eventually the mass of contaminant leaves the airspace, again by any number of removal mechanisms.

Conceptually, the information needed is:

- Where does the contaminant enter the room and at what rate?
- How does air move throughout the room and how does it transport the contaminant mass?
- Where and at what rate does the contaminant mass leave room air?

In practice, each of these conceptual aspects can be difficult to describe and quantify. The degree of complexity with which different models address each of these aspects also varies. However, there are a number of general principles that apply to most physical-chemical models. These principles will be discussed in this chapter.

2.1 Conservation of Mass

In models, mass is conserved. Contaminant mass enters the airspace by process release or is included with supply air. The contaminant mass moves throughout the room and eventually is removed from the airspace via local or general exhaust ventilation. The contaminant can also leave the airspace by non-ventilatory removal such as gas and vapor adsorption or particle deposition onto surfaces. In some instances, the contaminant can be degraded chemically within the air or on or in surfaces after absorption or deposition.

Air mass is also conserved as it moves through a room. When pressure and temperature are relatively constant, one can assume the volume of air and volumetric flow rate (Q) are conserved.

> ***Example 2.1***
>
> Air enters a room through a diffuser system at 20°C (293°K) and 760 mmHg. Processes in the room heat the air 10°C (up to 303°K) by the time it reaches the return register where the air pressure there is 755 mmHg. What is the percent change in the air volume?
>
> We write the ideal gas law (PV = nRT) for a unit mass of air twice: once for the entry conditions and once for the exit conditions:
>
> $P_{in} \bullet V_{in} = n \bullet R \bullet T_{in}$ and $P_{out} \bullet V_{out} = n \bullet R \bullet T_{out}$
>
> Rearranging and then setting the quantities as equal results in
>
> $$\frac{P_{in} \bullet V_{in}}{T_{in}} = n \bullet R \qquad \frac{P_{out} \bullet V_{out}}{T_{out}} = n \bullet R$$
>
> $$\frac{P_{in} \bullet V_{in}}{T_{in}} = \frac{P_{out} \bullet V_{out}}{T_{out}}$$
>
> $$\frac{V_{out}}{V_{in}} = \frac{P_{in}}{P_{out}} \bullet \frac{T_{out}}{T_{in}}$$
>
> $$\frac{V_{out}}{V_{in}} = \frac{760 \text{ mmHg}}{755 \text{ mmHg}} \bullet \frac{303°K}{293°K} = 1.04$$
>
> This illustrates that the change in the air volume from the diffuser to the return register is only 4%.

2.2 Parameters of the Indoor Air System

The following defines a number of parameters that are incorporated into many of the different models.

2.2.1 Room Volume

The room volume can be determined using basic geometry. For a typical rectangular box room, length • width • height = volume. For more complex rooms, the volume can be subdivided into smaller solid volumes and summed to obtain the room volume.

Sometimes room geometry suggests that there are actually two or more virtual air spaces. Consider an L-shaped room, with air supply and exhaust provided for each part of the "L." In this scenario, air and contaminants may mix in each leg of the "L" without appreciable mixing between the zones.

A potential complication in determining room volume is the volume displaced by the content of the room. At times the displaced volume might be significant, but often is not considered. Figure 2.1 shows a very crowded lab space. This room has 16% of the air volume displaced by the contents in the room.

One could imagine a 10 m × 5 m × 2.5 m marine biology lab with a room volume of 125 m^3. In this lab there could be two 5,000 gallon tanks displacing a total volume of 38 m^3, or 30% of the room volume.

2.2.2 Ventilation Configuration

The relative positions of air inlets and exhausts can greatly alter the transport of air contaminants

A room with diffuser and return locations, as illustrated in Figure 2.2, could result in two virtual airspaces in a room with little air mixing or contaminant transport between the halves.

Figure 2.3 illustrates how the relative location of air supply and return could affect the transport of contaminant in a room. In the left hand configuration, airflow through the work area is likely to be slow resulting in higher concentrations near the contaminant source. In the right hand configuration, supply air would move through the contaminant generation area and transport the contaminant mass away from the work area, diluting it throughout the workspace, until it were captured in the exhaust system on the opposite side of the room.

Figure 2.1 — A crowed lab with 16% of volume displaced.

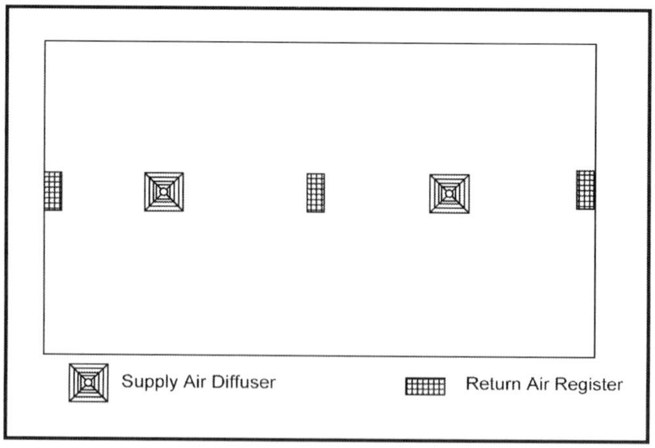

Figure 2.2 — A ventilation configuration that could result in two virtual air spaces.

2.2.3 Air Flow

The air entering room can be mechanically supplied or can infiltrate the room if it is under negative pressure relative to the surrounding air spaces. If the air is supplied as part of a HVAC system it is important to be aware of the recirculation of air (and potentially contaminants) and what percent of supplied air is fresh make-up air. A room is under negative pressure when air is mechanically exhausted at a faster rate than it is mechanically supplied; when this happens, air can enter the room through open doors, windows, drop ceilings, etc. Many models assume this make-up air has no or insignificant levels of contaminants. However re-circulated, make-up and infiltrating air can all carry air contaminants.

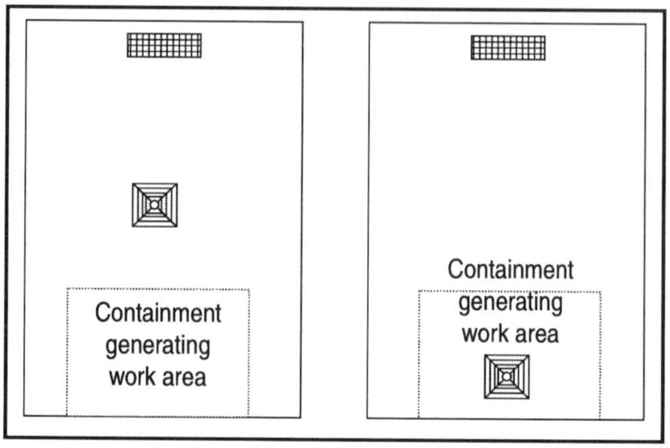

Figure 2.3 — Ventilation configurations can affect contaminant transport in rooms.

Air leaves a room through three general mechanisms. Air can be exhausted through the air return registers for the mechanical general ventilation. When this is part of an HVAC system, this air is often re-circulated. Air can also be removed from a room via local exhaust ventilation (LEV). LEV is process- or task-focused extraction system intended to prevent air contaminants from entering the general workspace air. LEV is normally exhausted outside of the workplace, unless effectively cleaned. It is important that these exhaust systems are not located near fresh-air intakes to the workplace, as the exhausted air could serve to contaminate the fresh-air that is supplied to the building. When a room is under positive pressure, air is mechanically supplied at a faster rate than it is exhausted: in this condition, air can leave one working environment and enter others through open doors, windows, drop ceilings, etc.

The room airflow rate (Q_{room}) is important for a number of models. Two main methods are used to determine this value. Design specifications can be used to extract performance criteria for supply and exhaust systems within the work environment. However, ventilation systems are notoriously altered from as-built designs or changed more frequently than design drawings, so in many cases, these may not be useful. Another more common approach is to identify all the points of air entry and exit from the room and take face velocity measurements across the surface of each supply and exhaust location within the system. Measuring face velocity, V_{face} (length/time) and face area, A_{face} (length^2), will allow the determination of individual flow rates, Q (length^3/time), into / out of the room by using simple algebra: $Q_1 = V_{face1} \bullet A_{face1}$. Then, the room airflow rate (Q_{room}) is computed by adding up all the flow rates for exhaust locations or, alternatively, for all supplied-air locations.

Within single family dwellings, the only mechanism typically available for fresh air ventilation or air exchange with outdoor air comes from the pressure differential of wind and the "stack effect" of warm air rising from a heated home. The rate of infiltration is typically in the range of 0.3–0.8 mixing airchanges/hr in a home with closed windows and doors and about 10 times this level with windows open.

As mentioned previously, the principle of conservation of mass and the ideal gas law dictate that the volumetric flow rate *into* a room must be equal to the volumetric flow rate *out of* a room. Ideally, the sum of the flow rates entering the room will equal the sum of all the exhaust flows. However, it is sometimes difficult to measure the flows through all air entry or exit points, especially when air enters or leaves the building through non-ducted systems, such as through gaps in the doors or windows or through left-open doors. These are typically low velocity flowrates and most velocity equipment is incapable of measuring values below 50 to 75 fpm, indicating some loss in certainty in the total room airflow rate. This author uses the following rules of thumb:

- Q_{in} and Q_{out} within 10% is excellent agreement and pretty much within the measurement errors you might expect, use the average of the two for Q_{room}
- Q_{in} and Q_{out} differ by 10% to 25%, average the two to estimate Q_{room} but be aware of the increased uncertainty
- Q_{in} and Q_{out} differ by greater than 25%, consider which side of the air balance you have the most confidence in and use that as estimate of Q_{room}.

For example, air enters the room through open doors and supply diffusers but exits only through two LEV systems and the room is clearly under negative pressure. If you can get reliable measurements on the two LEV exhaust rates, you should be able to estimate Q_{room} with reasonable confidence.

Airflow through a room and effective ventilation rates at locations in a room can also be determined using tracer gas techniques based on the well-mixed room model. Where there are no obvious sources of input or exhaust (*e.g.*, residential spaces) this is the only practical way of measuring the ventilation rate. This application will be discussed along with well-mixed room models.

2.2.4 Mixing Characteristics

Both the ventilation configuration and the amount of air/contaminant mixing are the key determinants to selecting which model to apply to a situation.

A well-mixed room is often the easiest approach to modeling air concentrations. This is a useful approach if the room air is, indeed, well-mixed or if the contaminant is released at multiple locations throughout a room. The obvious drawback to the well-mixed model is that concentration is higher near the source, and this model does not estimate this concentration gradient between the source and the rest of the room. Farther from the contaminant source, however, the well-mixed approach is still useful at approximating concentrations. The ACGIH® *Industrial Ventilation Manual* approach for designing general ventilation uses a safety/mixing factor to account for incomplete mixing in rooms. Note that when the safety/mixing factor approach is used for modeling concentrations rather than designing ventilation rates, the assumption of conservation of mass is violated.

One way to begin to account for the concentration gradients that occur in a room is to divide the airspace into two or more "zones." These zones may be virtual zones that divide the room into near, middle, and far areas. Alternatively, the ventilation configuration and geometry in the room itself may create distinct zones of air mixing. Inside each zone, the air is treated as if well-mixed. This has the advantage of accounting for the "positional variability" in concentration by using relatively simple mathematical approaches. Yet, when interested in concentrations close to contaminant releases, these zone models do not fully represent concentration gradients.

Furthermore, defining the actual air transfer rate and pattern between the zones within a room is more difficult than determining the net ventilation rate for the room.

To adequately capture steep concentration gradients often found near contaminant sources, continuous concentration gradient models can be used. However, these models often require estimates of turbulent diffusion parameters in the airspace, and these are not often readily available and require much more sensitive velocity equipment than the industrial hygienist is used to using.

2.2.5 Contaminant Sources

Generation rate determinants will be covered in detail in Chapter 3. General considerations for modeling the contaminant source are briefly discussed here. But in general, sources can be described by their spatial and temporal release pattern.

Sources can be considered as "point" or "area." Point sources refer to a contaminant generated from a localized area, such as a specific process or machine. Area sources refer to a spread-out source from a large area, such as contaminant off-gassing from wall coatings or from products stored in large quantities. Remember that the make-up air supplied to a room may also contain contaminants: if it exists, this source should be incorporated into modeling equations.

Contaminant generation may be continuous, intermittent, or even continuous but at changing rates. Continuous releases tend to be easier to model, as a steady-state concentration maximum may be computed easily, with often simpler mathematical solutions than time-varying models. Later chapters will describe how to model emission rates that may be decreasing due to back pressure from vapors already in room air, shrinking spill sizes and decreased rates of off-gassing. Intermittent releases will produce a time-varying concentration. A challenge to modeling these concentrations is using the appropriate mathematics to determine average concentrations over time, critical to computations of time-weighted exposure estimation.

2.2.6 Air Concentrations and Worker Exposures

The air concentration is typically the desired result of a model. While ppmv is often used to express concentrations of gases and vapors, mg/m^3 is often (though not always) the resulting units for model predictions. This is because the generation rate from sources is often conveniently described in terms of mass per time (mg/min, mg/sec).

Air concentrations are only part of what makes up a workers exposure. Worker exposure results from activity patterns within a room, where concentrations may vary in time and space. The duration that a worker spends in each concentration area is as important as the concentrations in the areas. Various levels of complexity can be incorporated into accounting for workers movement throughout the concentration field. If the modeling is performed to make decisions on risk acceptability, then a quick look at a reasonable worst-case scenario might quickly reveal low risk. When a more precise reconstruction of exposure is needed, as may be the case for epidemiology, litigation or when worst-case scenarios estimate an unacceptable exposure, more specific worker-activity data information and time-weighting approaches will be needed. Depending on the goal of the modeling, different approaches may be needed to consider the relationship between concentration and exposure.

2.3 Zero Ventilation Models

This approach models the average concentration resulting in unventilated rooms following a discrete release of contaminant into the air. Note that this method yields the average concentration in room; however, a concentration gradient may exist. Conceptually these models are simple: divide the mass released by the volume of the unventilated room to obtain an average room concentration:

Mass (mg) / volume (m^3) = average room concentration (mg/m^3)

When mass is released over time and there is ventilation, this approach can represent a reasonable worst-case scenario.

Example 2.2

A 6.0 gram ampule of ethylene oxide breaks in a 2.5 m^3 unventilated storage closet. What is the resulting concentration? (IDLH = 1,440 mg/m^3)

6.0 gram = 6.0 × 10^3 mg

6.0 x 10^3 mg / 2.5 m^3 = 2,400 mg/m^3

The IDLH level for ethylene oxide is 1,440 mg/m^3

High risk!

Example 2.3

15 mL of a cleaning solvent (density 1.2 g/mL) evaporates in an essentially unventilated 4.0 m × 4.0 m × 3.0 m storage closet. What is the resulting concentration? How much of a spill would result in concentrations exceeding an OEL of 2400 mg/m^3?

The mass of contaminant in the air space is

15 mL x 1.2 g/mL x 1,000 mg/g = 1,800 mg.

The closet volume is 4.0 m × 4.0 m × 3.0 m = 48 m^3.

The average concentration in the closet is

1,800 mg / 48 m^3 = 38 mg/m^3.

OEL = 2400 mg/m^3

Low risk

To compute spill volume to reach 2400 mg/m^3

X mg spilled / 48 m^3 = 2400 mg/m^3

X = 115,200 mg

Given the density above, 115,200 mg / (1.2 g/mL) x 1g / 1000 mg = 96 mL

If spill 96 mL, room concentration may be in the range of the OEL.

Example 2.4

A size K cylinder containing a calibration gas (150 ppm NO_2 in air) has a gauge pressure of 2,250 psi. The regulator on the cylinder fails and all of the gas rapidly escapes into a 12 m x 24 m x 3 m lab. If the room ventilation is off, estimate the resultant average concentration? If the ceiling limit is 5 ppm, what does this mean?

The room volume is 12 m • 24 m • 3 m = 864 m^3 = 864,000 L

A size K gas cylinder has an internal volume of 49.9 L.

If atmospheric pressure is 14.7 psi, the volume of compressed gas once released to environmental conditions is 49.9 L • 2250 psi / 14.7 psi = 7,640 L

If the 7,640 L of released calibration gas has 150 ppm NO_2

7,640 L • 150/10^6 = 1.1 L of NO_2 is released into the room

$$\frac{1.1\ L_{NO2}}{864{,}000\ L_{room\ air}} = \frac{1.3 \cdot 10^{-6}\ L_{NO2}}{L_{room\ air}} \cdot \frac{10^6}{10^6} = \frac{1.3\ L_{NO2}}{10^6\ L_{room\ air}} = 1.3\ \text{ppm}$$

The 5 ppm ceiling is not exceeded; however, we know that this model underestimates concentrations near the source.

2.4 Solution Approaches

The above examples illustrate the use of deterministic models to estimate room or exposure concentration. That is, application of equations developed from first principles to specific physical and chemical characteristics of the space under evaluation will yield an estimate of concentration. Given one estimate for generation rate or room airflow rate, the modeler will generate a point estimate for the room (or sub-volume of the room) concentration. However, we know that changes in generation rates, ventilation rates, and even room dimensions (with moving equipment) will result in variations to the estimate. To account for these variations, use of stochastic techniques can be applied to the range of determinants of concentration to evaluate the variability in chemical concentrations using probabilistic or "Monte Carlo" approaches. Chapter 10 provides more details on these methods, but an overview is provided here.

To illustrate probabilistic modeling consider the simple well mixed model:

$$C = \frac{G}{Q} \qquad (2\text{-}1)$$

Where

C: concentration (mg/m^3)
G: contaminant generation rate (mg/min)
Q: ventilation rate (m^3/min)

If a typical G for a process is 150 mg/min and a typical Q for the room containing the process is 6.0 m^3/min, then the typical concentration will be 25 mg/m^3. This is a deterministic solution.

But the contaminant generation rate and ventilation rate, in fact, vary with time. Now say that it is known that the room's ventilation rate (Q) can actually range from 4.0 to 8.0 m^3/min. For this example consider that 4.0 m^3/min is just as likely as 6.0 or 8.0 m^3/min. This is called a uniform distribution. And say that 150 mg/min is an average generation rate, but it varies with a standard deviation of 30 mg/min and is normally distributed.

Using Monte Carlo techniques, we can now use the equation, above, but instead of assigning only one value to G and one to Q, we allow the selection of real values within the range and distribution of real values and compute many values of concentration (C), repeating the computation for thousands to tens or hundreds of thousands of times. From this, we derive an estimate of the mean value and standard deviation for the room concentration values. There are computer programs that can be purchased to perform these simulations, but a spreadsheet can also be used. Simple Monte Carlo models can be run using the ubiquitous Excel spreadsheet program. Be warned that Excel has a very poor random number generator. It is being used here for illustrative purposes and should be used at most for screening applications of Monte Carlo model. The steps described here use Excel 2007. Other than initiating the add-in and locating the data analysis tool, the process using the actual tool is the same in Excel 2003.

To activate the "Analysis ToolPak" add-in follow these steps. Select the top left Office button and select "Excel Options" (Figure 2.4). When the options window comes up select "Add-ins"(Figure 2.5). Once the "Add-ins" menu is available, set the bottom "Manage" window to "Excel Add-ins" (Figure 2.6) and select "Go". The "Available Add-ins"

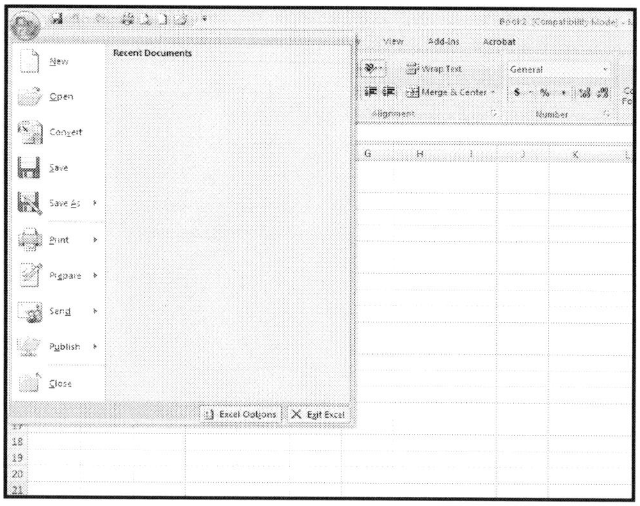

Figure 2.4 — Getting to Excel Options.
Microsoft product screen shots reprinted with permission from Microsoft Corporation.

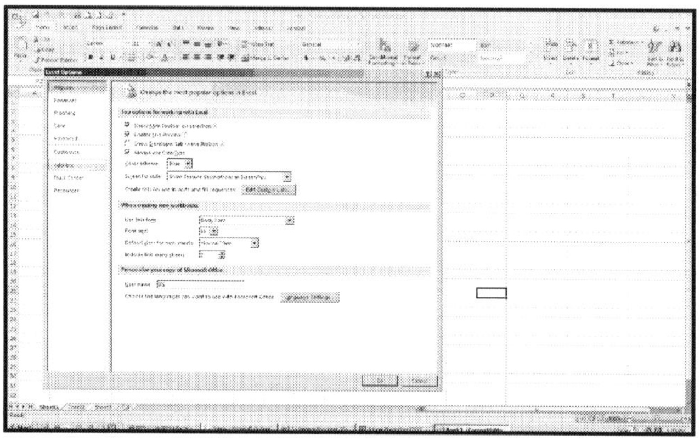

Figure 2.5 — Getting to Add-ins.
Microsoft product screen shots reprinted with permission from Microsoft Corporation.

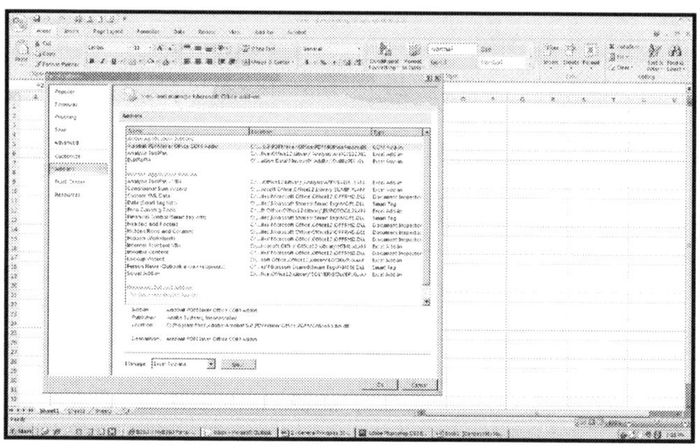

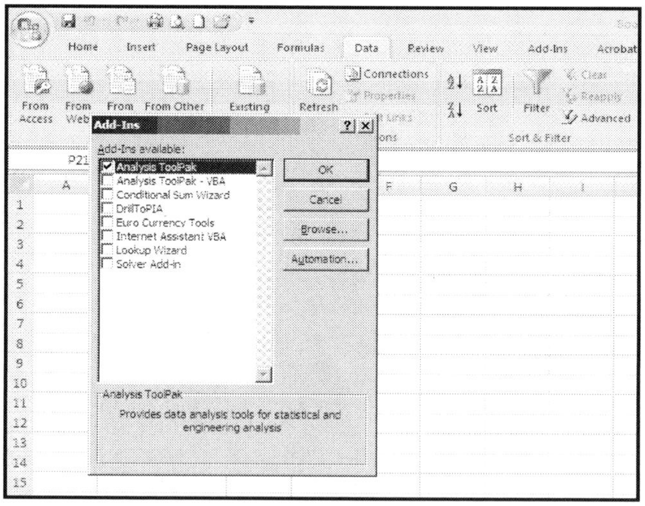

Figure 2.7 — Add-ins Dialogue Box.
Microsoft product screen shots reprinted with permission from Microsoft Corporation.

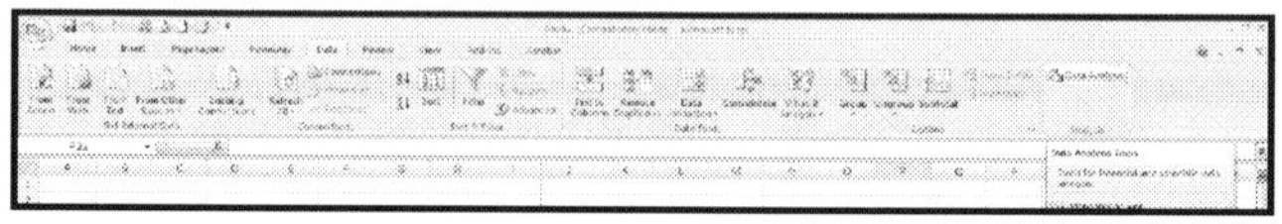

Figure 2.8 — Location of Data Analysis Tool.
Microsoft product screen shots reprinted with permission from Microsoft Corporation.

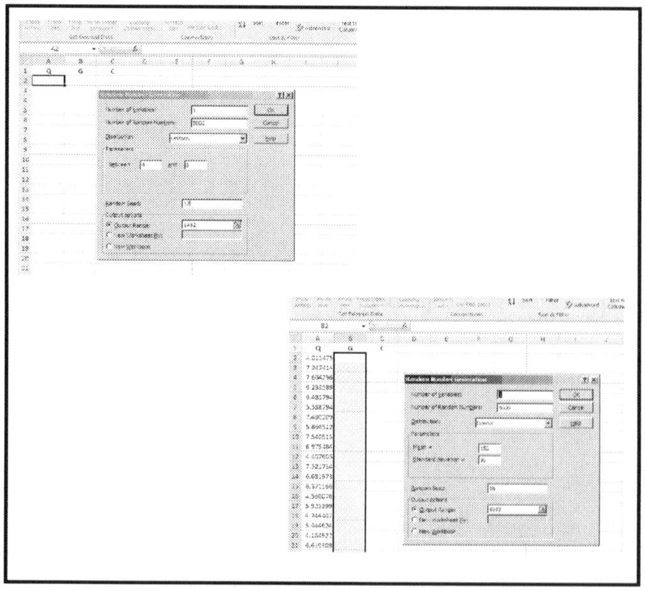

Figure 2.9 — With the random number generator, you can define the distribution associated with your input variables and generate a random sample from that distribution.
Microsoft product screen shots reprinted with permission from Microsoft Corporation.

dialogue box will appear (Figure 2.7.). Check "Analysis ToolPak" and select "OK". With the "Analysis ToolPak" add-in selected, the "Data Analysis" tool is available now at the far right of the Data Tab Menu (Figure 2.8).

By selecting the "Random Number Generator" tool in "Data Analysis", you can define the distribution and create random values. Figure 2.9 has a screen shot of preparing the variables for Q, uniformly distributed from 4.0 to 8.0 (for Column A). Figure 9 also shows the process used to create the G variables, normally distributed with a mean of 150 and a standard deviation of 30 (for column B). In this computation, 5000 values are selected (Rows 2 through 5001). It is ***vitally*** important that you put in different random seed numbers for each input variable in the model.

The values of associated concentrations (C) are computed in column C by programming in the well mixed model equation, C = G/Q (Figure 2.10). In this example column B (defined as being G) is divided by column A (defined as Q). In Excel this looks like =B2/A2 where the 2 represents the row 2 cells in each of the columns. If this equation is copied down the column, the row references will change so that cell C3 will calculate B3/A3, etc. From this simple math, 5000 values for C are calculated based on the distribution of input parameters (Figure 2.10).

These 5000 random samples from the range of reasonable values of Q and G resulted in a mean concentration of 26 mg/m^3 with a standard deviation of 7.4 mg/m^3, with a slight skewness to the right, although additional simulations are needed to confirm the trend. A histogram (also available from the Analysis Tool Pack) is generated and provided in Figure 2.11 along with an approximation of a normal distribution curve for a mean of 26 and S of 7.4. To complete this analysis, the simulation would be performed for another 5000 iterations, mean and standard deviation computed from the new and previous iterations (total of 10,000), and comparison of the results to the first 5000 iterations. This would continue until the addition of more random samples no longer changed the mean and standard deviation and the resulting histogram shape was stable.

With results from this simple Monte Carlo simulation combined with a simple steady-state, well mixed model, you can see that much more information than a simple point estimate of concentration could be available. By knowing the range of values of key parameters in the environment you are modeling, you can estimate percentiles and the exceedance fraction (percent of time that concentrations might be expected to be above a given level) and other estimates of exposure upon which we base our decisions of risk acceptability. These stochastic tools are presented only briefly here to illustrate the additional tools available to improve the information available from models and improve the inferences that can be drawn from your modeling efforts. These techniques are discussed more fully in Chapter 10, after we exhaust the basic models for estimating indoor contaminant concentrations.

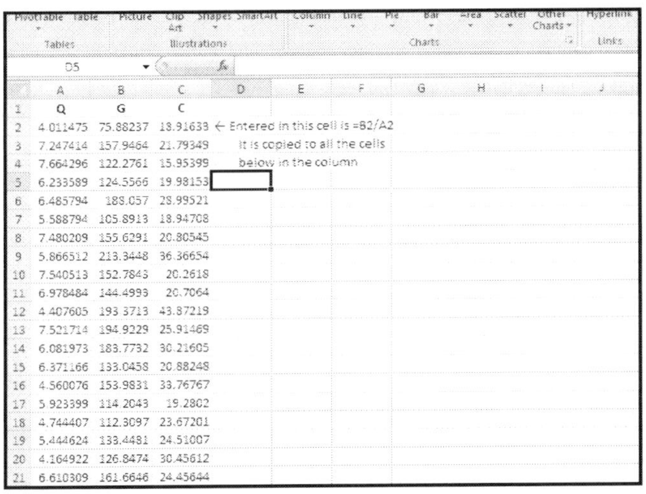

	A	B	C	D
1	Q	G	C	
2	4.011475	75.88237	18.91633	← Entered in this cell is =B2/A2
3	7.247414	157.9464	21.79349	It is copied to all the cells
4	7.664296	122.2761	15.95399	below in the column
5	6.233589	124.5566	19.98153	
6	6.485794	188.057	28.99521	
7	5.588794	105.8913	18.94708	
8	7.480209	155.6291	20.80545	
9	5.866512	213.3448	36.36654	
10	7.540513	152.7843	20.2618	
11	6.978484	144.4993	20.7064	
12	4.407605	193.3713	43.87219	
13	7.521714	194.9229	25.91469	
14	6.081973	183.7732	30.21605	
15	6.371166	133.0458	20.88248	
16	4.560076	153.9831	33.76767	
17	5.923399	114.2043	19.2802	
18	4.744407	112.3097	23.67201	
19	5.444624	133.4481	24.51007	
20	4.164922	126.8474	30.45612	
21	6.610309	161.6646	24.45644	

Figure 2.10 — Calculating C for each pair of random inputs.
Microsoft product screen shots reprinted with permission from Microsoft Corporation.

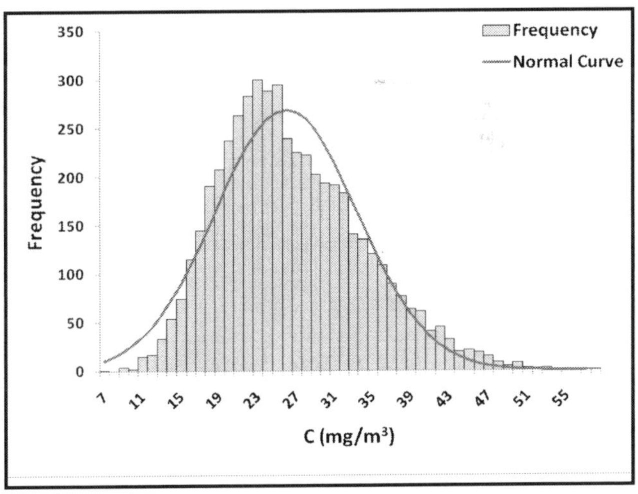

Figure 2.11 — Results of the example Monte Carlo solution.

Modeling Pollutant Generation Rates

By
Charles B. Keil, PhD, CIH

3.0 Overview

As mentioned in the General Principles chapter, characterizing pollutant sources in terms of their emissions is a basic parameter of indoor air systems. The chemicals released into the air interact with the room geometry and ventilation conditions to produce the concentrations in a room that result in worker exposures. The zero ventilation model examples in Chapter 2 present a straightforward pollutant generation scenario: a fairly instantaneous release of a known quantity of a chemical into a fixed volume of air. In many cases, the release of pollutant occurs over time and needs to be characterized in terms of a generation rate or emission rate. There are a number of approaches to arriving at a generation rate for use in concentration modeling. This chapter will introduce some of these approaches.

3.1 Mass Balance

Since mass is conserved in manufacturing processes, it is possible to do mass balances on the materials used in a process. By determining the amount of a chemical entering the process, the amount of the chemical incorporated into the product and the amount of the chemical accounted for in the waste the amount of chemical released (emitted) into indoor air spaces can be calculated.

$$\text{Mass}_{\text{into process}} - \text{Mass}_{\text{incorporated into product}} - \text{Mass}_{\text{collected as waste}} = \text{Mass}_{\text{released in room}} \qquad (3\text{-}1)$$

The mass released into the room is associated with the time scale for which the information is available, per day, per week, per month. Therefore, the mass released divided by the time it took to occur is the average emission rate over that timeframe. This simple average emission rate will not account for intermittent release cycles, where there may be periods of pollutant release at higher or lower rates, than described by the average.

Also note that the generation rates determined with this method represent total emissions and may not be a direct estimate of the concentration of the chemical in the air in the room. The mass of chemical removed by local exhaust ventilation or the HVAC system for the room must be considered separately.

Example 3.1

An offset lithography operation uses approximately 1 L of cleaning solvent every 8-hour shift. In this process, the solvent entirely evaporates into the air. The solvent density is 0.78 g/mL.

$$1 \text{ L/8 hr} \times \text{hr/60 min} \times 1000 \text{ mL/L} \times 0.78 \text{ g/mL} \times 1000 \text{ mg/g}$$

$$= 1625 \text{ mg/min}$$

This is an average emission rate for the 8-hour period. Clearly there would be periods of higher and lower emission rates. The averaging time of the emission rate is limited by the time resolution of the material use data.

Purchasing and inventory data can be used along with production data and disposal data to obtain an estimate of the fugitive indoor release. However, this information tends to have an even longer time scale so the results may be a longer-term average emission rate.

Example 3.2

A metals manufacturing facility has a chrome plating process. The gassing from the electroplating process produces a chromic acid mist. The purchasing department provides information that approximately 1750 lbs (795 kg) of flake CrO_3 (MW 100) are purchased each year. The production engineers estimate that the process plates about 13,500 m^2 (1.35×10^8 cm^2) of surface area each year. The plating thickness of the metallic chromium (MW 52) is 0.150 mil (0.000381 cm). The density of metallic chromium is 7.14 g/cm^3.

795 kg CrO_3 × 52 kg Cr / 100 kg CrO_3 = 413 kg Cr brought into the plant each year

Volume of metallic chromium applied to product = 1.35×10^8 cm^2 × 0.000381 cm = 51,435 cm^3

Mass of metallic chromium applied to product = 51,435 cm^3 × 7.14 g/cm^3 = 367,246 g = 367 kg

$Mass_{Cr\ released}$ = 413 kg – 367 kg = 46 kg

413 kg of Cr is brought into the plant each year; 367 kg leaves on parts. The balance, 46 kg, must be released into the air. If the plating operation is run 16 hours a day, 5 days a week, 50 weeks a year, the average emission rate during the process operation is 46 kg/4000 hrs or 192 mg/min. Note that this represents the total emission rate; much of this may be captured by local exhaust ventilation.

3.2 Emission Factors

Emission factors are expressions of pollutant releases in terms of some process variable. By using the rate of the process variable, the emission rate can be estimated. For example, if there is 0.015 mg of pollutant released for every widget made, and 22 widgets are made each minute the emission rate would be 0.33 mg/min. Emission factors have the advantage of being scalable based on the size of the operation in the workplace. A small shop may produce 15 widgets a minute, but a large factory may produce hundreds of the same type of widget each minute. The same emission factor can be applied to each to get the appropriately scaled emission rate. Similarly within a facility there may be times of higher or lower production volume. The emission factor allows the varying emission rate associated with the process to be estimated for each rate of production. The implicit assumption is that the emission factor is a function of the process. Therefore, if the process changes that emission factor will most likely change.

Emission factors have been used for years by the EPA for estimating ambient air emissions and are collected in AP-42 *Compilation of Air Pollution Emission Factors.*[1] Examples of ambient emission factors include mass of CO emitted per vehicle mile driven and mass of SO_2 emitted per ton of coal burned for electricity production.

Some emission factors have been developed specifically for workplace pollutant sources.[2–10] These emission factors describe workplace emissions in terms such as:

- µg Cr per area of part plated or ampere hour
- mg Trichloroethylene per basket of parts degreased
- mg VOC per cleaning of a printing press
- formaldehyde emission per active dissection table

Example 3.3

A company has decided to look retrospectively at worker exposures. One process under consideration is a CFC 113 degreasing process. There is a published emission factor for this operation of 37 g/m^2 of parts degreased.[5] A review of production data indicates that the average rate of operation was around 0.6 m^2/hr, though it could peak at 1.8 m^2 per hour. What was the average and peak emission rate (G)?

Average G = 0.6 m^2/hr x 37 g/m^2 = 22 g/hr

Max G = 1.8 m^2/hr x 37 g/m^2 = 67 g/hr

Even certain EPA ambient emission factors can be applied in some occupational situations.

Example 3.4

What emission rate of lead could be expected from an abrasive blasting operation that must be accomplished before a highway bridge can be repainted? The paint being used contains 12% lead by weight. A 5/16 inch nozzle will be used at a pressure of 80 psi.

AP-42 lists total particle (TP) emission factors this abrasive blasting operation as a function of wind speed. Higher wind speeds increase emissions into the environment through enhanced ventilation of the process and by decreasing coarse particle deposition. At a wind speed of 5 mph the emission factor for total particulate is 27 lb / 1000 lb of abrasive used. The amount of abrasive used is a function of both the blasting nozzle size and the air pressure used. AP-42 provides this information in a tabular format. A 5/16th inch nozzle operating at 80 psi has an abrasive flow rate of 420 lb/hr.

$$\frac{420\ lb_{abrasive}}{hr} \bullet \frac{27\ lb_{TP}}{1000\ lb_{abrasive}} \bullet \frac{453.6\ g_{TP}}{lb_{PM}} \bullet \frac{1000\ mg_{TP}}{g_{TP}} \bullet \frac{hr}{3600\ sec} = \frac{1430\ mg_{TP}}{sec}$$

The TP is 12% lead by weight giving a lead emission rate of 170 mg/sec. This is within the range reported by Conroy, et al. in a field study.[7]

3.3 Vapor Pressure Models

3.3.1 Saturation Vapor Pressure Concentrations

Vapor pressure is a physical property of substances, primarily liquids, that describes the tendency of molecules of the substance to move to the gas phase. In a closed system, when the rate of molecules moving into the gas phase equals the rate of molecules returning to the liquid phase are the same, the system is said to be at equilibrium. When at equilibrium, the pressure exerted by the gas phase chemical is the chemical's vapor pressure. The vapor pressure is the partial pressure of the gas phase portion of the chemical in the system.

Vapor pressure is very temperature dependent. Also, how vapor pressure applies to mixtures deserves special notice. Appendix II covers some of the chemistry of vapor pressure.

The ideal gas law tells us that pressure is proportional to volume so the ratio of the saturation vapor pressure of a chemical to the total pressure of system is proportional to the gas phase volume of the chemical to the volume of the entire gas phase. Volume to volume ratios are used in expressions of concentrations for gas phase pollutants: percent (parts per hundred) for high concentrations and parts per million (ppm) and parts per billion (ppb) for low concentrations. The volume fraction of a chemical (A) and concentration can be calculated in a system at equilibrium knowing the vapor pressure of a chemical and the atmospheric pressure.

$$\frac{VP_A}{P_{atm}} = \frac{V_A}{V_{Tot}} = \text{vapor fraction of A} \qquad (3\text{-}2)$$

Where

VP_A: vapor pressure of chemical A
P_{atm}: atmospheric pressure
V_A: volume of chemical A
V_{Tot}: total volume of the system

For example, if chemical A has a vapor pressure of 15 mmHg, in equilibrium with a normal atmospheric pressure the pressure volume relationship can be expressed in the following manner:

$$\frac{15\text{ mm Hg}_A}{760\text{ mm Hg}} = \frac{16\text{ L}_A}{760\text{ L}_{air}} = \frac{0.020\text{ L}_A}{\text{L}_{air}}$$

By convention volume fractions are often converted into percent values by multiplying them by 100 or into ppm or ppb by multiplying by 10^6 or 10^9 respectively.

$$\frac{15\text{ mm Hg}_A}{760\text{ mm Hg}} = \frac{16\text{ L}_A}{760\text{ L}_{air}} = \frac{0.020\text{ L}_A}{\text{L}_{air}}$$

$$\frac{0.020\text{ L}_A}{\text{L}_{air}} \bullet \frac{100}{100} = \frac{2.0\text{ L}_A}{100\text{ L}_{air}} = 2.0\%$$

$$\frac{0.020\text{ L}_A}{\text{L}_{air}} \bullet \frac{10^6}{10^6} = \frac{20{,}000\text{ L}_A}{10^6\text{ L}_{air}} = 20{,}000\text{ ppm}$$

Simplified the saturation vapor pressure model can be expressed.

$$\frac{VP_A}{P_{atm}} \bullet 10^6 = \text{ppm} \qquad (3\text{-}3)$$

If the concentration estimate is desired in units of mg/m^3 the standard conversion from ppm to mg/m^3 at NTP can be incorporated.

$$\frac{VP_A}{P_{atm}} \bullet 10^6 \bullet \frac{MW}{24.45} = mg/m^3 \qquad (3\text{-}4)$$

This saturation vapor pressure model can be used to estimate the headspace concentration in sealed containers and other scenarios where equilibrium conditions may be reached. These concentrations are clearly represent the worst case scenario and are not achievable in an open system such as an occupied work room and would be well over any applicable OEL. These calculations are useful determining the concentration limits in air for materials with known or estimated vapor pressures.

Example 3.5

The lid is left off a bottle of n-heptane in a small chemical storage cabinet. A researcher opens up the cabinet and is startled by the strong odor and quickly steps back out of the plume of n-heptane vapors which quickly dissipate. What is the maximum concentration that the brief exposure might have been in ppm and mg/m^3? Assume NTP.

n-Heptane data (NIOSH *Pocket Guide to Chemical Hazards*): VP (72°F): 40 mmHg, MW: 100.2, PEL: 500 ppm, immediately dangerous to life and health (IDLH): 750 ppm, Lower Explosive Limit (LEL): 1.05%)

$$\frac{40\text{ mm Hg}}{760\text{ mm Hg}} \bullet 10^6 = 53{,}000\text{ ppm}$$

$$53{,}000\text{ ppm} \bullet \frac{100.2}{24.45} = 220{,}000\text{ mg/m}^3$$

The saturation vapor pressure concentration is far higher than these health indices. From a safety standpoint this concentration (53,000 ppm = 5.3%) is above the lower explosive limit presenting a safety hazard as well.

Example 3.6

Your company's grounds maintenance people are considering using nicotine (VP 0.08 mmHg, MW 162.2) as an insecticide. What concentration of gas phase nicotine can you expect in the headspace of the storage containers? Assume NTP.

$$\frac{0.08 \text{ mm Hg}}{760 \text{ mm Hg}} \bullet 10^6 = 105 \text{ ppm}$$

$$105 \text{ ppm} \bullet \frac{162.2}{24.45} = 700 \text{ mg/m}^3$$

The NPG lists the IDLH concentration as 5 mg/m^3. Thus, any decision to use this compound should include appropriate protective measures.

At temperatures and pressures differing from NTP, the molar volume of an ideal gas will not be equal to 24.45 L/mole. In those cases the saturation vapor pressure can be calculated using the molar volume of an ideal gas

$$\frac{\text{vol}}{\text{mol}} = \frac{R \bullet T}{P_{atm}} \quad (3\text{-}5)$$

Substituting this into the concentration model above

$$\frac{VP_A}{P_{atm}} \bullet 10^6 \bullet \frac{MW \bullet P_{atm}}{R \bullet T} = \text{mg/m}^3 \quad (3\text{-}6)$$

$$VP_A \bullet 10^6 \bullet \frac{MW}{R \bullet T} = \text{mg/m}^3$$

3.3.2 Liquid Container Filling Releases

The saturation vapor pressure concentration model has been presented in this chapter because of its utility in a category of pollutant generation rate models, liquid container filling.

As a container is being filled with a liquid, the headspace air is being displaced into the room air at the same rate at which filling is taking place. So the fill rate (F) multiplied by the concentration in the headspace air (C_{hs}) is the chemical mass emission rate (G) from the container filling process.

$$F \bullet C_{hs} = G \quad (3\text{-}7)$$

Dimensionally this is

$$\frac{\text{vol}}{\text{time}} \bullet \frac{\text{mass}}{\text{vol}} = \frac{\text{mass}}{\text{time}}$$

If the headspace air is in equilibrium with the liquid phase, C_{hs} can be calculated using the equations for C_{sat} at NTP or other conditions.

$$F \bullet \frac{VP_A}{P_{atm}} \bullet 10^6 \bullet \frac{MW}{24.45} = G \text{ (mg/time)} \quad (3\text{-}8)$$

$$F \bullet VP_A \bullet 10^6 \bullet \frac{MW}{R \bullet T} = G \text{ (mg/time)} \quad (3\text{-}9)$$

Where

R: 62.36 L mmHg mol^{-1} K^{-1}
F: m^3/min, if F is in L/time the constant becomes 10^3

Matthiessen[11] suggests that this fill rate can be modified to represent conditions that are different than saturation. When the filling hose is submerged in the liquid level a factor of 0.50 to 1.00 may be applied to results of the calculations using Equation 3-9. In the case where splashing occurs during the filling operation mists as well as vapors may be generated. In this case a factor of 1.00 to 1.45 may be applied.

Example 3.7

Four 55 gallon drums of methyl ethyl ketone (CAS No. 78–93–3) are filled each hour in the bulk material transfer area of your facility. Use the drum-filling model to estimate the release rate of methyl ethyl ketone (MEK) into the air. MEK has a molecular weight of 72.12 and a vapor pressure of 71.2 mmHg at 20°C (68°F). Room temperature in this example is 20°C (68°F). Filling is conducted with the hose submerged, saturation is assumed. Ambient pressure is 1 atmosphere (760 mmHg).

$$F = \frac{4 \text{ drums}}{\text{hr}} \bullet \frac{55 \text{ gal}}{\text{drum}} \bullet \frac{\text{m}^3}{264.2 \text{ gal}} \bullet \frac{\text{hr}}{60 \text{ min}} = \frac{0.0139 \text{ m}^3}{\text{min}}$$

$$G = F \bullet C_{hs}$$

$$G = \frac{0.0139 \text{ m}^3}{\text{min}} \bullet 71.2 \text{ mmHg} \bullet 10^6 \bullet \frac{72.12 \text{ g}}{\text{mol}} \bullet$$

$$\frac{\text{mol K}}{62.36 \text{ L mm Hg}} \bullet \frac{}{293°\text{K}} = 3{,}900 \frac{\text{mg}}{\text{min}}$$

3.4 Evaporation from Open Surfaces

A number of models have been proposed for evaporation from open surfaces such as open tanks and spilled liquids. Lennert, et al., evaluated a number of these models.[12] One that performed well under a variety of conditions was developed by Gray.[13]

$$G = 1300 \bullet D^{1.9} \bullet \nu^{-0.9} \bullet \left[\frac{100 \bullet \nu \bullet V}{D}\right]^{0.625 \bullet \left[\frac{\nu}{D}\right]^{0.3}}$$

$$\bullet L^{-0.11} \bullet \sqrt{\frac{P_{atm}}{P_{atm} - P_V}} \bullet \frac{P_V}{R \bullet T} \bullet MW \bullet A \qquad (3\text{-}10)$$

where

- G: evaporation rate (g/min)
- D: molecular diffusion coefficient (m^2/sec)
- ν: kinematic viscosity of air ($1.56 \bullet 10^{-5}$ m^2/sec at NTP)
- V: air speed (m/sec)
- L: surface length of liquid pool (m)
- P_v: vapor pressure of evaporating chemical (Pa)
- P_{atm}: atmospheric pressure (Pa)
- R: ideal gas constant (8.3144 Pa m^3 $°K^{-1}$ $mole^{-1}$)
- MW: molecular weight (g/mole)
- A: surface area of liquid pool (m^2)
- T: temperature (°K)

The values of molecular diffusion coefficients for many gas phase compounds in air are available in chemical property reference manuals. For chemicals for which molecular diffusion data are not available, there are some estimation methods.

One such method based on Chapman-Enskog kinetic theory is[14]

$$D = \frac{4.14 \bullet 10^{-4} \bullet T^{1.9} \bullet \sqrt{\frac{1}{29} + \frac{1}{MW}} \bullet MW^{-0.33}}{P} \qquad (3\text{-}11)$$

Where

- D: molecular diffusion coefficient (m^2/sec)
- T: temperature (°K)
- MW: molecular weight (g/mole)
- P_{atm}: atmospheric pressure (Pa)

Another method for estimating molecular diffusion based on atomic diffusion volumes is the Fuller method.[15] The EPA website http://www.epa.gov/athens/learn2model/part-two/onsite/estdiffusion.htm provides an on-line calculator for molecular diffusion coefficients in air.

A more recent model has been developed by Hummel and Fehrenbacker and has been evaluated as being useful for predicting concentrations as a results of evaporation of chemicals from small spills.[14,16,17]

$$G = \frac{0.166 \bullet MW^{0.833} \bullet P_V \bullet \left[\frac{1}{MW} + \frac{1}{29}\right]^{0.25} \bullet A}{T^{0.05}} \bullet \sqrt{\frac{V}{L \bullet P_{atm}}} \qquad (3\text{-}12)$$

This equation uses the same units as the Equation 3-10.

Example 3.8

A drum of waste toluene ruptures, leaks and its contents flow into a 1.0 meter by 2.0 meter containment area surrounded by a curb (A: 2.0 m^2). Air is flowing across the length (L: 2.0 m) of the surface at a rate of 0.075 m/sec (V). Room conditions are at NTP. What evaporation rate can be expected?

Toluene Properties

MW: 92.1
P_v: 21 mmHg = 2800 Pa
D: $8.04 \bullet 10^{-6}$ m^2/sec

Using Gray's equation

$$G = 1300 \bullet (8.04 \bullet 10^{-6})^{1.9} \bullet (1.56 \bullet 10^{-5})^{-0.9} \bullet$$

$$\left[\frac{100 \bullet 1.56 \bullet 10^{-5} \bullet 0.075}{8.04 \bullet 10^{-6}}\right]^{0.625 \bullet \left[\frac{1.56 \bullet 10^{-5}}{8.04 \bullet 10^{-6}}\right]^{0.3}}$$

$$\bullet 2.0^{-0.11} \bullet \sqrt{\frac{101325}{101325 - 2800}} \bullet \frac{2800}{8.3144 \bullet 298}$$

$$\bullet 92.1 \bullet 2.0 = 8.8 \frac{g}{min}$$

Using Hummel's equation

$$G = \frac{0.166 \bullet 92.1^{0.833} \bullet 2800 \bullet \left[\frac{1}{92.1} + \frac{1}{29}\right]^{0.25} \bullet 2.0}{298^{0.05}} \bullet$$

$$\sqrt{\frac{0.075}{2.0 \bullet 101325}} = 8.5 \frac{g}{min}$$

3.5 Releases from Pressurized Vessels

Reactors, process lines, storage tanks and other types of equipment contain gas phase pollutants at elevated pressures. When a leak develops in a pressurized system, how can the pollutant release rate be estimated? The concept of critical flow is useful for this scenario. Critical flow is a complex phenomenon. A basic treatment of the subject for normal room conditions and example scenarios are presented as an introduction to the approach.

When fluids flow through an orifice the flow rate increases as the pressure differential/ratio across the orifice increases. Once the pressure ratio, low (downstream) pressure / higher (upstream) pressure reaches 0.53 the flow rate through the orifice is sonic, i.e., the gas/vapor is traveling at the speed of sound. Sonic flow rates have been demonstrated to be maximum values therefore, the flow rate through the orifice will not increase with further increases in pressure differences. The ratio of pressures that creates this maximum flow rate is called the critical pressure ratio (at the given conditions). The composition of the flowing gas and the temperature can both affect the critical pressure ratio. Table 3.1 shows the critical pressure ratios for various gases at approximately 20°C. For most all gasses and gas mixtures, when the upstream pressure is twice the pressure downstream of the orifice, flow will be sonic.

Table 3.1. Example Critical Flow Ratios for Various Gasses

Gas	Critical flow ratio ($P_{downstream}/P_{upstream}$)
Krypton	0.484
Xenon	0.487
Air (dry)	0.528
Benzene	0.528
Methane	0.544
Sulphur dioxide	0.553
Ethylene	0.555
Acetylene	0.557
Ethane	0.565
Ethyl chloride	0.567
Nitrogen peroxide	0.570
Propane	0.577
Chloroform	0.583
Ethyl ether	0.601

Figure 3.1 illustrates the effect of temperature on the critical flow ratio for ethylene. Even for this gas, which happens to be particularly sensitive to temperature, the critical flow ratio is affected by only ~8% over a relatively wide range of temperatures.

At 20°C the speed of sound in air is approximately 343 m/sec. If the area through which the critical flow is occurring is known or can be estimated, the volumetric flow rate through the orifice can be estimated.

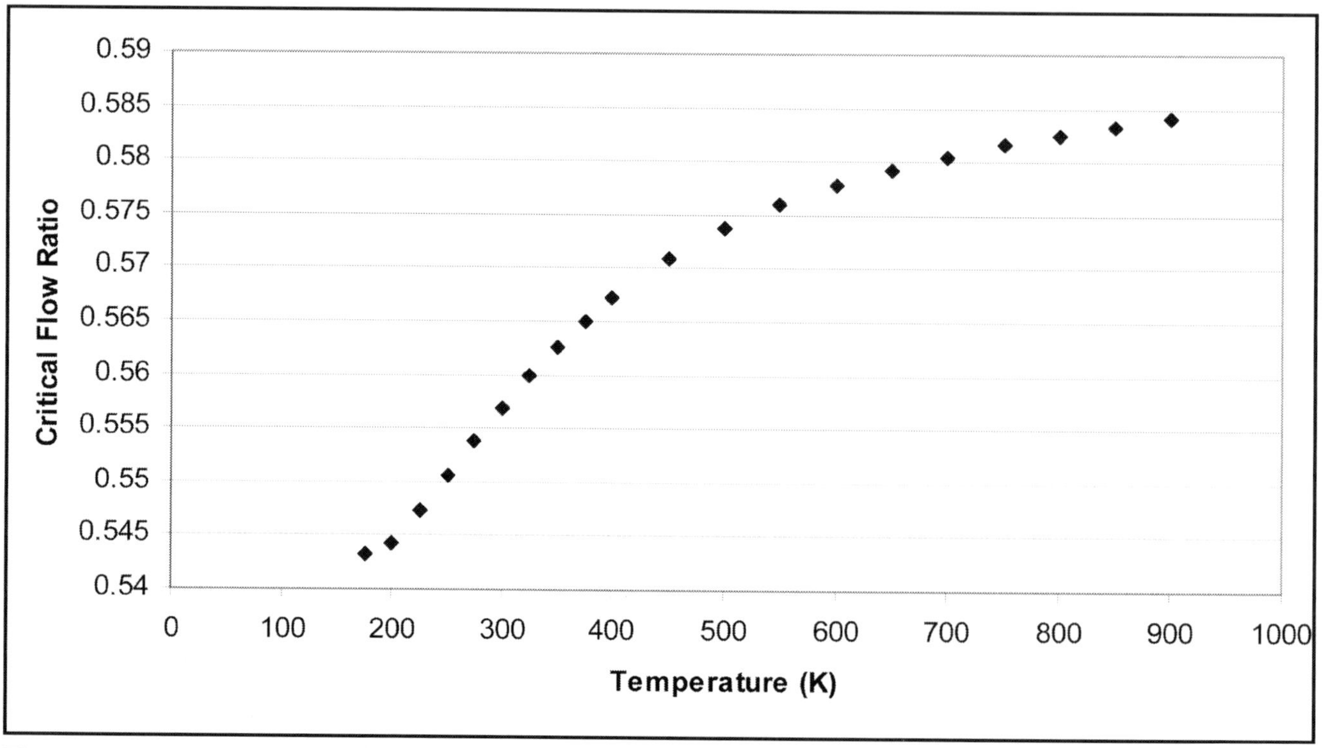

Figure 3.1 — Effect of Temperature on Critical Flow Ratio for Ethylene.

$$Q_{critical} = c \bullet A_{orifice} \qquad (3\text{-}13)$$

Where

$Q_{critical}$: volumetric flow rate (Length^3/time)
$A_{orifice}$: orifice area (Length^2)
c: speed of sound (Length/time)

Equation 3-13 can be used to calculate the volumetric flow rate of the compressed gas being released through an orifice at sonic velocity. To get the mass release rate, we first need to calculate the density of the pressurized gas from its molecular weight and the pressure at which it is kept in the vessel.

$$\rho = \frac{P \bullet MW}{R \bullet T} \qquad (3\text{-}14)$$

Where

ρ: pressurized gas density (g/m^3, mg/L)
P: absolute gas pressure (Pa), add atmospheric P to gauge P
R: ideal gas constant (8.314 m^3 Pa mol^{-1} K^{-1})
T: pressurized gas temperature (K)

The product of the flow rate (L/sec) and the gas density (mg/L) provides an estimate of the mass release rate (mg/sec) from a leaking pressurized vessel.

In the case of a storage container, such as a compressed gas cylinder, the mass release rate will decrease as pressure inside the container decreases and the compressed gas density decreases. This is true even while the volumetric release rate based on sonic flow stays the same as long as the critical pressure ratio is maintained. This can be accounted for with more advanced modeling, but the initial release rate can be useful as a maximum value. So for typical room conditions the mass release rate, G (mg/sec) for a continuously pressurized container (or initial release rate from a storage vessel) can be estimated (using appropriate units) as follows:

$$G = \frac{c \bullet A_{orifice} \bullet P \bullet MW}{R \bullet T} \qquad (3\text{-}15)$$

Example 3.9

A pressurized transfer line carries pure HCl (MW: 36.5). The operating gauge pressure is 5.0 Atmospheres (507 kPa gauge, 608 kPa absolute) and the temperature is approximately the same as the room, 25°C (298°K). A crack 1.0 cm long by 1.0 mm wide forms in the transfer line. Estimate the mass release rate in mg/sec from the crack.

The flow ratio is 1 Atm/6 Atm = 0.17 which is below the critical flow ratio and thus flow is sonic.

$$A_{orifice} = 1.0\text{ cm} \bullet 0.10\text{ cm} = 0.10\text{ cm}^2$$

$$G = \frac{c \bullet A_{orifice} \bullet P \bullet MW}{R \bullet T}$$

$$G = \frac{343\text{ m}}{\text{sec}} \bullet \frac{0.10\text{ cm}^2}{} \bullet \frac{\text{mol K}}{8.3145\text{m}^3\text{ Pa}} \bullet$$

$$\frac{608000\text{ Pa}}{} \bullet \frac{}{298°\text{K}} \bullet \frac{36.5\text{ g}}{\text{mol}} \bullet \frac{\text{m}^2}{10000\text{ cm}^2} \bullet \frac{1000\text{ mg}}{\text{g}}$$

$$G = 31000\text{ mg/sec}$$

If the pressurized gas in question is not a pure substance but a mixture, the molecular weight of the mixture should be used to determine the mass release rate of the mixture. The mass fraction of the chemical of interest should then be applied to get the mass release rate for that particular chemical.

3.6 Off – gassing models

Another source of indoor volatile organic compounds is off-gassing from finished product. Treated woods, surface treatments, paints, adhesives and other materials release associated volatile compounds into the air. Semivolatile compounds are released from other products. Volatile compound off-gassing can be relatively short term and of high intensity while semivolatiles may be released at a lower rate, but over a longer period of time. Until recently the approach to characterizing these emissions has been empirical based on chamber studies.[18,19] New initiatives are underway to develop and validate fundamental emission models that link parameters such as the initial material-phase concentration, the material-air partition coefficient, and the material-phase diffusion coefficient.[20]

3.7 A Note on Particle Emissions

In practice, estimating the generation rate of particles is often challenging. Many of the detailed and theory based examples in this chapter have been for gases and vapors. A challenge to estimating particle releases from a mass balance approach is that particle mobility in the air is highly dependent on the particle size. Particles with an aerodynamic diameter of > 0.1 mm generally settle out of the air quickly. For example, a mass balance on a grinding operation using the area of the surface that is ground and the thickness removed (area • thickness = volume) and the density of the material (mass/volume) can provide information on the total mass removed from the part but using the resulting mass emission rate is complicated by the fact that the emitted mass is composed of particles of varying size, composition and settling velocities. Additionally the tendency to the particles to agglomerate or absorb water vapors also will affect the proportion of the particles released that can remain suspended in the air. Without information on the size and mass distribution created in the particles that are suspended in the air, airborne release estimates can only be worst case.

Much additional work is needed from a theoretical as well as emission factor approach to particle generation rates.

3.8 References

1. **U.S. Environmental Protection Agency (EPA):** Office of Air Quality Planning and Standards. In *Compilation of air pollutant emission factors [microform].* Research Triangle Park, NC: Office of Air Quality Planning and Standards, Office of Air and Radiation, U.S. Environmental Protection Agency, 1995.
2. **Franke, J.E. and R.A. Wadden:** Indoor contaminant emission rates characterized by source activity factors. *Environ. Sci. Technol. 21*:45–51 (1987).
3. **Wadden, R.A., J.E. Franke, and P.A. Scheff:** Emission factors for trichloroethylene vapor degreasers. *Am. Ind. Hyg. Assoc. J. 50*:496–500 (1989).
4. **Wadden, R.A., J.L. Hawkins, and P.A. Scheff:** Characterization of emission factors related to source activity for trichloroethylene degreasing and chrome plating processes. *Am. Ind. Hyg. Assoc. J. 52*:349–356 (1991).
5. **Scheff, P.A., R.L. Friedman, and J.E. Franke:** Source Activity Modeling of Freon Emissions for Open-Top Degreasers. *Appl. Occup. Env. Hyg. 7*:127–134 (1992).
6. **Wadden, R.A., D.I. Baird, and J.E. Franke:** Ethanol emission factors for glazing during candy production. *Am. Ind. Hyg. Assoc. J. 55*:343-351 (1994).
7. **Conroy, L.M., R.M.M. Lindsay, and P.M. Sullivan:** Lead, Chromium, and Cadmium Emission Factors During Abrasive Blasting Operations by Bridge Painters. *Am. Ind. Hyg. Assoc. J. 56*:266 (1995).
8. **Wadden, R.A., J.E. Franke, and P.A. Scheff:** Determination of VOC emission rates and compositions for offset printing. *Am. Ind. Hyg. Assoc. J. 56*:368–376 (1995).
9. **Keil, C.B.:** The development and evaluation of an emission factor for a toluene parts-washing process. *Am. Ind. Hyg. Assoc. J. 59*:14–19 (1998).
10. **Keil, C.B., F. Akbar-Khanzadeh, and K.A. Konecny:** Characterizing Formaldehyde Emission Rates in a Gross Anatomy Laboratory. *Appl. Occup. Env. Hyg. 16*:967 (2001).
11. **Matthiessen, R.C.:** Estimating Chemical Exposure Levels in the Workplace. *Chem. Eng. Prog. 16*:30–34 (1986).
12. **Lennert, A., F. Nielson, and N.O. Breum:** Evaluation of Evaporation and Concentration Distribution Models — A Test Chamber Study. *Ann. Occ. Hyg. 41*:625–641 (1997).
13. **Gray, D.C.:** Solvent Evaporation Rates. *Am. Ind. Hyg. Assoc. J. 35*:675–710 (1974).
14. **Hummel, A.A., K.O. Braun, and M.C. Fehrenbacher:** Evaporation of a Liquid in a Flowing Airstream. *Am. Ind. Hyg. Assoc. J. 57*:519–525 (1996).
15. **Fuller, E.N. K. Ensley, and J.C. Giddings:** Diffusion of halogenated hydrocarbons in helium. The effect of structure on collsion cross sections. *Ind. Eng. Chem. 5*:18–27 (1969).
16. **Fehrenbacher, M.C. and A.A. Hummel:** Evaluation of the Mass Balance Model Used by the Environmental Protection Agency for Estimating Inhalation Exposure to New Chemical Substances. *Am. Ind. Hyg. Assoc. J. 57*:526 (1996).
17. **Keil, C.B. and M. Nicas:** Predicting Room Vapor Concentrations Due to Spills of Organic Solvents. *Am. Ind. Hyg. Assoc. J. 64*:445–454 (2003).
18. **Jayjock, M.A. D.R. Doshi, E.H. Nungesser, and W.D. Shade:** Development and Evaluation of a Source/Sink Model of Indoor Air Concentrations from Isothiazolone Treated Wood Used Indoors. *Am. Ind. Hyg. Assoc. J. 56*:546 (1995).
19. **Que, Z. and T. Furuno:** Formaldehyde emission from wood products: relationship between the values by the Chamber method and those by the Desiccator test. *Wood Sci. & Tech. 41*:267–279 (2007).
20. **Kephalopoulos, S.A., M.A. Arvanitis, and M.A. Jayjock (eds.):** Global CEM Net Report of the Workshop no. 2 on "Source Characterization, Transport and Fate." Intra, Italy, June 20–21, 2005.

Well-Mixed Box Model

4

By
Patricia H. Reinke
Charles B. Keil, PhD, CIH

4.0 Overview

An indoor air model has two main components: a source (or generation) term and a dispersion term. In the zero ventilation models presented in previous chapters, discrete pollutant releases into a fixed volume of air were considered. This chapter describes the simplest representation of continuous pollutant release into a ventilated air spaces, the well-mixed box (WMB) model. The well-mixed box models a room as a large box with highly turbulent internal airflow (see Figure 4.1). Equations describing contaminant concentrations in the box are developed using a mass balance and a number of simplifying assumptions. These equations are described and compared, and examples are presented.

4.1 Model Development

We are concerned with the concentration of a specific airborne contaminant (A) in the box in Figure 4.1. Air enters the box with a flow rate Q_{in} and a contaminant concentration $C_{A,in}$. Air exits the room with an airflow rate Q_{out}. The box is perfectly mixed (that is, air is moving turbulently as if there were a number of fans in the room blowing air around). This mixing creates a uniform contaminant concentration throughout the room including at the exhaust locations. The outgoing air concentration is therefore equal to the room air concentration $C_{A,room}$.

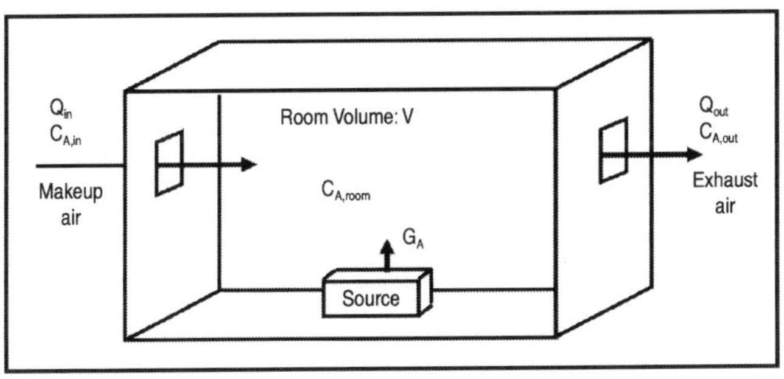

Figure 4.1 — Conceptual Model of the Well Mixed Box.

Now, we place a contaminant source in the room. The source can generate gas, vapor, or aerosol. It is intuitive that if a gas or vapor is released in a room, the volume of contaminated air flowing out of the room would be greater than that into the room, that is, $Q_{out} > Q_{in}$. If not, the room would build up pressure. In developing the WMB model, however, it is assumed that generation rates are low when compared with room airflow rates, so $Q_{in} = Q_{out} = Q$.

4.1.1 Mass Balances

To develop an equation that describes the time course of room concentration in the idealized box, we use a mass balance. A mass balance is a mathematical expression of the law of conservation of mass on a given system. The law of conservation of mass states that mass is neither created nor destroyed in a process or non-nuclear chemical reaction. When chemical reactions occur in a system, strict application of the law of conservation of mass requires accounting for all of the mass in a system before and after a reaction. This chapter will cover only systems in which reactions are not occurring. When no reactions occur, a mass balance is a simple accounting of any substance within the system. In words, a mass balance requires that the rate of chemical accumulation within any closed system is equal to the rate of mass into the system minus the rate of mass out of the system. The mass into the system can be divided into the mass into the system through the system boundaries and the mass generated inside the system. The general mass balance equation for a system is:

$$\text{Mass}_{accumulated} + \text{Mass}_{out} = \text{Mass}_{generated} + \text{Mass}_{in} \qquad (4\text{-}1)$$

4.1.2 Room Air Mass Balance

To develop a mass balance on contaminant in the box shown in Figure 4.1, one should first choose the closed boundary system as the air within the room at a given time. The rate of accumulation of mass of contaminant A in the air is written

$$\frac{dmass_A}{dt} = V \cdot \frac{dC_{A,room}}{dt} \quad (4\text{-}2)$$

Where

V: volume of air in the room ($length^3$)
$C_{A,room}$: uniform room concentration of A ($mass/length^3$)
t: time

Because concentration has dimensions of mass per volume and air volume has units of volume, the rate of accumulation in Equation 4-2 has dimensions of mass per unit time.

The rate of contaminant mass leaving the room air has two components: the mass that leaves the room through the exhaust and the contaminant that leaves room air when it is adsorbed onto or absorbed into room surfaces (walls, floors, clothing, equipment). The latter is referred to as sink or "non-ventilatory" losses. The ventilatory loss can be described as:

$$\frac{-dmass_A}{dt} = Q \cdot C_{A,room} \quad (4\text{-}3)$$

Sinks are not well understood or characterized. Data indicate, however, that the rate of contaminant loss through sinks is proportional to the concentration of contaminant in the room.[1–3]

$$\text{Mass removal rate via sinks} = K_{sink} \cdot C_{A,\,room} \quad (4\text{-}4)$$

Where

K_{sink}: rate constant for the sink ($length^3/time$)

The proportionality constant, K_{sink}, is a property of the room that depends on the surfaces in the room, their area, and their affinity for the contaminant.

So, the rate that pollutant mass is removed from the room air becomes:

$$\frac{-dmass_A}{dt} \; Q \cdot C_{A,room} + K_{sink} \cdot C_{A,room} \quad (4\text{-}5)$$

Although it is commonplace to assume that sinks are negligible in modeling (due primarily to lack of good estimates of K_{sink}), this term is included as a reminder that, without considering sinks, overestimates of exposure potential can result. As pointed out in *A Strategy for Assessing and Managing Occupational Exposures*, 3rd edition, "This is especially true of high molecular weight or high boiling point contaminants and aerosols, which are more likely to adhere to surfaces."[4]

The next term in the mass balance equation is the rate of mass generated (or consumed, although that is less of a concern for the industrial hygienist) within a system. The generation term is designated as G_A and has dimensions of mass per unit time.

$$\text{Rate of Generation: } G_A \text{ (mass/time)} \quad (4\text{-}6)$$

Finally, the rate of contaminant coming into the system through its boundaries is:

$$\frac{dmass_A}{dt} = Q \cdot C_{A,in} \quad (4\text{-}7)$$

Where

$C_{a,in}$: the concentration of A in the incoming air ($mass/length^3$)

Combining the equations, we obtain the complete differential equation for air concentration in the well-mixed room:

$$V\frac{dC_{A,room}}{dt} = Q \cdot C_{A,in} + G_A - K_{sink}C_{A,room} - Q \cdot C_{A,room} \quad (4\text{-}8)$$

4.1.3 Common WMB Model Assumptions

To integrate this differential equation, several assumptions must be made. In this section, the WMB model is integrated using the most common assumptions. More complicated forms of the model are discussed later. The assumptions are as follows:

- *No Sinks.* As indicated earlier, sinks are difficult to characterize. Therefore, with the understanding that this may result in overestimation of room concentrations, we eliminate the sink term.
- *Constant Generation Rate, G_A.* This assumption states that the mass generation rate is constant with time. Although this is often the case, it is important to note that in some situations, such as in a small spill that shrinks as it evaporates or in a cyclic process, the generation rate is not constant. Models presented later better characterize such situations.
- *Low Generation Rates.* It is assumed that the generation rates are low, so the airflow into and out of the box is nearly equal. A relatively low generation rate also means that the resulting concentration within the volume is not a significant proportion of the saturation concentration of a vaporizing source. This assures that backpressure is not an issue. Note that modeling scenarios where backpressure is an issue (i.e., relatively high generations rates) are treated in Chapter 5.
- *Constant Incoming Contaminant Concentration, $C_{A,in}$.* We assume that the concentration of contaminant in the incoming air is constant with time at $C_{A,in}$.

4.1.4 Closed Form Solution of the WMB Model

Using the assumptions stated above, the differential equation of the mass balance on the room is integrated to obtain the general WMB model.

$$C_{A,room,t} = \left[C_{A,room,0} - C_{A,in} - \frac{G_A}{Q} \right] \bullet e^{\frac{-Q \bullet (t-t_0)}{V}} + C_{A,in} + \frac{G_A}{Q} \quad (4\text{-}9)$$

Where

$C_{A,room,t}$: Concentration of A in room air at time t (mass/length^3)
$C_{A,room,0}$: Concentration of A in room air at time t_0 (mass/length^3)
$C_{A,in}$: Concentration of A in supply air (mass/length^3)
G_A: Generation rate of A (mass/time)
V: Room volume: (length^3)
Q: Ventilation rate (length^3/time)

This equation can be used to estimate the concentration of contaminant A in a room after a period of contaminant generation. It is important to review the assumptions inherent in the equation before using it. Constant and low generation rates are often good assumptions. Perfect and instantaneous mixing is usually not a good representation of mixing indoors; chapters of this book investigate improved representations of mixing and dispersion. The following example uses this form of the model to predict room concentration changes with time.

Example 4.1

After lunch, workers are scheduled to run a small roll-coating experiment. The toluene based coating will be applied in a 5.0 m × 6.0 m × 2.0 m room, with 0.10 m^3/s airflow. The toluene concentration in the room in the morning was constant at 10.0 mg/m^3. Thus, you estimate the incoming air also has a concentration of 10 mg/m^3. The research chemist tells you to expect a worst-case emission rate of 50.0 mg/s for this small experiment. Estimate the room concentration 20 minutes after the start of coating.

Assumptions:

- Room has perfect mixing.
- No sinks.
- Incoming air contains 10.0 mg/m^3 toluene (A).
- $C_{A,in}$ = 10.0 mg/m^3.
- Initial concentration is 10.0 mg/m^3
- $C_{A,room,0}$ = 10.0 mg/m^3
- Generation Rate G_A = 50.0 mg/s.

Room volume can be calculated from the room dimensions. V = 5.0 m × 6.0 m × 2.0 m = 60 m^3

Now the equation for room concentration at 20 minutes (1200 seconds) can be solved:

$$C_{A,room,t} = \left[10\frac{mg}{m^3} - 10\frac{mg}{m^3} - \frac{50.0\frac{mg}{sec}}{0.10\frac{m^3}{sec}} \right] \bullet e^{\frac{-0.10\frac{m^3}{sec}(1200\ sec)}{60\ m^3}} + 10\frac{mg}{m^3} + \frac{50.0\frac{mg}{sec}}{0.10\frac{m^3}{sec}} = 440\frac{mg}{m^3}$$

Programs such as EXCEL can be used to calculate room concentrations at a number of times. Figure 4.2 shows the concentration build up from 0 to 100 minutes. Note that room concentration rises quickly and approaches a steady value of 510 mg/m^3. This concentration is termed the steady state concentration and, as is show below, is equal to G_A/Q.

4.1.5 Simplifications of the WMB Time-Dependent Model

Simplifications of the time dependent model can be made to the model if conditions permit. These simplifications include the following:

1. Initial room air concentration is zero, $C_{A,room} = 0$. If there is no contaminant in room air when the generation begins:

$$C_{A,room,t} = \left[- C_{A,in} - \frac{G_A}{Q} \right] \bullet e^{\frac{-Q \bullet (t-t_0)}{V}} + C_{A,in} + \frac{G_A}{Q} \quad (4\text{-}10)$$

2. No contaminant in the incoming ventilation air, $C_{A,in} = 0$:

$$C_{A,room,t} = \left[C_{A,room,0} - \frac{G_A}{Q} \right] \bullet e^{\frac{-Q \bullet (t - t_0)}{V}} + \frac{G_A}{Q} \quad (4\text{-}11)$$

3. No contaminant in incoming air and initial room concentration is zero (1 and 2 combined):

$$C_{A,room,t} = \frac{G_A}{Q} \bullet \left[1 - e^{\frac{-Q \bullet (t - t_0)}{V}} \right] \quad (4\text{-}12)$$

These simplified forms of the WMB mass balance equation provide a fair first estimate of average room air concentration changes with time. The models require that the GA is constant and known, and that sinks are negligible. The WMB model does not, however, provide information on the spatial dispersion of air contaminants.

4.2 Steady State Solutions

Figure 4.2 shows the change in room concentration over time for the WMB described in Example 4.1. After approximately 50 minutes, room concentration becomes nearly constant at 510 mg/m^3. This is called the steady state concentration, and it remains constant as long as the system conditions remain the same.

Steady state is defined as a condition in which the physical properties of a system do not change with time. Attainment of steady state requires stabilization time. This time required before a room or any system comes to steady state can differ greatly depending on the situation.

Equations for steady state room concentrations can be obtained from the unsteady state equations by assuming the system is allowed to operate unchanged for a very long period of time. That is, $(t - t_0)$ gets very large. Thus, $Q\,(t - t_0)\,/\,V$ gets very large, and $e^{-Q(t-t_0)/V}$ approaches zero. The WMB equation becomes:

$$C_{A,room,SS} = C_{A,in} + \frac{G_A}{Q} \quad (4\text{-}13)$$

This can be further simplified when there is no contaminant in the incoming air stream ($C_{A,in} = 0$). Then:

$$C_{A,room,SS} = \frac{G_A}{Q} \quad (4\text{-}14)$$

Example 4.2 shows how steady state equations are used.

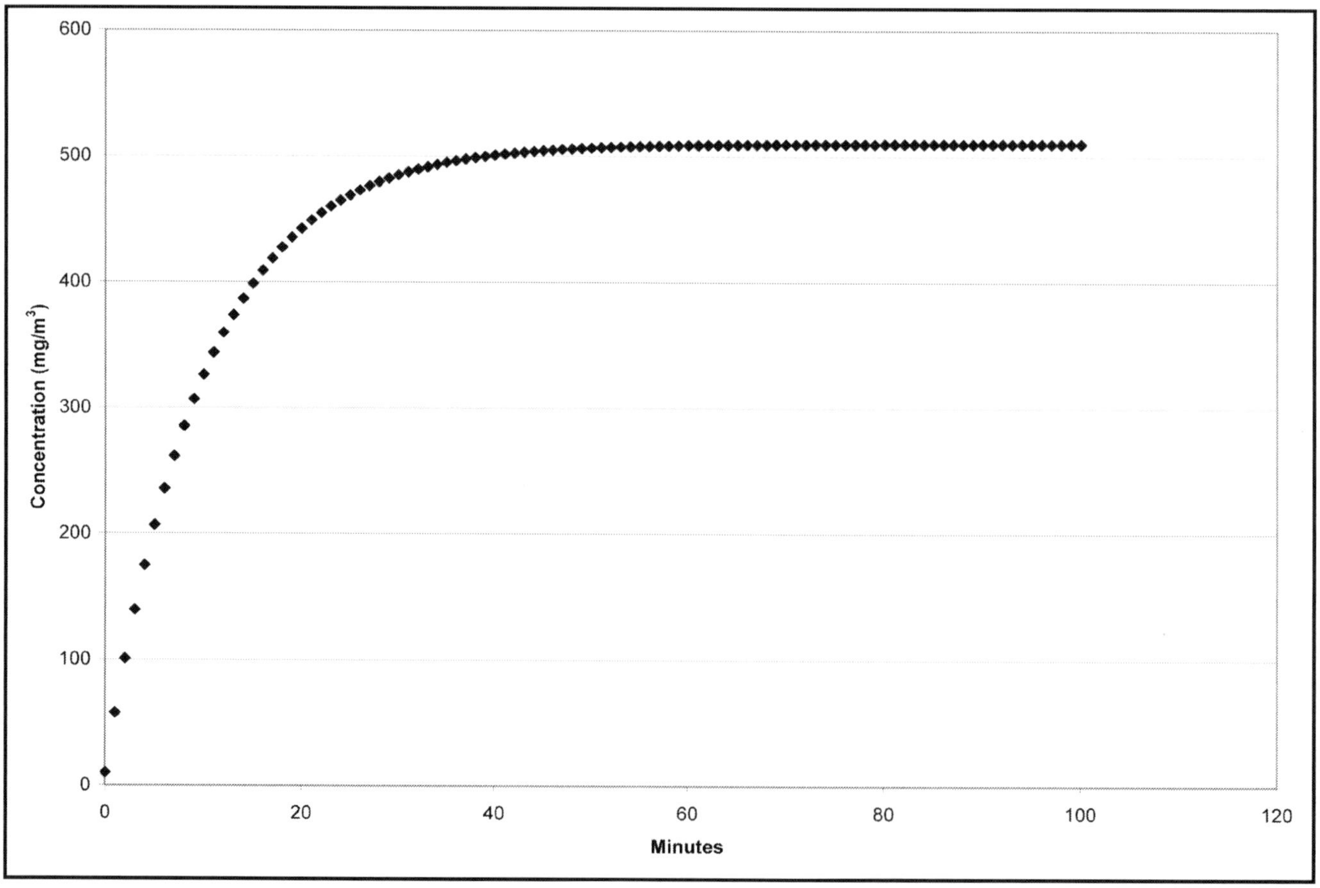

Figure 4.2 — Build-up of Concentration in the WMB Example 4.1.

Example 4.2

Using the same conditions as Example 4.1, estimate the steady state concentration in the room

$$C_{A,room,SS} = 10\frac{mg}{m^3} + \frac{50\frac{mg}{sec}}{0.10\frac{m^3}{sec}} = 510\frac{mg}{m^3}$$

Note that this is the same as the steady state concentration shown in Figure 4.2. The solution in Example 4.2 does not use $C_{A,room,0}$ or V. The steady state room concentration *does not depend on either the initial room concentration or on the room volume. (This is not always intuitive.) These factors influence how quickly a system reaches steady state, but not the steady state concentration.*

4.2.1 Time to Reach Steady State

The exponential time term in the WMB equation never actually reaches zero, so theoretically steady state is never definitively reached. One can, however, calculate the time to reach some concentration less than that of steady state.

The time to reach a percentage of the steady state room concentration can be estimated by algebraic manipulation of the WMB equation. This can be solved to determine the time to reach some value less than the steady state concentration.

$$t - t_0 = \frac{-V}{Q} \bullet \ln\left[1 - \frac{C_{A,room,t} - C_{A,room,0}}{C_{A,room,SS} - C_{A,room,0}}\right] \tag{4-15}$$

Example 4.3

Develop an equation to calculate the time it takes to reach 95% of steady state concentration when the initial concentration is zero.

$$t - t_0 = \frac{-V}{Q} \bullet \ln\left[1 - \frac{C_{A,room,t} - 0}{C_{A,room,SS} - 0}\right]$$

$$= \frac{-V}{Q} \bullet \ln\left[1 - \frac{C_{A,room,t}}{C_{A,room,SS}}\right]$$

$$= \frac{-V}{Q} \bullet \ln\,(1 - 0.95)$$

$$= \frac{-V}{Q} \bullet \ln(0.05)$$

$$= \frac{-V}{Q} \bullet (-\,3.0)$$

$$= 3.0\,\frac{V}{Q}$$

4.3 Concentration Decay

When a concentration exists in a well mixed room and the generation rate term goes to zero, the concentration in the room will decay according to the following form of the WMB model.

$$C_{A,room,t} = (C_{A,room,0} - C_{A,in}) \bullet e^{\frac{-Q \bullet (t - t_0)}{V}} + C_{A,in} \tag{4-16}$$

When there is no chemical A in the incoming air the decay equation is:

$$C_{A,room,t} = C_{A,room,0} \bullet e^{\frac{-Q \bullet (t - t_0)}{V}} \tag{4-17}$$

Example 4.4

A production process that emits a chemical into a well mixed room ceases. At the time of the shut down of the process the room concentration was 127 mg/m^3. The room volume is 620 m^3. The room airflow rate is 31 m^3/min. What is the concentration 25 minutes after the process stops? How long does it take for the concentration to decay to 15 mg/m^3?

$$C_{A,room,t} = 127\frac{mg}{m^3} \bullet e^{\left[\frac{-31\frac{m^3}{min} \bullet 25\,min}{620\,m^3}\right]}$$

$$= 127\frac{mg}{m^3} \bullet e^{-1.25}$$

$$= 127\frac{mg}{m^3} \bullet 0.29$$

$$= 36\frac{mg}{m^3}$$

Rearranging the decay equation to solve for time yields an equation similar to that presented for determining the time to reach a certain concentration during build-up.

$$t - t_0 = \frac{-V}{Q} \bullet \ln\left[\frac{C_{A,room,t}}{C_{A,room,0}}\right]$$

$$t - t_0 = \frac{-620\ m^3}{31\ \frac{m^3}{min}} \bullet \ln\left[\frac{15\ \frac{mg}{m^3}}{127\ \frac{mg}{m^3}}\right] = 43\ min$$

As alluded to in the general principles chapter, the decay equation can be used to get an estimate of the ventilation in a room or the effective ventilation at a location in a room.

To determine the ventilation rate for a room a tracer gas needs to be well mixed in a room and then the release stopped. The tracer can be from a process existing in the room or selected specifically for the ventilation study. Sulfur hexafluoride is a common tracer gas. For an estimate of the room ventilation rate, good mixing needs to be maintained throughout the decay. Mixing is often induced with the use of fans. If the tracer gas is mixed throughout the room, but mixing is not good during the decay, the calculated ventilation rate is the effective ventilation rate for that location of concentration measurements.

Rearranging the decay equation and defining $t - t_0$ as the time elapsed (t) yields a useful form of the equation.

$$C_{A,room,t} = C_{A,room,0} \bullet e^{\frac{-Q \bullet t}{V}}$$

$$\frac{C_{A,room,t}}{C_{A,room,0}} = e^{\frac{-Q \bullet t}{V}}$$

$$-\ln\left[\frac{C_{A,room,t}}{C_{A,rom,0}}\right] = \frac{Q}{V} \bullet t \qquad (4\text{-}18)$$

Which is in the form of a linear relationship, $y = m \bullet x$. If the left hand side of the equation is plotted on the y-axis and t on the x-axis, the slope of the line is $-Q/V$. If V is known, Q can be determined. Consider the following example.

Example 4.5

Fans are used to induce good mixing in a 294 m^3 room. SF_6 is released and the concentration builds up to 18 ppm. The release is stopped and the concentration decay is measured with an infrared analyzer. A plot of the decay is illustrated below.

The data in tabular form is shown below with additional columns that calculate C_t/C_0 and -$\ln(C_t/C_0)$. The plot of time vs. -$\ln(C_t/C_0)$ is presented below the data along with the equation of the line through the data points.

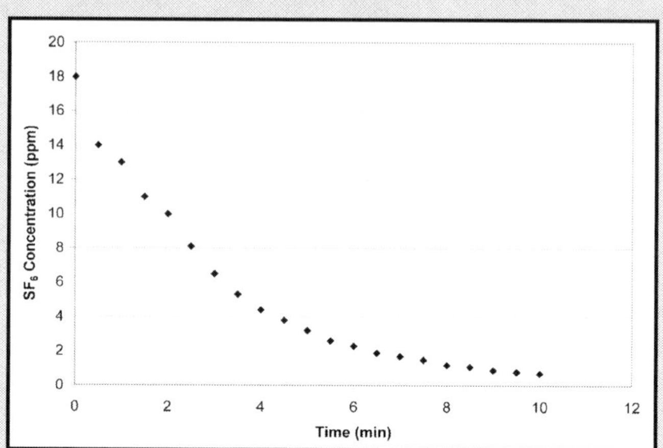

Figure 4.3 — Concentration decay over time.

Example 4.5 Data

Time (min)	C (ppm)	C_t/C_0	$-\ln(C_t/C_0)$
0	18	1.00	0.00
0.5	14	0.78	0.25
1	13	0.72	0.33
1.5	11	0.61	0.49
2	10	0.56	0.59
2.5	8.1	0.45	0.80
3	6.5	0.36	1.02
3.5	5.3	0.29	1.22
4	4.4	0.24	1.41
4.5	3.8	0.21	1.56
5	3.2	0.18	1.73
5.5	2.6	0.14	1.93
6	2.3	0.13	2.06
6.5	1.9	0.11	2.25
7	1.7	0.09	2.36
7.5	1.5	0.08	2.48
8	1.2	0.07	2.71
8.5	1.1	0.06	2.80
9	0.9	0.05	3.00
9.5	0.8	0.04	3.11
10	0.7	0.04	3.25

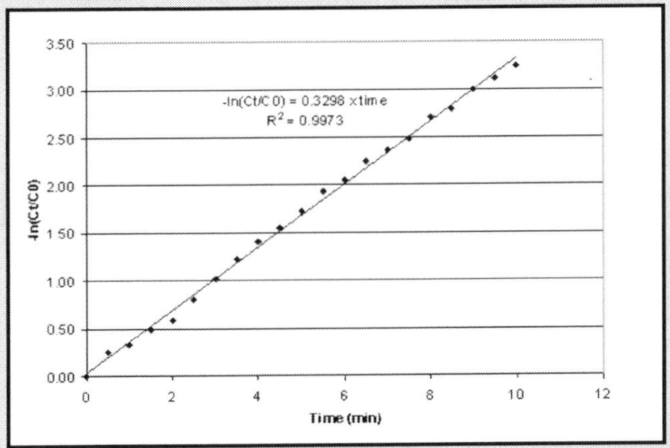

Figure 4.4 — Slope computation to determine Q/V from experimental decay data.

If the slope (Q/V) is 0.33/min and the volume of the room is 294 m^3 the fresh air ventilation rate (Q) of the room is 0.33/min • 294 m^3 = 97 m^3/min. Linear regression analysis of this type is easily done in programs such as EXCEL.

4.4 Mixing Factors

The WMB models presented thus far describe behavior in perfect and instantaneously mixed rooms. Since perfect mixing does not exist in the real world, a mixing factor is often used to correct for deviation from perfection. A common representation of the mixing factor, m, is a factor by which the actual ventilation rate is multiplied to obtain a lower "effective" ventilation rate. That is:

$$Q_{eff} = m \bullet Q \tag{4-19}$$

Where

- Q_{eff}: effective ventilation in and out of the box
- Q: actual ventilation rate of the box
- m: mixing factor 0–1.0 (dimensionless)

The mixing factor can also be considered as the fraction of the total ventilation air that effectively ventilates a location in the room. If the tracer gas decay approach to determining ventilation rates is done without fans inducing mixing during decay, the ventilation rate determined is the effective ventilation rate.

A number of studies have been done to predict mixing factors, the value of m is typically set at 0.1–0.33. According to one study, m varied from 0.3 to 0.7 in small rooms without fans.[5]

WMB model equations can be used with the mixing factor approach by substituting Q_{eff} for Q in the equations. Although these equations may provide a fair

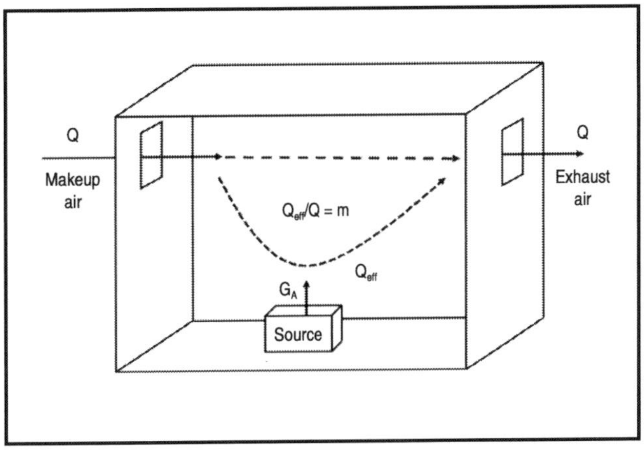

Figure 4.5 — Illustration of inadequate mixing via short-circuiting.

representation of average air concentration in the occupied zone of a room, it is important to note that they defy the law of conservation of mass of applied to the entire room. That is, if room concentration is multiplied by room volume, over time the total contaminant in the space will exceed the total mass generated. This is the result of considering only an effective ventilation rate and not an effective volume. In actuality, the entire room is not at this corrected concentration, but a portion of the room, perhaps a small portion with a very high airflow rate, will have a much lower concentration than that calculated using the mixing factor equations above. This is often referred to as a short-circuiting, and is shown in Figure 4.5. The air that short circuits occupies a volume, and, although the $C_{A,room}$ calculated using the mixing factor equations may provide a good estimate of concentration at a point in the room, there are still variations in concentration in the room.

Because mixing factors defy the law of conservation of mass, and because using a mixing factor with a well-mixed room model does not describe nonideality well, some experts discourage their use.[2] Mixing factors, however, are still used in ventilation and industrial hygiene literature and therefore addressed here. Alternatives to using mixing factors to account for less than complete mixing in rooms are addressed in later chapters.

4.5 Average Exposures over Time

Since airborne concentrations build up and decay, a technique is needed to determine an average concentration over time. The general WMB equation can be integrated to obtain the time average exposure to a contaminant. More information on integrating concentration functions to find average concentrations is addressed in Appendix A. Integrating the WMB equation from time t = 0 to time t = t gives

$$C_{avg} = C_{A,in} + \frac{G_A}{Q} + \frac{V}{Q \bullet (t - t_0)}$$

$$\bullet \left[C_{A,room,0} - C_{A,in} - \frac{G_A}{Q} \right]$$

$$\bullet \left[e^{\frac{-Q \bullet (t - t_0)}{V}} - 1 \right] \quad (4\text{-}20)$$

4.6 The Effect of Airflow other than Fresh Air Mechanical Ventilation

Figure 4.1 is a highly idealized representation of an indoor room. In reality, Q, the airflow into a room, is made up of a number of separate airflows from different locations and with different contaminant concentrations.

Three modes of air exchange are described in the *ASHRAE 1993 Fundamentals Handbook*: 1) forced ventilation; 2) natural ventilation; and 3) infiltration.[7] Therefore, the Q_{in} entering the box in Figure 4.1 has three components, Q_F, the forced ventilation component; Q_N, the natural ventilation component; and Q_{INF}, infiltration.

4.6.1 Natural Ventilation

Q_N is natural ventilation through intentional openings such as windows and doors. Many older facilities and most residences rely on natural ventilation, which is driven by pressure differences from wind and indoor–outdoor temperature differences. Because natural ventilation is difficult to measure and control, it is not included in the models in this section. If doors and windows are open, however, natural ventilation contributes to Q and should be considered when evaluating a model.

4.6.2 Infiltration and Exfiltration

Infiltration airflow, $Q_{INF,}$ is the uncontrolled flow of air through cracks or other unintentional openings in a building envelope. The same forces that influence natural ventilation influence infiltration. These include wind, temperature differences, and mechanically induced pressure differences. Because these forces can change with weather conditions, they are often difficult to characterize and control.

The infiltration/exfiltration flow rate Q_{INF} depends on the pressure difference inside and outside of a wall and on the area of openings within the wall. Estimates of the infiltration rate (and the natural ventilation rate) are difficult to make without experimental data on a given building. Some data and mathematical correlations exist for housing and other residences.[8] A rule of thumb for ordinary home construction (not airtight for energy efficiency) is an infiltration rate of 1 air change per hour. Reference data also exist for shopping malls, schools, and office buildings.[8]

In modeling contaminant concentrations in workplaces with forced flow ventilation, the infiltration rate is generally much smaller than the forced ventilation rate, Q_F. When mechanical ventilation is not functioning, however, the infiltration term may become very important.

4.6.3 Forced Ventilation

The forced ventilation component of incoming airflow, $Q_{F,}$ is relatively easy to control and measure, especially in a well-designed ventilation system. Q_F can be further divided into that portion that is recycled and returned to the room through a filter, Q_R, and that portion that is fresh makeup air, Q_{MU}. These separate flows are shown schematically in Figure 4.6. When there is no natural ventilation and the infiltration rate is equal to the exfiltration rate.

The forced ventilation air leaving the room can also be divided. When infiltration and exfiltration are equal, the exiting airflow is the sum of the air exhausted outdoors (Q_E) and the air that is recirculated. So:

$$Q_F = Q_{MU} + Q_R = Q_E + Q_R \quad (4\text{-}21)$$

Figure 4.7 also includes a filtration system for contaminant removal. Filters can be on make-up air F_{MU} and recirculated air F_R. Each filter has efficiency, symbolized by F, and defined as:

$$F = \frac{\text{Concentration out of filter}}{\text{Concentration into filter}} \quad (4\text{-}22)$$

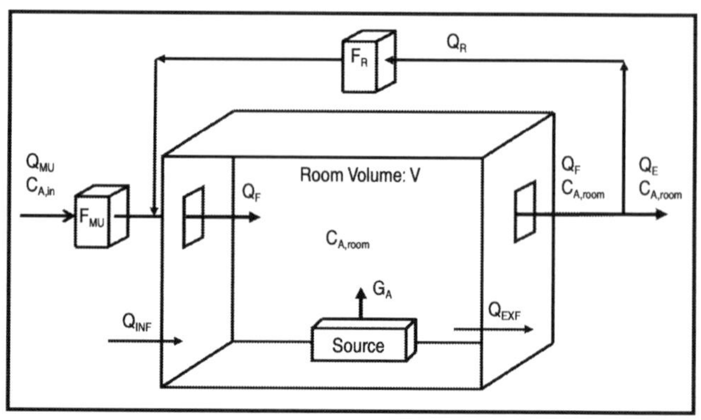

Figure 4.6 — Well-mixed model with recirculation and filtration.

4.7 WMB Model with Multiple Air Flows

A WMB model for the more realistic but more complicated representation of an indoor space has been developed, as shown in Figure 4.6. The analysis does not consider a separate natural ventilation term and is best applied to rooms where mechanical ventilation and infiltration airflows predominate. The analysis assumes that infiltration and exfiltration are equal. It also assumes that all Qs, Fs and G_A and $C_{A,in}$ are not changing with time.

A mass balance on the room in Figure 4.6 gives:

$$C_{A,room,t} = \frac{[Q_{MU} \bullet (1 - F_{MU}) + Q_{INF}] \bullet C_{A,in} + G_A}{Q_{MU} + Q_R \bullet F_R + Q_{INF}} \bullet \left[1 - e^{\frac{-(Q_{MU} + Q_R \bullet F_R + Q_{INF})}{V} \bullet (t - t_0)}\right] + C_{A,room,0} \bullet e^{\left[\frac{-(Q_{MU} + Q_R \bullet F_R + Q_{INF})}{V} \bullet (t - t_0)\right]} \quad (4\text{-}23)$$

Where

$C_{A,room,t}$: Concentration of A in room at time t (mg/m³)
$C_{A,room,0}$: Concentration of A in room at time 0 (mg/m³)
Q_{MU}: Volumetric flow rate of mechanical supplied fresh make up air (m³/min)
Q_{INF}: Volumetric flow rate of infiltrating fresh air (m³/min)
Q_R: Volumetric flow rate of recirculated air (m³/min)
G_A: Pollutant A generation rate (mg/min)
F_{MU}: Efficiency of filter on make up air (dimensionless)
F_R: Efficiency of filter on recirculated air (dimensionless)
t: Time of concern (min)
t_0: Start time (min)

At steady state this is equal to:

$$C_{A,room,t} = \frac{[Q_{MU} \bullet (1 - F_{MU}) + Q_{INF}] \bullet C_{A,in} + G_A}{Q_{MU} + Q_R \bullet F_R + Q_{INF}} \quad (4\text{-}24)$$

4.8 References

1. **Tichenor, B.A., L.A. Sparks, J.B. While, and M.D. Jackson:** Evaluating Sources of Indoor Air Pollution. *J. Air Waste Manag. Assoc. 40*:487–492 (1990).
2. **Heinsohn, R.J.:** General Ventilation Well-Mixed Mode (Chapter 5). In *Industrial Ventilation Engineering Principles*. New York: John Wiley & Sons, Inc., 1991.
3. **Sparks, L.A.:** Modeling Indoor Concentrations and Exposure. *Ann. N.Y. Acad. Sci. 641*:102–111 (1992).
4. **Jayjock, M.A.:** Estimating Airborne Exposure with Physical–Chemical Models. In *A Strategy for Assessing and Managing Occupational Exposures,* 2nd edition. Mulhausen, J. and J. Damiano (eds.). Fairfax, VA: American Industrial Hygiene Association, 1998.
5. **Drivas, P.J., P.G. Simmonds, and F.H. Shair:** Experimental Characterization of Ventilation Systems. *Environ. Sci. Technol. 6*:609–614 (1972).
6. **Perry, R.H.:** Mathematics (Section 2). In *Perry's Chemical Engineers' Handbook*. Green, D.W. and J.O. Maloney (eds.). New York: McGraw-Hill, 1997.
7. **American Society of Heating, Refrigerating and Air-Conditioning Engineers (ASHRAE):** Infiltration and Ventilation (Chapter 23). In *ASHRAE Handbook: Fundamentals*. Atlanta, GA: ASHRAE, 1997.
8. **Wadden, R.A. and P.A. Scheff:** Air Quality Models (Chapter 6). In *Indoor Air Pollution: Characterization, Prediction, and Control*. New York: John Wiley & Sons, Inc., 1983.

Well Mixed Rooms with Changing Conditions

By
Patricia H. Reinke
Michael Jayjock, PhD, CIH
Mark Nicas, PhD, CIH

5.0 Overview

The preceding chapter described modeling a well mixed room in which the conditions are relatively unchanging. This however is not always the case in situations where concentrations need to be modeled. This chapter will touch on some of the situations in which emission rates change.

5.1 Numerical Solution of the Differential Mass Balance

In many situations, the simplifying assumptions used to integrate Equation (4-8) in the previous chapter into the closed form solution are not accurate. G, Q, and/or $C_{A,in}$ may vary with time in such a way that makes integration of the differential mass balance very difficult or impossible. In such cases, one can calculate room air concentrations using the differential form of the well-mixed box (WMB) mass balance, and numerical integration methods for first order equations. One of these methods is the first order Runge–Kutta method described below.

The WMB mass balance can be put in the form:

$$\frac{dC_{A,room}}{dt} = C_{A,room} \bullet A(t) + B(t) \qquad (5\text{-}1)$$

A(t) and B(t) are the functions of time in the mass balance equation:

$$A(t) = \left[\frac{-(Q + K_{sink})}{V}\right] \qquad (5\text{-}2)$$

$$B(t) = \frac{Q_A}{V} \bullet C_{A,in} + \frac{G_A}{V} \qquad (5\text{-}3)$$

In the first order Runge–Kutta method, a small interval of time, Δt, is selected. Contaminant concentration is then calculated for each time increment. For example, concentration is calculated first at $t = t_0$, then at $t = t_0 + \Delta t$, then at $t = t_0 + 2\Delta t$, and continued until the end of the time period of interest. The Runge–Kutta equations used to calculate $C_{A,room}$ are the following:

At $t = t_0$, $C_{A,room} = C_{A,room,0}$

At $t = t_1 = t_0 + \Delta t$,

$$C_{A,room,1} + \frac{C_{A,room,0} \bullet \left[1 + \frac{\Delta t}{2} \bullet A(t_0)\right] + \Delta t \bullet B(t_0)}{\left[1 - \frac{\Delta t}{2} \bullet A(t_0)\right]} \qquad (5\text{-}4)$$

Then when $t = t_2 = t_1 + \Delta t$

$$C_{A,room,2} + \frac{C_{A,room,1} \bullet \left[1 + \frac{\Delta t}{2} \bullet A(t_1)\right] + \Delta t \bullet B(t_1)}{\left[1 - \frac{\Delta t}{2} \bullet A(t_1)\right]} \qquad (5\text{-}5)$$

The method continues through the time period of interest. More precise results are obtained with third and forth order Runge–Kutta methods, but for most industrial hygiene applications the first order method is adequate. Details on the more involved methods can be found in *The Chemical Engineer's Handbook.*(1)

Example 5.1

A new parts coater is to be installed in a well-mixed room in your facility. The room has a volume of 25.0 m^3 and an airflow rate (Q) of 1.0 m^3/min. All of the makeup air entering the room is outside air containing minimal contaminant. The manufacturer's specifications for the coater state that, at full speed and 21°C (70°F), the toluene emission rate is 27 g/hr. Your plant is not climate controlled, and you realize that many days the machine will run in temperatures significantly above 21°C. You contact the manufacturer for information on how the toluene generation rate, G_A, varies with temperature. The engineer who responds to your request gives you the following information:

$$G_A = 450 \bullet e^{\left[4660 \bullet \left[\frac{1}{294.3} - \frac{1}{T + 273.2}\right]\right]}$$

Where

T: operating temperature (°C)
G: toluene emission rate (mg/min)

You immediately recognize that this equation is based on the Clausius-Clapeyron equation and decide that the engineer has done his or her homework. To do your homework, you look up temperature logs for July and find that on the hottest days the temperature at the beginning of the day is about 23°C (≈73°F) and increases about 1°C/hour during the 8-hour shift. You develop the following equation to estimate room temperature as a function of time into the shift:

$$T = 23 + \frac{1}{hr} \bullet \left[\frac{t}{\frac{60\ min}{hr}}\right]$$

Where

t: time into the shift (minutes)

Use these two equations and the Runge-Kutta method to predict the time course of room concentration throughout a shift.

Solution:

Assumptions:

- No sinks
- No contaminant in the incoming air, $C_{A,in} = 0$.
- At the start of the day, there is no contaminant in the air, $C_{A,room,0} = 0$.

Calculations:

The calculations for this problem are very lengthy and involved, but with the aid of a spreadsheet the problem can be worked out easily. With the stated assumptions, the Runge-Kutta A(t) and B(t) can be written:

$$A(t) = \frac{-Q}{V}$$

$$B(t) = \frac{G_A}{V}$$

Since Q and V do not change with time in this problem, A(t) is constant

$$A(t) = \frac{-Q}{V} = \frac{1.0 \frac{m^3}{min}}{25.0\ m^3} = -\frac{-0.040}{min}$$

A Δt of 10 minutes was chosen

$$\left[1 + \frac{\Delta t}{2} \bullet A(t)\right] = \left[1 + \frac{10\ min}{2} \bullet \frac{-0.040}{min}\right] = 0.8$$

$$\left[1 - \frac{\Delta t}{2} \bullet A(t)\right] = \left[1 - \frac{10\ min}{2} \bullet \frac{-0.040}{min}\right] = 1.2$$

At t_0, $T_0 = 23°C$, $C_{A,\,room,0} = 0$ and:

$$G_{A,0} = 450 \bullet e^{\left[4660 \bullet \left[\frac{1}{294.3} - \frac{1}{23.0 + 273.2}\right]\right]}$$

$$= 498 \frac{mg}{min}$$

$$B_0 = \frac{G_A}{V} = \frac{498 \frac{mg}{min}}{25\ m^3} = 19.9 \frac{mg}{m^3\ min}$$

For the first time, $t_1 = 10$ min, the calculations are made as follows

$$T_1 = 23 + \frac{1}{hr} \bullet \frac{10\ min}{\frac{60\ min}{hr}} = 23.17°C$$

$$G_{A,1} = 450 \bullet e^{\left[4660 \bullet \left[\frac{1}{294.3} - \frac{1}{23.17 + 273.2}\right]\right]}$$

$$= 502 \frac{mg}{min}$$

$$B_1 = \frac{G_{A,1}}{V} = \frac{502 \frac{mg}{min}}{25\ m^3} = 20.1 \frac{mg}{m^3\ min}$$

So the concentration at 10 minutes can be calculated using:

$$C_{A,room,1} = \frac{C_{A,room,0} \bullet \left[1 + \frac{\Delta t}{2} \bullet A_0\right] + \Delta t \bullet B_0}{\left[1 - \frac{\Delta t}{2} \bullet A_0\right]}$$

The concentration at t_0 was assumed to be 0 so C at 10 minutes is:

$$C_{A,room,1} = \frac{0 \bullet 0.8 + 10\ min \bullet 19.9 \frac{mg}{m^3\ min}}{1.2} = 166 \frac{mg}{m^3}$$

At time t_2 = 20 min, the calculations are made as follows:

$$T_2 = 23 + \frac{1}{hr} \bullet \left[\frac{20\ min}{\frac{60\ min}{hr}}\right] = 23.33°C$$

$$G_{A,2} = 450 \bullet e^{\left[4660 \bullet \left[\frac{1}{294.3} - \frac{1}{23.33 + 273.2}\right]\right]}$$

$$= 507 \frac{mg}{min}$$

$$B_2 = \frac{G_{A,2}}{V} = \frac{507 \frac{mg}{min}}{25\ m^3} = 20.3 \frac{mg}{m^3\ min}$$

The concentration at 20 minutes can be calculated using:

$$C_{A,room,2} = \frac{C_{A,room,1} \bullet \left[1 + \frac{\Delta t}{2} \bullet A_1\right] + \Delta t \bullet B_1}{\left[1 - \frac{\Delta t}{2} \bullet A_1\right]}$$

$$C_{A,room,2} = \frac{166 \bullet 0.8 + 10\ min \bullet 20.1 \frac{mg}{m^3\ min}}{1.2} = 278 \frac{mg}{m^3}$$

These and additional calculations were made for each time increment using a personal computer spreadsheet. The results are shown in Figure 5.1 and in Table 5.1.

Table 5.1 – Results for Example 5.1

Time (minutes)	Concentration (mg/m³)	Time (minutes)	Concentration (mg/m³)
0	0	250	604
10	166	260	609
20	278	270	614
30	354	280	620
40	407	290	625
50	443	300	630
60	469	310	636
70	488	320	641
80	502	330	647
90	513	340	652
100	522	350	658
110	529	360	664
120	536	370	669
130	542	380	675
140	547	390	681
150	553	400	687
160	558	410	693
170	563	420	698
180	568	430	704
190	573	440	710
200	578	450	716
210	583	460	722
220	588	470	729
230	593	480	735
240	599		

5.2 Cyclic Processes

Many generation processes are not constant but cyclical, in which contaminant generation occurs for a period of time and is turned off for a period of time. Contaminant concentrations in rooms containing these types of processes can be estimated using the unsteady state WMB equations. When the generation rate is cyclical, the time period of interest is divided into smaller periods during which the generation rate is constant. The WMB model is then used to predict concentrations during each individual time period. The endpoint of the previous time period is used as the starting conditions for the next. This method is illustrated in Example 5.2.

Example 5.2

The schedule for a spray coater has been modified to adopt a must-in-time manufacturing approach. The machine now runs 40 minutes and then is turned off for 30 minutes throughout the day. From a mass balance on the process, you estimate a xylene (o,m,p mixture) generation rate of 40 mg/s while the machine is running. The room volume is approximately 40 m^3; the airflow rate, Q, is 0.075 m^3/s; and there is no xylene in the air at the start of the day. Xylene in the incoming air steam is also assumed to be 0 ppm. Predict the room concentration history from startup until break time, 3 hours later.

Solution:

Assumptions:

- The room is well-mixed.
- Incoming air contains no xylene, $C_{A,in} = 0$.

Calculations:

Equation 4-11 is the simplest form of the WMB model that can be used for this problem. The time from startup until break time is divided into time increments with constant generation rate, GA. Equation 4-11 is simplified for each time period, and concentration versus time data are calculated and plotted.

Time period 1: 0 minutes to 40 minutes

$C_{A,room,0}$: 0 mg/m^3
G_A : 40 mg/sec = 2400 mg/min
Q: 0.075 m^3/sec = 4.5 m^3/min

$$C_{A,room,t} = \left[C_{A,room,0} - \frac{G_A}{Q} \right] \bullet e^{\frac{-Q \bullet (t-t_0)}{V}} + \frac{G_A}{Q}$$

Using the constant terms in the equation

$$C_{A,room,t} = \left[0 - \frac{2400 \frac{mg}{min}}{4.5 \frac{m^3}{min}} \right] \bullet e^{\frac{-4.5 \frac{m^3}{min} \bullet (t-t_0)}{40\ m^3}}$$

$$+ \frac{2400 \frac{mg}{min}}{4.5 \frac{m^3}{min}}$$

$$C_{A,room,t} = -533 \frac{mg}{m^3} \bullet e^{\frac{-0.1125 \bullet (t-t_0)}{min}} + 533 \frac{mg}{m^3}$$

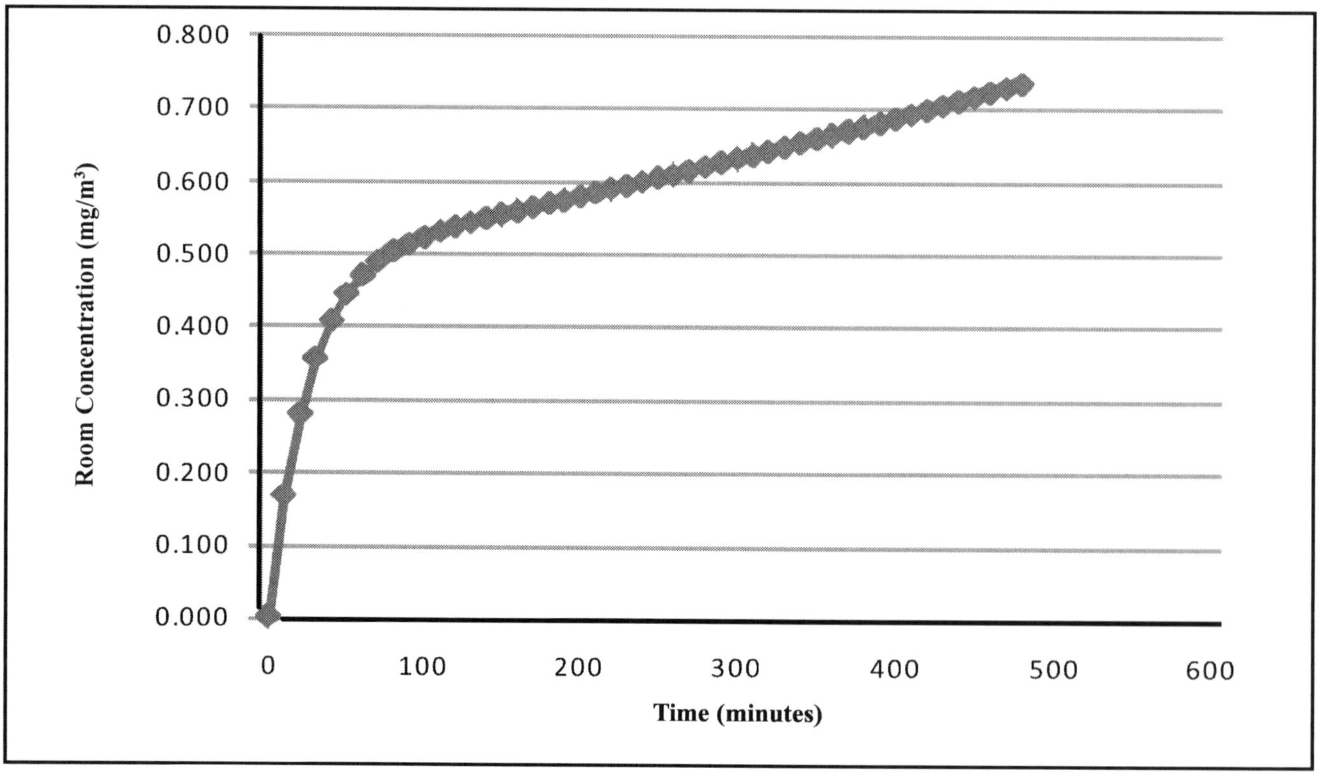

Figure 5.1 — Concentrations resulting from Example 5.1.

t is in minutes.
The proceeding equation is used to find $C_{A,room}$ at various times during the first 40 minutes.

Time (min)	Room Concentration (mg/m³)
0	0
5	229
10	359
15	434
20	477
25	501
30	515
35	523
40	527

Time Period 2: 40 minutes to 70 minutes

Process is off and concentration decays

$C_{A,room,0}$: 527 mg/m³
G_A : 0 mg/min
Q: 0.075 m³/sec = 4.5 m³/min

$$C_{A,room,t} = \left[C_{A,room,0} - \frac{G_A}{Q}\right] \bullet e^{\frac{-Q \bullet (t-t_0)}{V}} + \frac{G_A}{Q}$$

Using the constant terms in the equation

$$C_{A,room,t} = \left[527\frac{mg}{m^3} - 0\right] \bullet e^{\frac{-4.5\frac{m^3}{min} \bullet (t-40\ min)}{40\ m^3}} + 0$$

$$C_{A,room,t} = 527\frac{mg}{m^3} \bullet e^{\frac{-0.1125 \bullet (t-40\ min)}{min}}$$

Time (min)	Room Concentration (mg/m³)
40	527
45	300
50	171
55	97
60	56
65	32
70	18

Time Period 3: 70 minutes to 110 minutes

Process is on and concentration builds up, but from an initial concentration this time.

$C_{A,room,0}$: 18 mg/m³
G_A : 2400 mg/min
Q: 0.075 m³/sec = 4.5 m³/min

$$C_{A,room,t} = \left[18\frac{mg}{m^3} - \frac{2400\frac{mg}{min}}{4.5\frac{m^3}{min}}\right] \bullet e^{\frac{-4.5\frac{m^3}{min} \bullet (t-t_0)}{40\ m^3}} + \frac{2400\frac{mg}{min}}{4.5\frac{m^3}{min}}$$

$$C_{A,room,t} = -513\frac{mg}{m^3} \bullet e^{\frac{-0.1125 \bullet (t-70\ min)}{min}} + 533\frac{mg}{m^3}$$

Time (min)	Room Concentration (mg/m³)
70	18
75	239
80	366
85	438
90	479
95	503
100	515
105	523
110	527

These procedures can be repeated for each interval, producing the concentration pattern illustrated in Figure 5.2. The sawtooth pattern is typical of cyclical processes.

5.3 Backpressure

This model deals exclusively with evaporating sources with relatively large surface areas. The rate of evaporation from any source (G) is driven by the gradient in concentration between the source and the receiving volume. Just as water evaporates fastest on a day with low relative humidity, the evaporation rate of any compound is maximized when there is little or no partial pressure of that compound in the air above it. As the amount of compound in the air increases toward its saturation level (C_{sat}), G, decreases until it is zero when the air is saturated. Thus G is *inversely* proportional to the airborne concentration in the room over the source. We call this effect "back pressure."(2) This factor is ignored in the discussion on box models; however, modeling large evaporating sources indoors requires the inclusion of back pressure.
A two-film theory of volatilization has been developed.(3) From this a model for indoor evaporating sources was developed.(4)

$$G = 1000 \bullet \frac{K_t \bullet MW \bullet A \bullet (P_V - P_{VB})}{R \bullet T_L} \quad (5\text{-}6)$$

Where

G: generation rate (mg/min)
K_t: mass transfer rate (m/min)
MW: molecular weight (g/mole)
A: area of evaporating pool (m^2)
P_V: saturation vapor pressure (atm)
P_{VB}: time-dependent actual vapor pressure (atm) i.e. backpressure
R: gas constant (8.205 x 10^{-5} atm m^3/(mole °K))
TL: temperature of the volatilizing liquid (°K)

For any particular volatilizing species at constant temperature, the constant terms can be collected into the following simplified relationships:

$$G = K \bullet (P_V - P_{VB}) \quad (5\text{-}7)$$

$$K = 1000 \bullet \left[\frac{K_t \bullet MW \bullet A}{R \bullet T_L}\right], \text{ a constant} \quad (5\text{-}8)$$

Determining the mass transfer coefficient (K_t) for the evaporating liquid is not easy. A simple approach is to use the following estimation developed by the U.S. Environmental Protection Agency[5]:

$$K_T = 0.5 \bullet \left[\frac{18.0}{MW}\right]^{1/3} \quad (5\text{-}9)$$

It is the authors' experience that this may result in the overestimation of the evaporation rate, but other approaches are available.[4,6]

Thus, given a relatively large and constant evaporating surface, we have the following sequence of events. When evaporation begins (time = 0), the airborne concentration C equals zero. At this point the evaporation rate (*i.e.*, the generation rate) is at its *maximum* because there is no concentration in the air and the concentration gradient is at maximum. After any time elapses, there is contaminant in the air. Which results in some finite *reduction* of the gradient (*i.e.*, some back pressure is now acting to slow down the evaporation rate, G). Ultimately, the system will reach steady-state at an evaporation rate that is lower than the maximum rate that occurred at time zero. In situations where the steady-state concentration is close to the saturation concentration (*i.e.*, large surface areas of evaporation in rooms with poor ventilation), the evaporation rate will be lowered to near zero.

The ideal gas law can be used to convert airborne concentration in units of mg/m^3 to partial pressure (i.e., back pressure) using the following relationship:

$$P_{V,B} = \frac{C \bullet T_{air} \bullet 8.2 \times 10^{-8}}{MW} \quad (5\text{-}10)$$

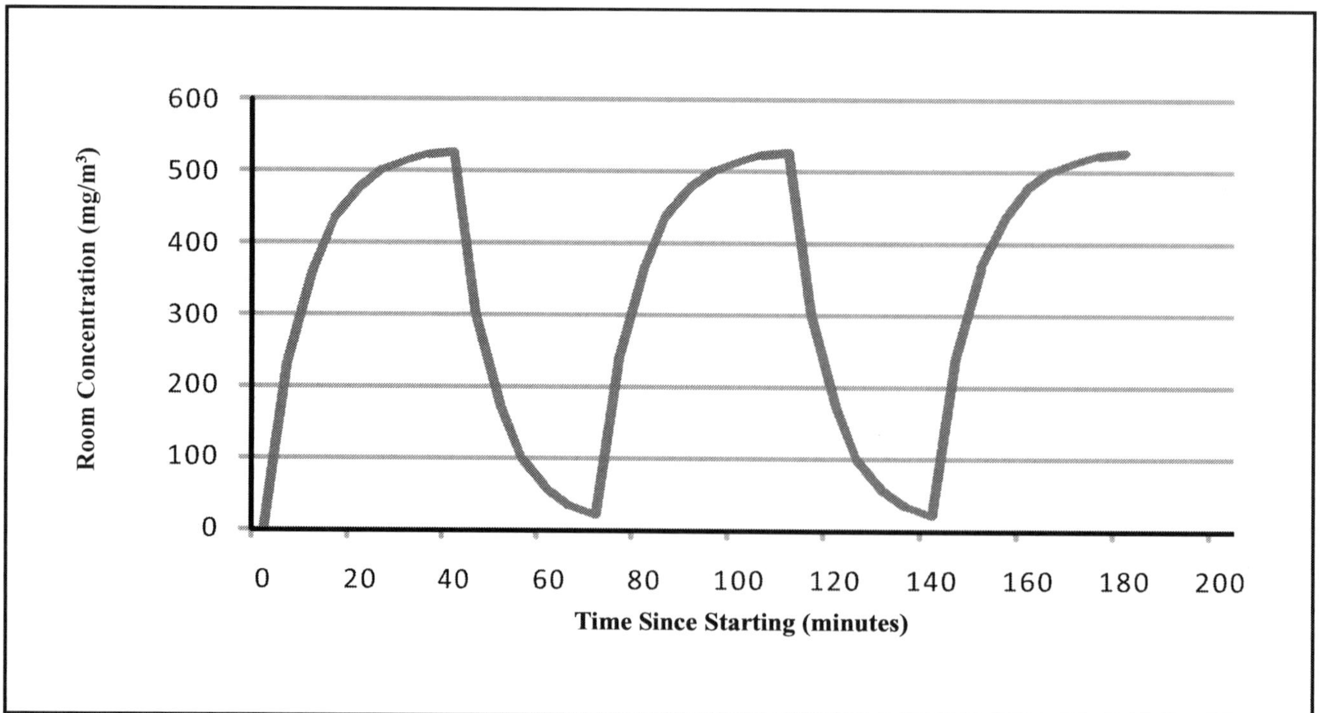

Figure 5.2 — Concentrations resulting from Example 5.2.

Where

C: airborne concentration (mg/m^3)
T_{air}: air temperature (°K)
MW: g/mole

The constant, 8.2×10^{-8}, combines the universal gas constant (R) and units of conversion.

For our purposes, we set $T_{air} = T_L$, as the temperature of the volatilizing liquid (°K).

Our approach to modeling this scenario is to use a personal computer to do stepwise solutions as follows:

Step 1. At t = 0 minutes, set concentration (C) = C_0 = 0 and P_{VB} = 0

Step 2. Calculate the airborne concentration (C) using Equation 10 from Chapter 4 for the time interval from t = 0 to t = 1 minute (i.e., t = 1 minute).

Step 3. Calculate the back pressure (P_{VB}) using Equation 5–10 above.

Step 4. Calculate G using Equation (5-1) above.

Step 5. Calculate the airborne concentration (C) using Equation 4-9 from Chapter 4 by setting C_0 equal to the C that was calculated in Step 2 and the slightly reduced G calculated in Step 4. Again, the interval is 1 minute (i.e., the period for the second interval is between t = 1 and 2 minutes).

Repeat Steps 3 through 5 for subsequent 1-minute periods (i.e., t = 2 to 3 minutes, t = 3 to 4 minutes, … t = n to n + 1 minutes) until the airborne concentration does not change (i.e., steady-state).

Note that a standard form mathematical solution for airborne concentration at any point in time for this specific model has been determined.[1]

$$C = \frac{1000 \bullet K_t \bullet MW \bullet A \bullet P_V}{R \bullet T_L \bullet (K_t \bullet A + Q)} \bullet \left[1 - e^{\frac{-(K_t \bullet A + Q) \bullet t}{V}} \right] \qquad (5\text{-}11)$$

The agreement between the stepwise model and the exact solution is remarkable for fine-time increments of one minute or less. An advantage to using the stepwise approach is its versatility in allowing the simple substitution and use of other expressions for the generation rate, which may be experimentally derived. One also might expand the computer code to account for a time-dependent vapor pressure (VP) — from a time-dependent change in concentration for a binary solution to an increasing (growing puddle) or decreasing (shrinking) evaporating source — without resorting to calculus.

Example 5.3 — Effect of Backpressure Adjustments

A ruptured vessel results in a 100 m^2 spill of 2-butoxyethanol (MW: 118, P_V: 0.6 mmHg). The room is 20 m x 30 m x 3 m (1800 m^3) with an airflow rate of 15 m^3/min (0.5 air changes per hour). Room temperature is 25°C.

Compare the resulting concentrations 90 minutes after the spill, taking into account whether back pressure is or is not considered.

First:

P_V in ATM = 0.6 mmHg/760 mmHg = 0.00079 ATM

Using the EPA method (Eq 5-9).

$K_t = 0.5 \cdot (18.0/118)^{1/3} = 0.267$ m/min

If back pressure is not considered:

$$G = 1000 \bullet \frac{K_t \bullet MW \bullet A \bullet P_V}{R \bullet T_L}$$

$$G = \frac{1000\ \text{mg}}{\text{g}} \bullet \frac{0.267\ \text{m}}{\text{min}} \bullet \frac{118\ \text{g}}{\text{mole}} \bullet 100\ \text{m}^2 \bullet$$

$$0.00079\ \text{atm} \Big/ \left[8.205 \times 10^{-5} \frac{\text{m}^3 \bullet \text{ATM}}{\text{mole} \bullet \text{K}} \bullet 298\text{K} \right]$$

$$G = \frac{102{,}000\ \text{mg}}{\text{min}}$$

If the initial C is zero and incoming air has no pollutant in it, the appropriate well mixed box model (Chapter 4, Equation (4-12) is

$$C_{A,t} = \frac{G_A}{Q} \bullet 1 - e^{\frac{-Q \bullet (t-t_0)}{V}}$$

$$C_{90\ \text{min}} = \frac{102{,}000\ \text{mg}}{\text{min}} \bullet \frac{\text{min}}{15.0\ \text{m}^3}$$

$$\left[1 - e^{\frac{-15.0\ \text{m}^3}{\text{min}} \bullet \frac{90\ \text{min}}{1800\ \text{m}^3}} \right]$$

$$C_{90\,min} = \frac{3590\ mg}{m^3}$$

If back pressure is considered, use Equation 5-10 above.

Using appropriate units

$$C = \frac{1000 \bullet 0.267 \bullet 118 \bullet 100 \bullet 0.00079}{8.25 \times 10^{-5} \bullet 298 \bullet (0.267 \bullet 100 + 15)}$$

$$\bullet \left[1 - e^{\frac{-(0.267\ \bullet\ 100\ +\ 15)\ \bullet\ 90}{1800}}\right]$$

$$C = 2140\ \frac{mg}{m^3}$$

Figure 5.3 illustrates the concentration buildup in this case with (solid) and without (dashed) consideration of back pressure.

5.4 Exponentially Decreasing Emission Rate

In the section above, a well-mixed room (WMR) dispersion construct was used alternatively with a constant contaminant emission rate and with a decreasing net contaminant emission rate due to "backpressure." For this discussion, a constant mass emission rate (mg/min) can be written as:

$$G(t) = G_0 \qquad (5\text{-}12)$$

where G_0 is the initial instantaneous emission rate. Equation 5-12 simply means that the emission rate at some future time t equals the initial rate. However, common scenarios involve contaminant emission at a rate that decreases over time. For example, the mass rate of emission due to a small spill of an organic solvent, or due to offgassing of a sorbed contaminant, will tend to be highest at the outset and decrease thereafter. The physical reason for a decreasing rate in the small solvent spill scenario is easy to appreciate — evaporation cools the remaining solvent (decreasing its tendency to evaporate) and also decreases the surface area from which further evaporation can occur. If new contaminant mass is not added to the spill (source), a simple model applied to this circumstance is an exponentially decreasing emission rate described by:

$$G(t) = G_0 \times \exp(-\alpha\, t) \qquad (5\text{-}12a)$$

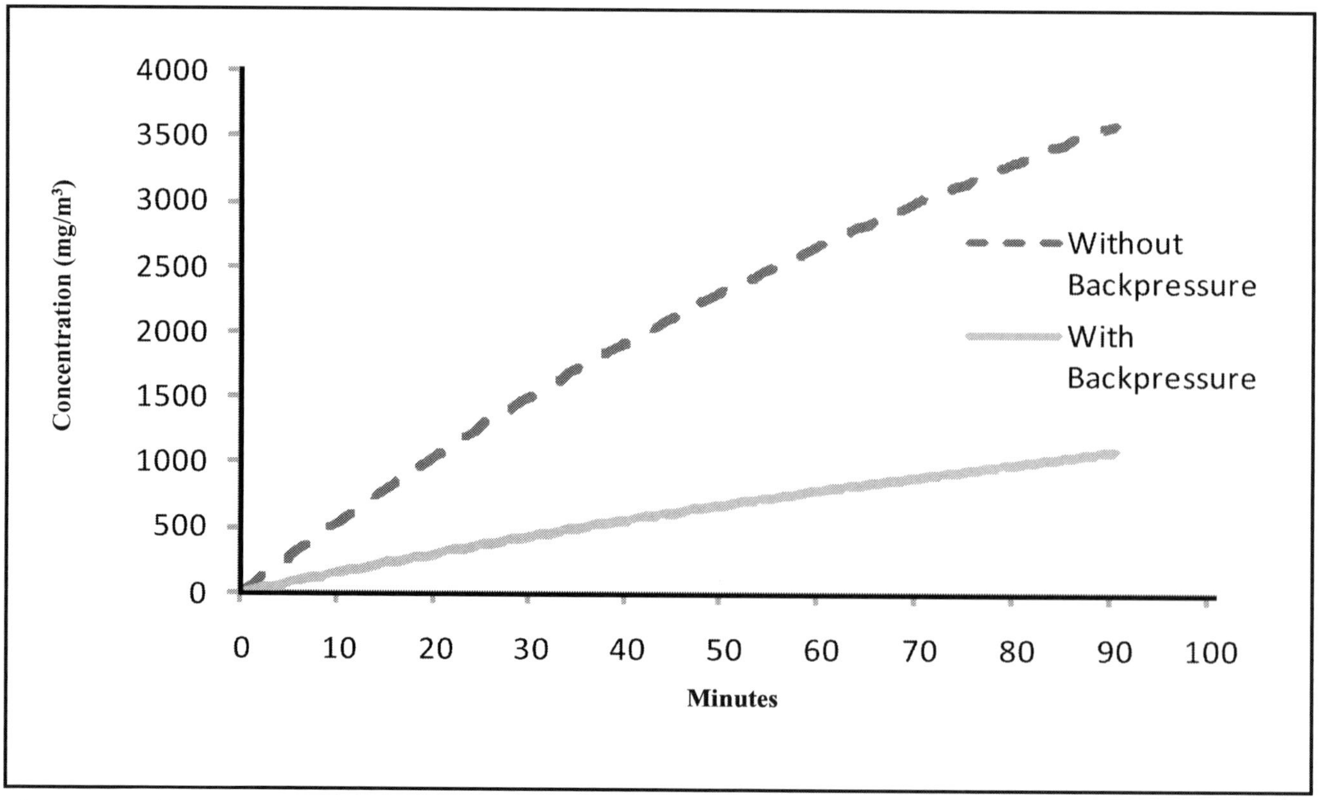

Figure 5.3 — Concentrations with and without backpressure for Example 5.3.

where G_0 is the initial mass emission rate, and α is the first-order emission rate constant (min^{-1}). If emission is from a liquid spill, α can be termed the first-order evaporation rate constant. If emission is due to offgassing of sorbed contaminant, α can be termed the first-order offgassing rate constant. If the initial contaminant mass (mg) in the source is denoted M_0, it can be shown that $G_0 = \alpha \times M_0$, in which case Equation 5-12a becomes:

$$G(t) = \alpha \times M_0 \times \exp(-\alpha t) \tag{5-12b}$$

In theory, the contaminant mass emission rate in Equation 5-12b never reaches zero, which is unrealistic in general. However, if 100% of a small solvent spill is known to evaporate in 15 minutes, but Equation 5-12b predicts that 95% evaporates, the degree of model error can be considered negligible.

If Equation 5-12b describes the contaminant mass emission rate in a WMR, and if the only mechanism of contaminant loss from room air is via the exhaust airflow, the following mass balance equation pertains:

$$V\,dC = \alpha\, M_0 \exp(-\alpha t)\,dt - Q\,C\,dt \tag{5-13}$$

C = contaminant concentration (mg/m^3)
M_0 = initial contaminant mass in the source (mg)
α = emission rate constant (min^{-1})
Q = air supply rate (m^3/min)
V = room volume (m^3)
dt = infinitesimal time interval (min)

The left-hand side of Equation 5-13 is the change in contaminant mass in room air in the interval dt. The mass gain term on the right-hand side is the mass emission rate function G(t) in Equation 5-12b multiplied by dt. The mass loss term involves mass leaving the room via the exhaust air flow, such that mass loss is the product of C, Q and dt.

If the initial chemical concentration in room air is zero, the general equation for the time-varying contaminant concentration in room air in mg/m^3 is the following:

$$C(t) = \frac{\alpha\, M_0}{\alpha V - Q} \times \left[\exp\left(-\frac{Q}{V}\,t\right) - \exp(-\alpha t)\right] \tag{5-14}$$

Several points are noted about this model equation.

- As t gets large, the terms $\exp(-(Q/V) \times t)$ and $\exp(-\alpha \times t)$ go to zero, in which case the right-hand side goes to zero. This signifies that no steady-state concentration (other than zero) is attained.
- There are no general published tables for α. We can expect that α will vary with temperature. Where offgassing of sorbed contaminant is being modeled, the α value may vary with the sorbing matrix. Where evaporation of a pure hydrocarbon solvent from a small spill is involved, an algorithm for estimating α will be offered.
- If the product $\alpha \times V$ were to equal Q exactly, the model equation would take a different form, because zero cannot be a denominator.
- The effect of backpressure is ignored. As a rule of thumb, Equation 5-14 can be used (i.e., backpressure can be ignored) when the worst-case solvent vapor concentration in the room (equal to $M_0 \div V$) is < 10% of the solvent's saturation vapor concentration. In other words, if all the solvent were to instantaneously evaporate, this worst-case concentration would be < 10% of C_{sat}.

According to Equation 5-14, the contaminant concentration will reach a maximum value at time t_{peak} given by the expression:

$$t_{peak} = \frac{V \times \ln \dfrac{\alpha V}{Q}}{\alpha V - Q} \tag{5-15}$$

Example 5.4

Consider the following hypothetical scenario. A spill of 50 mL acetone occurs in a room with volume $V = 100\ m^3$ which has a supply/exhaust air rate of $Q = 10\ m^3/min$ (corresponding to 6 ACH and $Q/V = 0.10\ min^{-1}$). Acetone has a liquid density of 0.778 g/mL, an evaporation rate constant $\alpha = 0.15\ min^{-1}$, and a saturation vapor pressure of 180 mm Hg (at 20°C). The initial acetone liquid mass is M_0 = (50 mL) × (778 mg/mL) = 38,900 mg.

Must back-pressure be considered? The worst-case acetone concentration that could develop would be if all the mass evaporated into the 100 m^3 room: (38,900 mg) ÷ (100 m^3) = 389 mg/m^3, or 164 ppm at NTP. In contrast, acetone's saturation vapor concentration is calculated from its vapor pressure: [(180 mm Hg) ÷ (760 mm Hg)] × 10^6 = 237,000 ppm. Because the worst-case concentration due to the 50 mL acetone spill is only 0.07% of the saturation vapor concentration, a backpressure effect can be ignored.

Compute the time-varying contaminant concentration in room air in mg/m^3. Given the specified parameter values, the acetone concentration function C(t) in mg/m^3 based on Equation 5-14 is as follows:

$$C(t),\ mg/m^3 = 1{,}167 \times [\exp(-0.10\,t) - \exp(-0.15\,t)]$$

Figure 5.4 depicts C(t) computed from the above equation over the interval $0 \le t \le 60$ min. The concentration increases to a maximum value around 8 minutes after the spill and decreases thereafter. The maximum concentration value is found by first computing t_{peak} based on Equation 5-15:

$$t_{peak} = \frac{100 \times \ln\left[\dfrac{0.15 \times 100}{10}\right]}{0.15 \times 100 - 10} = 8.11 \text{ min}$$

Next, the value of t = 8.11 min is inserted into the C(t) equation by which we find that C_{peak} = 73 ppm:

$$C(t_{peak}) = 1{,}167 \times [\exp(-0.10 \times 8.11) - \exp(-0.15 \times 8.11)] = 173 \text{ mg/m}^3 = 73 \text{ ppm}$$

Example 5.5: Fumigant Off-Gassing

Methyl bromide gas is still used for fumigating certain commodities such as tree fruits and cherries. During fumigation, the commodity sorbs the methyl bromide, which subsequently off-gasses with a first-order emission rate constant. Although a fumigated commodity must be aerated following fumigation and prior to handling by workers, the commodity can still retain an appreciable amount of methyl bromide which continues to off-gas. Therefore, workers handling the fumigated commodity can have inhalation exposure to methyl bromide gas.

Consider a hypothetical scenario in which 4×10^3 kg of fumigated cherries (about 5 m^3) are placed in a cold storage room with volume V = 100 m^3 and Q = 0.42 m^3/min (corresponding to 0.25 ACH). The cherries in the storage room initially contain 40 mg methyl bromide per kg (or 40 ppm weight by weight). The methyl bromide off-gassing rate constant for cherries at the cold storage temperature is .0083 min^{-1}. We want to know the time course of the methyl bromide concentration in storage room air.

The initial methyl bromide mass in the cherries is M_0 = 160,000 mg, or:

$$M_0 = (4 \times 10^3 \text{ kg}) \times (10^6 \text{ mg/kg}) \times (40 \times 10^{-6}) = 160{,}000 \text{ mg}$$

By inserting the values for M_0, α, Q and V into Equation 14, we obtain:

$$C(t), \text{mg/m}^3 = 3239 \times [\exp(-0.0042\, t) - \exp(-0.0083\, t)]$$

Figure 5.6 depicts C(t) over the interval $0 \leq t \leq 480$ min. The concentration increases to a maximum value around 2½ hours after the cherries are placed in the room, and decreases thereafter. The maximum concentration value is found by computing t_{peak} based on Equation 5-15:

$$t_{peak} = \frac{100 \times \ln\left[\dfrac{0.0083 \times 100}{0.42}\right]}{0.0083 \times 100 - 0.42} = 166 \text{ min}$$

This time value is inserted into the C(t) equation by which we find that C_{peak} = 205 ppm:

$$C(t_{peak}) = 3239 \times [\exp(-0.0042 \times 166) - \exp(-0.0083 \times 166)] = 796 \text{ mg/m}^3 = 205 \text{ ppm}$$

Federal OSHA has an instantaneous ceiling limit for methyl bromide of 20 ppm (Table Z1, 29 CFR 1910.1000). The peak exposure concentration of 205 ppm exceeds this by a factor of 10. The ACGIH® 8-hr TLV®-TWA for methyl bromide is 1 ppm. The 8-hr TWA concentration for the exposure time series in Figure 5.5 is 153 ppm, or 153-fold greater than the TLV®. This indicates that off-gassing poses a serious health risk to individuals in the cold storage space. The use of modeling to estimate methyl bromide concentrations in analogous fumigated commodity scenarios has been published.(7)

Similarly computations can be applied to estimating off-gassing from medical equipment sterilization. Non-published data suggest that at room temperature, the sterilant ethylene oxide offgasses from fumigated packages of medical equipment at a rate on the order of 0.00083 min^{-1}. This techniques illustrates that off-gassing is of concern in a distribution warehouse which can temporarily store more than a hundred pallets of fumigated medical equipment.

5.4.1 The Evaporation Rate Constant for a Small Solvent Spill

The value of α for a set of pure solvents was determined by timing the mass loss from 3 mL of liquid solvent placed in a watchglass on an electronic balance. For each solvent, the logarithm of the fraction of liquid mass remaining versus time (on an arithmetic scale) was an approximately linear relationship for the loss of the initial 75% of mass. A simple regression estimate of the absolute value of the slope of this line was set equal to α. Note that this is the same technique used to estimate an air exchange rate by timing the exponential decrease in a tracer gas concentration in a well-mixed room.

Table 5.2 shows the α values based on three replicate determinations for each of thirteen solvents. For the hydrocarbon solvents, α has a strong linear correlation with the solvent saturation vapor pressure P_V (mm Hg) at

20°C, while the α values for methylene chloride, chloroform, and carbon disulfide deviate substantially from this relationship. The value of α also depends on the initial size of the spill, or more precisely, on the initial surface area to volume ratio of the spill, SA/VOL in cm^{-1}. It was found that α linearly increases as SA/VOL increases. These two linear relationships are combined into the following prediction algorithm for hydrocarbon solvents.[8]

$$\alpha,\ min^{-1} = .000524\ P_V + .0108\ SA/VOL \qquad (5\text{-}16)$$

where

P_V = saturation vapor pressure in mm Hg at 20°C
SA/VOL = initial surface-area-to-volume ratio of the spill, cm^{-1}

Table 5.2 Estimated Evaporation Rate Constants for 3 mL Spills of Pure Solvent

Chemical	P_V (mm Hg at 20°C)	α (min^{-1})
ethyl ether	440	0.37
n-pentane	420	0.39
acetone	180	0.16
n-hexane	124	0.14
methanol	96	0.055
methyl ethyl ketone	78	0.055
2-propanol	33	0.028
toluene	21	0.025
n-butyl acetate	10	0.011
1-butanol	6	0.0046
methylene chloride	350	0.14
carbon disulfide	297	0.11
chloroform	160	0.082

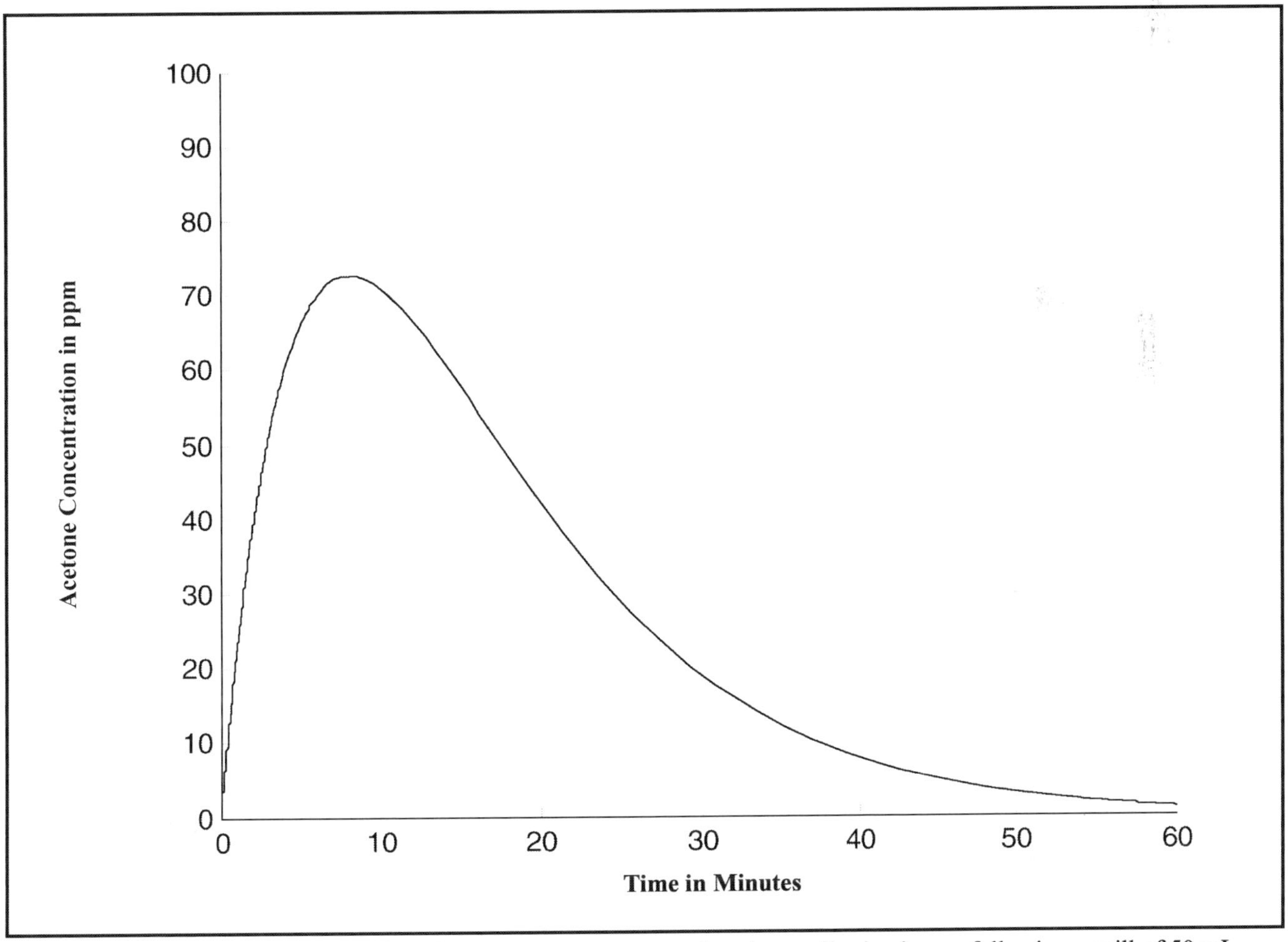

Figure 5.4 — The predicted time series of acetone vapor concentrations in a well-mixed room following a spill of 50 mL acetone. The room has volume V = 100 m^3 and a supply/exhaust air rate Q = 10 m^3/min. The temperature is 20°C, and the acetone evaporation rate constant is a = 0.15 min^{-1}. The peak concentration of 73 ppm occurs at 8.1 minutes following the spill.

5.4.2 Model Validation

A pertinent question is — how well does the model estimate exposure intensity? To the author's knowledge, there is only one published study that presents time series concentration measurements plus the background information necessary to make corresponding concentration estimations. In this study 15 mL n-pentane was transferred to a watch glass. Concentration measurements were made and compared to model estimates. Given a liquid density of 630 g/mL, the n-pentane "spill" had initial mass M_0 = (630 mg/mL) × (15 mL) = 9,450 mg. The initial spill surface area was measured to be 78.5 cm². Therefore, the initial surface-area-to-volume ratio was 5.2 cm^{-1}:

$$SA/VOL = \frac{78.5\ cm^2}{15\ cm^3} = 5.2\ cm^{-1}$$

For n-pentane at 20°C, P_V = 420 mm Hg. Thus, based on Equation (5-16):

$$\alpha = 0.000524\ (420\ mm\ Hg) + 0.0108\ (5.2\ cm^{-1}) = 0.28\ min^{-1}$$

This n-pentane spill occurred in a small laboratory with volume V = 22.7 m³ and Q = 18.6 m³/min. The room air was purposely mixed by operating several small fans positioned away from the watch glass to avoid promoting solvent evaporation. By inserting the values for M_0, α, Q and V into Equation 5-14, we obtain:

$$C(t),\ mg/m^3 = -216 \times [\exp(-0.819\ t) - \exp(-0.28\ t)]$$

The negative value for $\frac{\alpha M_0}{\alpha V - Q}$ at the front of the equation is not a problem, because the difference in the brackets is always negative for $t > 0$. The solid-line curve in Figure 5.6 depicts the modeled C(t) over the interval $0 \le t \le 15$ min. The model predicts that the concentration attains a peak value of 81 mg/m³ at 2 minutes following the "spill" and decreases thereafter. The open circles in Figure 5.6 show grab sample concentration measurements of n-pentane made at the room exhaust air outlet. The shape and concentration of this model's estimates fit the peak the measured concentrations reasonably well.

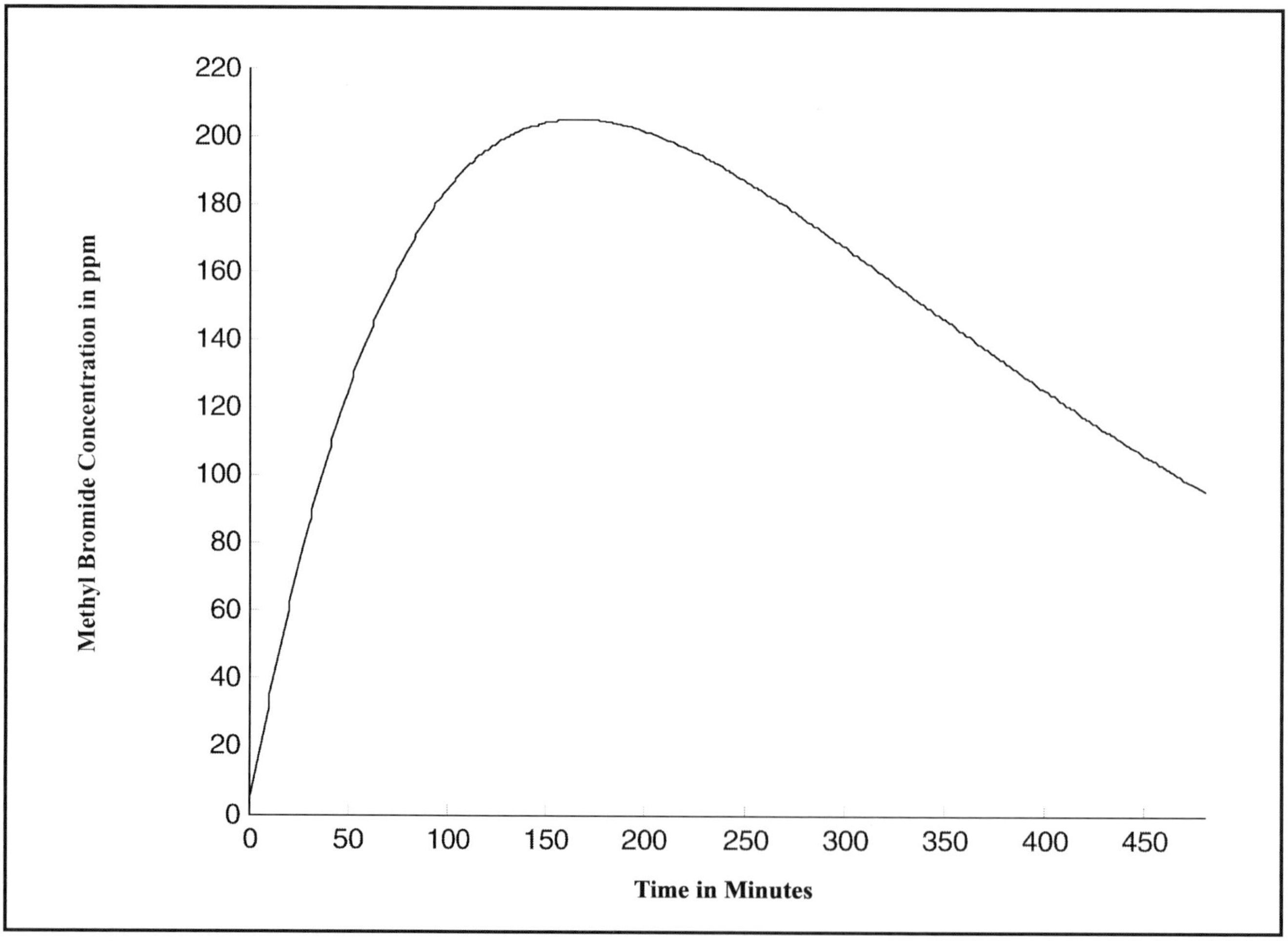

Figure 5.5 — The predicted time series of methyl bromide concentrations in a well-mixed cold storage room due to off-gassing from fumigated cherries. The initial methyl bromide mass is M_0 = 160,000 mg, and the off-gassing rate constant is α = 0.0083 min^{-1}. The room has volume V = 100 m³ and a supply/exhaust air rate Q = 0.42 m³/min. The peak concentration of 205 ppm occurs at 166 minutes following placement of the cherries spill in the storage room.

5.5 References

1. **Tuggle, M.:** "A Standard Mathematical Form Solution for Back Pressure Model." October 26, 1990. [Letter to M.A. Jayjock.]
2. **Jayjock, M.A.:** Back Pressure Modeling of Indoor Air from Evaporating Sources. *Am. Ind. Hyg. Assoc. J. 55*:230–235 (1994).
3. **Liss, P.S. and P.G. Slater:** Flux of Gases across the Air-Sea Interface. *Nature 247*:181–184 (1974).
4. **Thibodeaux, L.J.:** Exchange Rates between Air and Water (Chapter 4). In *Chemodynamics*. New York: Wiley Interscience, 1979.
5. **Braun, K.O. and K.J. Caplan:** "An Evaporation Rate of Volatile Liquids: (NTIS PB No. 92B232305). PACE Laboratories Inc. Project 890501.215/EPA744BRB92B001. Study conducted for the U.S. Environmental Protection Agency, 1989.
6. **Wu, J.M. and J.M. Schroy:** Emissions from Spills. In *Proceedings on Control of Specific Toxic Pollutants*. Pittsburgh, Pa.: Air Pollution Control Association, 1979.
7. **Nicas, M:** Estimating Methyl Bromide Exposure due to Offgassing from Fumigated Commodities. *Appl. Occup. Environ. Hyg. 18*:200–210 (2003).
8. **Keil, C. and M. Nicas:** Predicting Room Vapor Concentrations due to Spills of Organic Solvents. *Am. Ind. Hyg. Assoc. J. 64*:445–454 (2003).

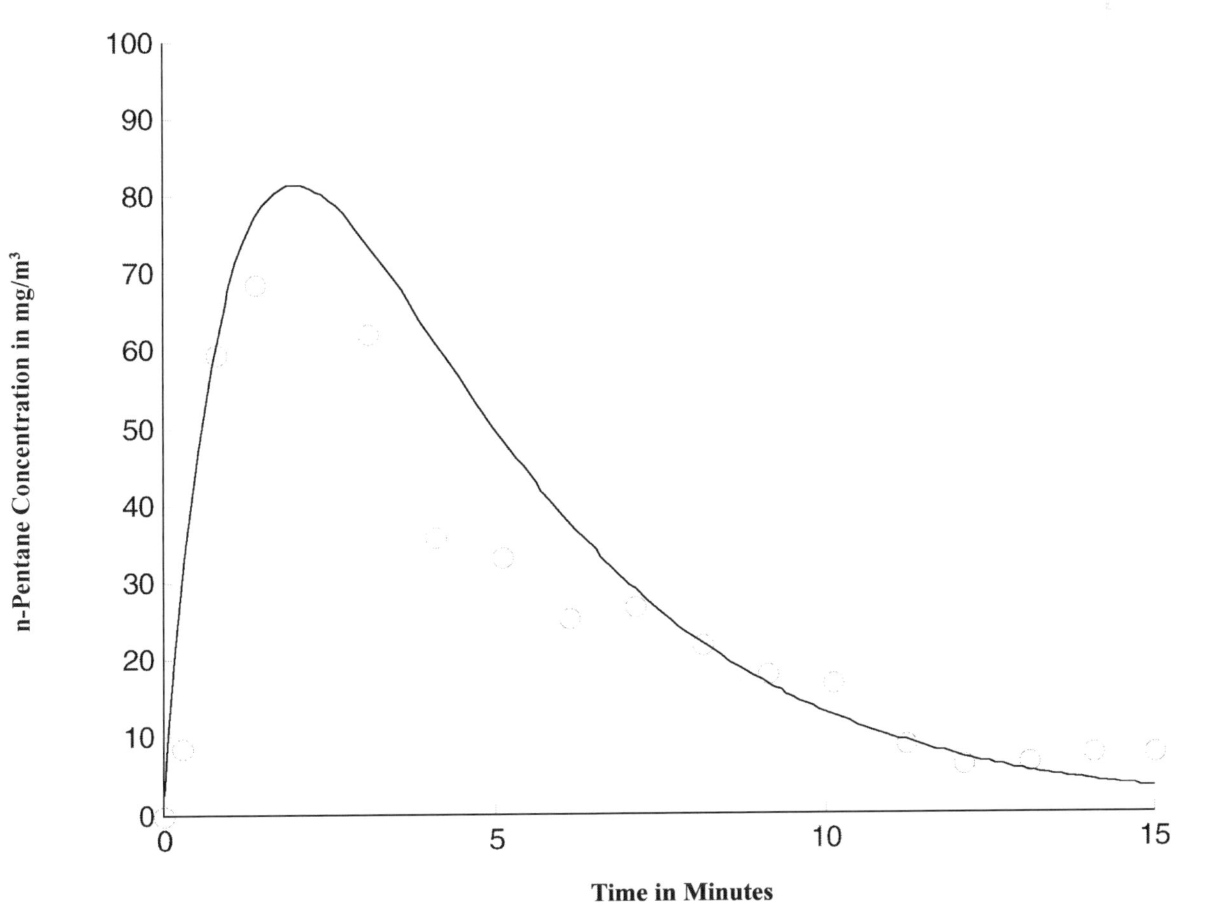

Figure 5.6 — The predicted time series of n-pentane vapor concentrations versus measured n-pentane vapor concentrations in a well-mixed room following the transfer of 15 mL n-pentane into a watch glass. The room had volume V = 22.7 m^3 and a supply/exhaust air rate Q = 18.6 m^3/min. The n-pentane evaporation rate constant is $\alpha = 0.28$ min^{-1}. The predicted peak concentration of 81 mg/m^3 occurs at two minutes following the transfer. The time profile of the measured concentrations matches the predicted concentrations reasonably well.

The Near Field/Far Field (Two-Box) Model with a Constant Contaminant Emission Rate

By
Mark Nicas, PhD, CIH

6.0 Overview

A well-mixed room (WMR) model can adequately estimate exposure intensity for individuals working at a distance (say, 3 m or more) from a contaminant emission source, but it usually underestimates exposure intensity close to the source. Intuition and air monitoring data indicate that exposure intensity is higher near the source than at distant points in a room, but a WMR model treats the concentration as uniform throughout the room.

If a WMR is considered a single box, an analogous model that accounts for spatial variability in exposure intensity is a two-box model. That is, the room is conceptually divided into two boxes. One box contains the emission source and is termed the "near field." Its dimensions are sized to include the breathing zone of the worker who is located near the source and whose exposure is to be estimated. The rest of the room is the conceptual second box and is termed the "far field." Figure 6.1 depicts this construct.

6.1 Model Development

This two-box model assumes the air within each box is perfectly mixed, but that there is limited air flow between the boxes. The inter-box air flow rate is denoted β with units of m^3/min. Air simultaneously moves into and out of the near field at rate β. Room mechanical supply air and/or infiltration air moves into and out of the far field at rate Q, also with units of m^3/min. Q is the same room supply air rate parameter used in the WMR model. In the two-box model described here, it is assumed that the contaminant mass emission rate is a constant value G in mg/min, and that the only way airborne chemical mass leaves a box is via air flow.

This description leads to the following coupled mass balance equations:

(Near Field)

$$V_N \, dC_N = G \, dt + \beta \cdot C_F \, dt - \beta \cdot C_N \, dt \qquad (6\text{-}1)$$

(Far Field)

$$V_F \, dC_F = \beta \, C_N \, dt - (\beta + Q) \cdot C_F \, dt \qquad (6\text{-}2)$$

V_N and V_F: the near-field and far-field volumes, respectively (m^3)
C_N and C_F: the near-field and far-field concentrations, respectively (mg/ m^3)
G: constant mass emission rate (mg/min)
β: air flow rate (m^3/min) between the near and far fields
Q: room supply air rate (m^3/min)
dt: an infinitesimal time interval (min)

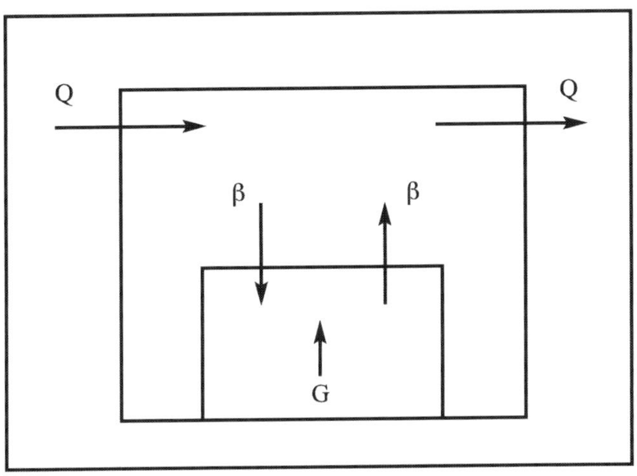

Figure 6.1 — A conceptual diagram of the two-box model. The smaller near field box contains the contaminant emission source (G) and the breathing zone of the individual whose exposure is to be estimated. The larger far field box represents the rest of the room. Air flows between the near field and the far field at rate β (m^3/min). Air flows into and out of the far field (the general room space) at rate Q (m^3/min).

Consider Equation 1. The left-hand side is the change in mass in the near field in the interval dt. Mass enters the near field in two ways: (1) it is emitted from the source (the G term), and (2) it is carried into the near field from the far field by the inter-box air flow β. Mass leaves the near field in the same inter-box air flow. Next, consider Equation 6-2. The left-hand side is the change in mass in the far field in the interval dt. The only mass entering the far field is that carried in from the near field by the inter-box air flow β. However, mass leaves the far field in two ways: (1) it is carried back into the near field by the inter-box air flow, and (2) it exits the room altogether in the exhaust air flow Q.

If the initial contaminant concentration in each box is zero, the general solutions for the time-varying concentrations in the near field, $C_N(t)$, and in the far field, $C_F(t)$, are the following:

$$C_N(t) = \frac{G}{\left[\frac{\beta}{\beta + Q}\right] Q} + G\left[\frac{\beta \cdot Q + \lambda_2 \cdot V_N(\beta + Q)}{\beta \cdot Q \cdot V_N(\lambda_1 - \lambda_2)}\right]\exp(\lambda_1 t) - G\left[\frac{\beta \cdot Q + \lambda_1 \cdot V_N(\beta + Q)}{\beta \cdot Q \cdot V_N(\lambda_1 - \lambda_2)}\right]\exp(\lambda_2 t) \tag{6-3}$$

$$C_F(t) = \frac{G}{Q} + G\left[\frac{\lambda_1 \cdot V_N + \beta}{\beta}\right]\left[\frac{\beta \cdot Q + \lambda_2 \cdot V_N(\beta + Q)}{\beta \cdot Q \cdot V_N(\lambda_1 - \lambda_2)}\right]\exp(\lambda_1 t) - G\left[\frac{\lambda_2 \cdot V_N + \beta}{\beta}\right]\left[\frac{\beta \cdot Q + \lambda_1 \cdot V_N(\beta + Q)}{\beta \cdot Q \cdot V_N(\lambda_1 - \lambda_2)}\right]\exp(\lambda_2 t) \tag{6-4}$$

where

$$\lambda_1 = 0.5\left[-\left(\frac{\beta \cdot V_F + V_N(\beta + Q)}{V_N \cdot V_F}\right) + \sqrt{\left(\frac{\beta \cdot V_F + V_N(\beta + Q)}{V_N \cdot V_F}\right)^2 - 4\left(\frac{\beta \cdot Q}{V_N \cdot V_F}\right)}\right]$$

$$\lambda_2 = 0.5\left[-\left(\frac{\beta \cdot V_F + V_N(\beta + Q)}{V_N \cdot V_F}\right) - \sqrt{\left(\frac{\beta \cdot V_F + V_N(\beta + Q)}{V_N \cdot V_F}\right)^2 - 4\left(\frac{\beta \cdot Q}{V_N \cdot V_F}\right)}\right]$$

The general solutions are complicated, but once numerical values are inserted for the various parameters, the resulting equations are relatively simple. The model parameters λ_1 and λ_2 correspond to air turnover rates (in units of min^{-1}) and do not depend on the contaminant emission rate. The absolute value of λ_1 is approximately equal to the air turnover rate in the far field (general room), or $|\lambda_1| \cong Q \div V_F$. The absolute value of λ_2 is approximately equal to the air turnover rate in the near field, or $|\lambda_2| \cong \beta \div V_N$. Because the parameters λ_1 and λ_2 are real, distinct, negative numbers for nearly all realistic scenarios, the terms $\exp(\lambda_1 \times t)$ and $\exp(\lambda_2 \times t)$ go to zero as t gets large, in which case the steady-state contaminant concentrations in the near field, $C_{N,SS}$, and in the far field, $C_{F,SS}$, are as follows:

$$C_{N,SS} = \frac{G}{Q} + \frac{G}{\beta} \tag{6-5}$$

$$C_{F,SS} = \frac{G}{Q} \tag{6-6}$$

If $\beta < Q$, then $C_{N,SS} > 2\, C_{F,SS}$, which signifies that exposure intensity is at least twice as high near the source as elsewhere in the room. On the other hand, if $\beta >> Q$, then $C_{N,SS} \cong C_{F,SS}$, which signifies there is no difference in exposure intensity close to versus far from the source, which is to say that the room is essentially well-mixed. Therefore, for a given Q value, the difference between $C_{N,SS}$ and $C_{F,SS}$ depends on the value of β, which makes intuitive sense. If β is small, contaminant lingers near the source and can build up to a high concentration. If β is large, contaminant is quickly dispersed away from the source into the general room, so the concentration near the source does not build up to a high level.

The inter-box air flow rate β equals one-half the free surface area, FSA, of the near field zone times the average random air speed, S, which is assumed to be the same throughout the room, or:

$$\beta = ½ \cdot FSA \cdot S \tag{6-7}$$

To explain the rationale for Equation 6-7, if the near field is a box with its base on the floor as in Figure 6.1, air is free to flow through the vertical sides and top surface of the box, but not through the base which is on the floor. Thus, for a box geometry, FSA = 2(L · H) + 2(W · H) + L · W, where L, W and H denote the respective length, width and height dimensions of the box. In an analogous fashion, if the near field has a hemispherical geometry with its base on the room floor, air is free to flow through the curved surface of the hemisphere, but not through the base which is on the floor. Thus, for a hemispherical geometry, $FSA = 2 \cdot \pi \cdot r^2$, where r is the hemisphere's radius. To maintain a constant mass or volume of air in the near field field zone, at any moment air flows *into* the near field through one-half the free surface area, and flows *out of* the near field through the other half. This is analogous to a room's supply air rate Q being equal to its exhaust air rate Q. Given some free surface area in m^2, the volumetric air flow rate into the near field (and out of the near field) is ½ FSA times the average air speed S in m/min. It is assumed here that there is no overall direction of air flow in the vicinity of the contaminant emission source.

Example 6.1

To illustrate applying the two-box model, consider the following hypothetical scenario. An individual sits at a work table and manually applies an adhesive containing toluene to small parts. Toluene evaporates at a constant rate G = 1,000 mg/min (or 1.15 mL per minute). The individual's breathing zone is at arm's length (approximately 2.5 feet) from the source of emission. We conceptualize the near field as a hemisphere with its base on the table and with radius 0.76 m (2.5 ft). The near field volume $V_N = (2/3) \times \pi \times (0.76)^3 = 0.93\ m^3$, and the near field $FSA = 2 \times \pi \times (0.76)^2 = 3.6\ m^2$. The random air speed is taken as S = 3.7 m/min (12 ft/min), which is the median air speed observed in a survey of indoor workplaces.(1) Based on Equation 6-7, $\beta = ½ \cdot (3.6\ m^2) \cdot (3.7\ m/min) = 6.7\ m^3/min$. The work operation is located in a room of volume 200 m^3 which receives Q = 20 m^3/min of air supply (corresponding to six nominal air changes per hour). Therefore, $V_F = 200\ m^3 - 0.93\ m^3 = 199\ m^3$. The initial toluene concentration is zero throughout the room.

Given these input values and based on Equations 6-3 and 6-4, the expressions for $C_N(t)$ and $C_F(t)$ in mg/m^3 are, respectively:

$$C_N(t) = 199.8 - 51.4 \exp(-0.10\,t) - 148.4 \exp(-7.23\,t)$$

$$C_F(t) = 50 - 50.7 \exp(-0.10\,t) + 0.7 \exp(-7.23\,t)$$

where:

$\lambda_1 = -0.10\ min^{-1}$, $\lambda_2 = -7.23\ min^{-1}$

Figure 6.2 depicts $C_N(t)$ and $C_F(t)$ over the interval $0 \le t \le 30$ min. There is a rapid increase in the near-field concentration, but a much more gradual increase in the far-field concentration. In this scenario, the near-field concentration attains 75% of its steady-state value in the first minute after the start of toluene emission, and 90% of its steady-state value by 10 minutes. In contrast, the far-field concentration reaches only 60% of its steady-state value by 10 minutes. Because the air turnover rate in the near field is high ($\beta \div V_N = 7.2\ min^{-1}$), direct toluene emission into the near field at mass rate G causes the near field concentration to attain the value $G \div \beta = 150$ mg/m^3 after the first minute. The more gradual increase toward the steady-state value thereafter reflects the slower increase in the far field concentration which adds the component $G \div Q = 50$ mg/m^3. The eventual near field steady-state concentration would be 200 mg/m^3, and the eventual far field steady-state concentration would be 50 mg/m^3. In this scenario, $C_{N,SS}$ is four-fold greater than $C_{F,SS}$.

Note that if the room were perfectly mixed such that the concentration were uniform throughout the room, the concentration curve would look like the $C_F(t)$ curve in Figure 6.2. This is the basis for the previous statement that a well-mixed room model gives a reasonable estimate of exposure intensity for individuals who are not positioned close to the emission source. However, in a room that is not perfectly mixed, the concentration near the source can be considerably higher, four-fold in this case.

6.2 Concentration Decay

Analogous to Equations 6-3 and 6-4 for the *de novo* buildup of contaminant concentration in the two zones, the following expressions describe the decay in contaminant concentration in the two zones upon cessation of emission (G = 0) and given starting concentrations $C_{N,0}$ and $C_{F,0}$:

$$C_N(t) = \frac{\beta(C_{F,0} - C_{N,0}) - \lambda_2 \bullet V_N \bullet C_{N,0}}{V_N(\lambda_1 - \lambda_2)} \exp(\lambda_1 t) + \frac{\beta(C_{N,0} - C_{F,0}) + \lambda_1 \bullet V_N \bullet C_{N,0}}{V_N(\lambda_1 - \lambda_2)} \exp(\lambda_2 t) \quad (6\text{-}8)$$

$$C_F(t) = \quad (6\text{-}9)$$

$$\frac{(\lambda_1 \bullet V_N + \beta)[\beta(C_{F,0} - C_{N,0}) - (\lambda_2 \bullet V_N \bullet C_{N,0})]}{\beta \bullet V_N(\lambda_1 - \lambda_2)} \exp(\lambda_1 t)$$

$$+ \frac{(\lambda_2 \bullet V_N + \beta)[\beta(C_{N,0} - C_{F,0}) + (\lambda_1 \bullet V_N \bullet C_{N,0})]}{\beta \bullet V_N(\lambda_1 - \lambda_2)} \exp(\lambda_2 t)$$

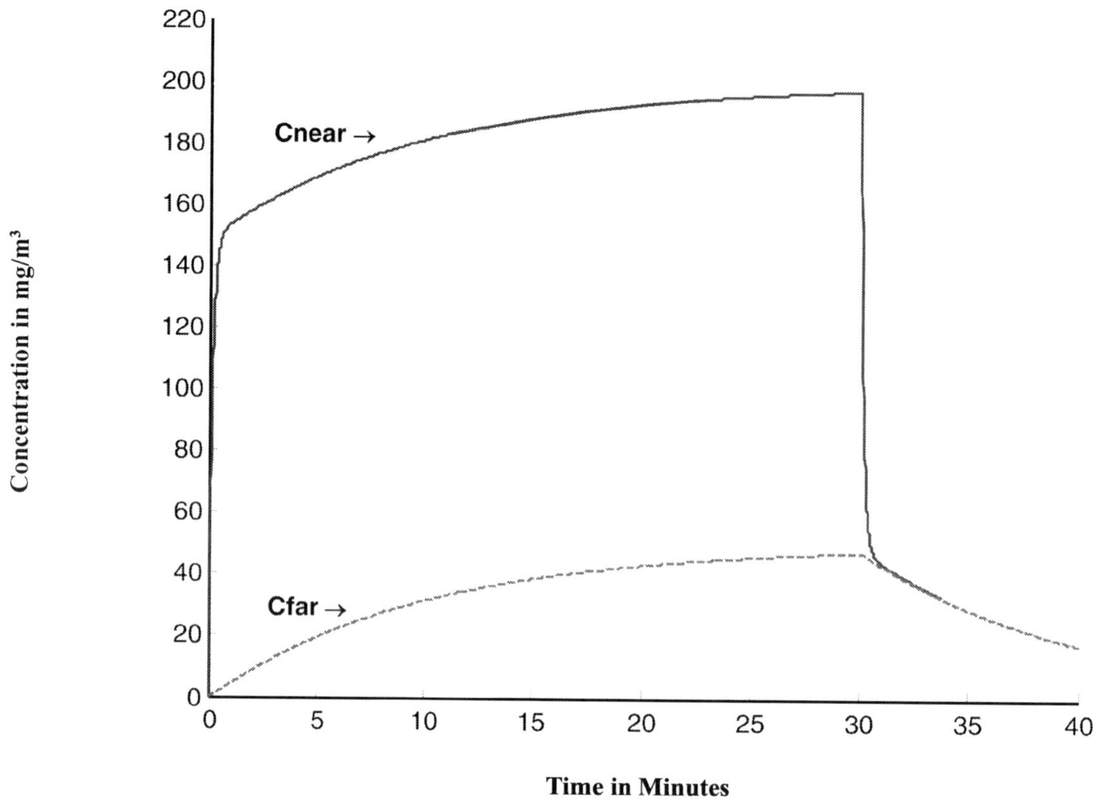

Figure 6.2 — The toluene vapor concentration (mg/m³) in the near field and the far field as predicted by the near field/far field model. From time 0 to 30 minutes, there is constant emission at G = 1,000 mg/min, and the concentration increases in both zones toward the theoretical steady-state values (200 mg/m³ for the near field and 50 mg/m³ for the far field). At time 30 minutes, emission ceases and there is a decrease in the toluene concentration in both zones. The model input values for the scenario are described in the text.

The parameters λ_1 and λ_2 are computed by the same expressions associated with Equations 6-3 and 6-4. Again, the general expressions are complicated, but once numerical values are inserted for the various parameters, the resulting equations are relatively simple.

Example 6.2

For the toluene evaporation scenario previously considered in which $V_N = 0.93$ m³, $V_F = 199$ m³, Q = 20 m³/min, and $\beta = 6.7$ m³/min, assume that emission ceases after 30 minutes. At this point, the near field and far field concentrations are, respectively, 197.2 mg/m³ and 47.5 mg/m³ (see Figure 6.2) To consider the decay functions, let $C_{N,0}$ = 197.2 mg/ and $C_{F,0} = 47.5$ mg/m³. Given these inputs and based on Equations 6-8 and 6-9, the expressions for $C_N(t)$ and $C_F(t)$ in mg/m³ for t > 0 are, respectively:

$$C_N(t) = 48.9 \exp(-0.10\,t) + 148.4 \exp(-7.23\,t)$$

$$C_F(t) = 48.2 \exp(-0.10\,t) - 0.7 \exp(-7.23\,t)$$

where

$\lambda_1 = -0.10\ \text{min}^{-1}$, $\lambda_2 = -7.23\ \text{min}^{-1}$

Figure 6.2 depicts $C_N(t)$ and $C_F(t)$ starting at 30 minutes on the horizontal axis time scale. There is a rapid decrease in $C_N(t)$ governed by the λ_2 parameter such that by the first minute after emission ceases, the near-field concentration decreases by 75% and essentially equals the far-field concentration. After the first minute, both $C_N(t)$ and $C_F(t)$ exhibit a gradual decrease governed by the λ_1 parameter.

6.3 Model Validation

A pertinent question is — how well does the model estimate exposure intensity? Although the model has not been "validated" in a comprehensive manner, published studies indicate that it generally provides TWA exposure

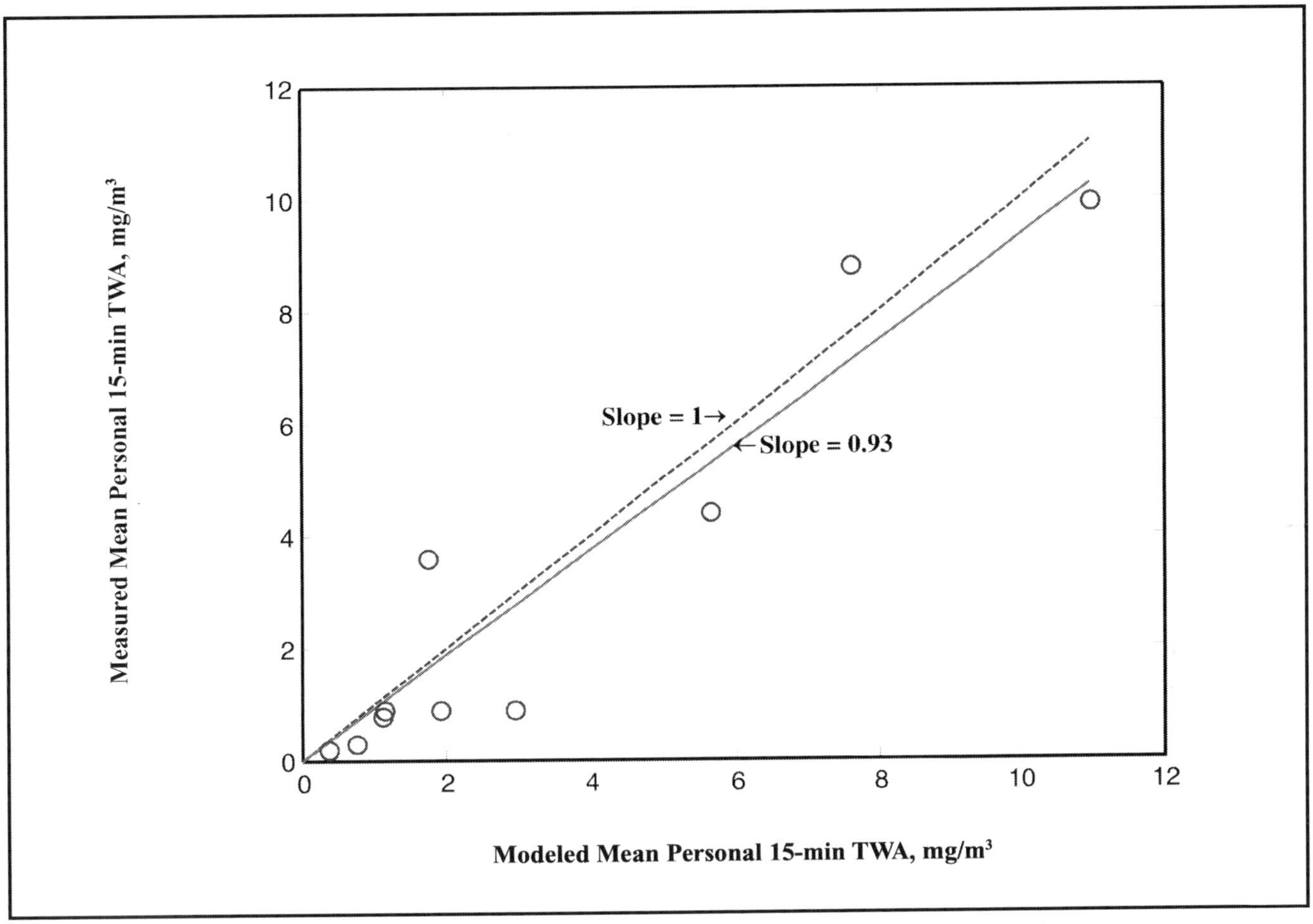

Figure 6.3 — Predicted versus measured 15-min TWA personal benzene vapor exposure concentrations across ten experimental combinations of factors (volume of liquid applied, benzene content, air exchange rate), taken from M. Nicas and J. Neuhaus, *J. Occup. Environ. Hyg.* *5*:599–608, 2008. Details of the experiments and measured values were taken from P. Williams, et al.: Airborne Concentrations of Benzene Associated with the Historical Use of Some Formulations of Liquid Wrench. *J. Occup. Environ. Hyg.* *4*:547–561, 2007.

estimates within a factor of ½- to 2-fold of measured values, which is to say, it reasonably estimates concentration values. In recent years, there have been three published studies that assessed the model's predictive ability. These are briefly described below.

In the first study, cyclohexane concentrations were measured while that solvent was applied to a valve being disassembled on a work table in a test room.[2] The cyclohexane was applied at a fairly uniform rate to mimic constant vapor emission, and a fan was used to create different air speeds near the work table. An 1-hr TWA breathing zone (near field) concentration was measured for the person working on the valve, and two 1-hr TWA area (far field) concentrations were measured in the test room. The simulation experiment was repeated six times for each of three air speeds generated (3.3 m/min, 23 m/min,
61 m/min). For the three air speeds, the best estimate of the near field concentration was 1.03-fold, 1.01-fold, and 0.72-fold the respective mean of the replicate breathing zone concentration measurements. Note that the investigators observed an approximate two-fold difference between the lowest and highest measured values for a nominally fixed set of test conditions. This natural variability in measured concentrations signifies that a model estimate should properly be compared against not a single measurement, but against the average of replicate measurements (preferably six or more) for the scenario being modeled.

In the second study, benzene concentrations were measured at parts washers in two different facilities.[3] At Facility A, one 1-hr TWA concentration was measured at breathing zone level above the part washer, and at Facility B, two 4-hr TWA concentrations were measured at breathing zone level above the part washer. No far field measurements were made. The benzene mass emission rate was slightly decreasing over the course of the monitoring, and the model was adjusted to account for this variable rate. For the three time periods monitored, the best estimate of the near field concentration was 0.60-fold, 1.0-fold, and 2.1-fold the respective single concentration measurement that was made.

In the third study, benzene concentrations were measured while a benzene-containing liquid was applied to metal parts on a work table in a test room, and the published values were compared against model estimates.(4) For each experiment, the mass of benzene applied and the room air exchange rate were reported, two 15-min TWA breathing zone (near field) concentrations were measured for the person working at the table, and two 1-hr TWA area (far field) concentrations were measured in the test room. Ten experimental combinations of factors (volume of liquid applied, benzene content, air exchange rate) were used, and the average values from two replicates for each combination were reported. The benzene mass emission rate was decreasing over the course of the monitoring, and the model was adjusted to account for this variable rate. Across the ten experimental combinations, the best estimate of the near field concentration ranged from 0.5-fold to 3.3-fold the respective average of the breathing zone measurements; in 8/10 combinations, this ratio was in the range 0.5-fold to 2.1-fold. Figure 6.3 shows a scatter plot of the average measured versus predicted near field concentrations across the ten experimental combinations. The estimated regression line slope was 0.93 with $r^2 = 0.94$, and the null hypothesis that the true slope equals one could not be rejected (p-value = 0.39).

6.4 The Local Purging Flowrate

The right-hand side of Equation 6-5 for the steady-state near field concentration can be rearranged to yield the following expression for $C_{N,SS}$:

$$C_{N,SS} = \frac{G}{\left[\dfrac{\beta}{\beta + Q}\right] Q} \qquad (6\text{-}10)$$

The denominator is variously termed the effective near-field ventilation rate or the "local purging flow rate." Because $\beta \div (\beta + Q) < 1$, the local purging flowrate is less than Q, the room supply air rate. In other words, given imperfect air mixing, the effective volumetric ventilation rate (m^3/min) in the near field will be less than the general room volumetric ventilation rate.

In traditional industrial hygiene texts, it is common to see a "mixing factor" m used to account for poor air mixing. The factor m is a fraction between 0 and 1, and the product of m and Q is termed the effective room ventilation rate. In some texts, the inverse of m is termed a "safety factor" to be applied in designing a dilution ventilation airfow rate. If contaminant is emitted into room air at constant rate G, and if $m \cdot Q$ is used instead of Q in the WMR model equation, the steady-state concentration in the room is given as:

$$C_{SS} = \frac{G}{m \bullet Q} \qquad (6\text{-}11)$$

Upon inspection of Equations 6-10 and 6-11, which are alternative attempts at accounting for imperfect air mixing in a room, the quotient $\beta \div (\beta + Q)$ appears to be equivalent to the mixing factor m. As it turns out, $\beta \div (\beta + Q) \neq m$, although the explanation is beyond the scope of this chapter. In general, if one wishes to estimate exposure intensity for an individual close to an emission source in an imperfectly-mixed room, it is preferable to use the two-box model described in this chapter, or a turbulent eddy diffusion model, rather than a WMR model with a ventilation rate adjusted by a mixing factor. A detailed discussion of the problem inherent in using a mixing factor is provided elsewhere.(5)

To close with a historical note, the first mention of the near field/far field (two-box) model as applied to industrial hygiene was by W.C. Hemeon in his book titled *Plant and Process Ventilation*, Second Edition.(6) The concepts underlying the model were described, and various geometries for the near field were discussed. However, the dynamic concentration equations, $C_N(t)$ and $C_F(t)$, were not provided, and the steady-state concentration equations, $C_{N,SS}$ and $C_{F,SS}$, were not as explicit as presented here. The derivation of the dynamic concentration equations is provided in the previously cited article in the 1996 *American Industrial Hygiene Association Journal*.(5)

6.5 References

1. **Baldwin, P. and A Maynard:** A Survey of Wind Speeds in Indoor Workplaces. *Ann. Occup. Hyg. 42*:303–313 (1998).
2. **Spencer, J.W. and M.J. Plisko:** A Comparison Study Using a Mathematical Model and Actual Exposure Monitoring for Estimating Solvent Exposures during the Disassembly of Metal Parts. *J. Occup. Environ. Hyg. 4*:253–259 (2007).
3. **Nicas, M., M.J. Plisko, and J.W. Spencer:** Estimating Benzene Exposure at a Solvents Parts Washer. *J. Occup. Environ. Hyg. 3*:284–291 (2006).
4. **Nicas, M. and J. Neuhaus:** Predicting Benzene Vapor Concentrations with a Near Field/Far Field model. *J. Occup. Environ. Hyg. 5*:599–608 (2008).
5. **Nicas, M.:** Estimating exposure intensity in an imperfectly mixed room. *Am. Ind. Hyg. Assoc. J. 57*:542–550 (1996).
6. **Hemeon, W.C.:** *Plant and Process Ventilation*, Second Edition. New York: Industrial Press, 1963.

7 Turbulent Eddy Diffusion Models

By
Mark Nicas, PhD, CIH

7.0 Overview

Although the near field/far field model has proven robust in estimating airborne contaminant exposures close to an emission source, an unrealistic aspect is the sharp discontinuity in the airborne contaminant concentration as one moves from the near field to the far field. Rather, it is expected that a continuous gradient of decreasing concentration will occur as one moves away from the emission source. In theory, three-dimensional contour surfaces of equal concentration could be mapped around an emission source, although all points on a surface need not be equidistant from the source. Turbulent eddy diffusion models account for a continuous concentration gradient (the term eddy is omitted hereafter for simplicity) and use two simplifying assumptions: (1) the equal concentration surfaces are symmetrical in shape, and (2) the turbulent diffusion coefficient is constant over time and space.

7.1 Initial Considerations

Turbulent diffusion involves the concept of an eddy or swirl of room air "grabbing" a parcel of contaminant at one location, and carrying the parcel to a nearby location in the room. This process leads to the dispersion of contaminant away from the source. Turbulent diffusion involves the *macroscale* random motion of parcels of air carrying numerous contaminant molecules (or particles). This dispersion mechanism is distinct from *microscale* molecular diffusion which involves the random motion of individual molecules in the gas phase (air) due to the kinetic energy of the molecules.

Both types of diffusion are characterized by a coefficient with the unit of length^2/time; this chapter uses m^2/min. The unit m^2/min reflects the rate or speed at which molecules (or particles) are dispersed away from a starting position. D_T denotes the turbulent diffusion coefficient (sometimes termed the eddy diffusivity coefficient), and D_M denotes the molecular diffusion coefficient. Whereas D_T values are on the order of 0.1 to 10 m^2/min, D_M values are on the order of 0.001 m^2/min. Because D_T values exceed D_M values by 2 to 4 orders of magnitude, turbulent diffusion is far more important than molecular diffusion as a means to disperse gas-phase contaminants. D_M is therefore typically ignored when modeling air transport of pollutant mass. There has been little research on predicting workplace values of D_T. The following algorithm has been suggested for assigning D_T in an isothermal room (no temperature gradients), although its overall validity has not been reported(1):

$$D_T,\ \text{m}^2/\text{min} = 60 \times (\text{ACS} \times U_{in}^2 \times H^4 \times 0.08)^{0.33} \qquad (7\text{-}1)$$

where ACS is the number of room air changes per second, U_{in} is the discharge speed (m/sec) of supply air into the room through air inlets, and H is the room height (m). Some experimentally determined values for D_T have been reported. These are summarized in Table 7.1.

Using a turbulent diffusion model requires specifying the worker's location relative to the emission source. That location can be expressed as a Cartesian coordinate position (x,y,z) relative to the source which is taken to be at the origin (0,0,0). However, if there is no advective air flow (the bulk movement of air in a given direction, on average), an equal concentration surface is predicted at a fixed distance from the emission point, and the individual's location can be expressed as a radial distance r (m) from the source, where

$$r = \sqrt{x^2 + y^2 + z^2} \qquad (7\text{-}2)$$

Next, consider that an advective airflow pattern exists in the vicinity of the emission source, which signifies that air is moving in a certain direction and at a certain speed, on average. For example, the average movement might be left-to-right along the x-axis. The molecules (particles) tend to move in the direction of the bulk air flow, but turbulent diffusion is superimposed on this movement. With both turbulent diffusion and advection involved,

Table 7.1 Some reported eddy diffusion coefficients[2–5]

Room Dimensions	Air changes per hour	D_T (m^2/min)	Other
150 ft x 120 ft x 16 ft 288,000 ft^3	0.3	11.5	
110 ft x 40 ft x 20 ft 88,000 ft^3	11	0.1–1.3	
180 ft x 120 ft x 20 ft 432,000 ft^3	5	0.1–0.8	
770,000 ft^3	5.7	0.085–0.885	mean 0.354, SD .29 median 0.279
289,000 ft^3	8.2	0.085–9.988	mean 1.26, SD 2.9 median 0.304
64.5 ft x 60.5 ft x 12 ft 47,000 ft^3	10.5	0.11–0.23	mean 0.16, SD .04 median 0.16

dispersion is characterized by D_T and by a mean air speed U (m/min) in a specified direction. That direction need not be along a room axis, and an average airflow vector can be decomposed into components along each of the three room axes. Because advective flow does not lead to an equal concentration surface at a fixed distance from the emission point, the worker's location must be expressed as a Cartesian coordinate position (x,y,z) relative to the emission source taken to be at the origin (0,0,0).

The most commonly used turbulent diffusion models do not account for contaminant removal from room air via the exhaust air flow or other mechanisms (for example, adsorption onto surfaces and particle settling). Therefore, the room volume V, the supply/exhaust air rate Q, and non-ventilation removal rate constants do not appear in the model equations. These factors can be included in a turbulent diffusion model, but a different mathematical approach is required as outlined later in this chapter. However, in the vicinity of the emission source (say, within 1 m), these removal mechanisms are relatively unimportant in reducing exposure intensity and, in general, turbulent diffusion models are used to estimate exposure intensity near the source.

7.2 Spherical Turbulent Diffusion without Advection following a Pulse Release

Consider that contaminant mass M_0 (mg) is released as a pulse at time zero at the origin. Assume that mass disperses away from the origin in a spherically symmetrical pattern via turbulent diffusion with coefficient D_T (m^2/min). Assume there is no advective air flow near the source that affects the pattern of dispersion, or U = 0. Let C(x,y,z,t) denote the contaminant concentration in mg/m^3 at position (x, y, z) relative to the pulse release position (0,0,0) at time t (min) after the pulse release. All distances are in m. The concentration equation is as follows:

$$C(x,y,z,t),\ mg/m^3 = \frac{M_0}{8(\pi D_T t)^{1.5}} \times \exp\left[-\frac{x^2+y^2+z^2}{4 D_T t}\right] \qquad (7\text{-}3)$$

Due to the symmetry in dispersion, there is a contour surface of equal concentration at the same radial distance r (m) from the origin. That is, at all points which are radial distance r from the pulse release position, the same contaminant concentration exists. Due to this symmetry, C(x,y,z,t) can be written as C(r,t), where *r* is radial distance (m) from the pulse release position:

$$C(r,t),\ mg/m^3 = \frac{M_0}{8(\pi D_T t)^{1.5}} \times \exp\left[-\frac{r^2}{4 D_T t}\right],\ \text{where}$$

$$r = \sqrt{x^2+y^2+z^2} \qquad (7\text{-}4)$$

Example 7.1

To illustrate applying the pulse release equation, consider that contaminant is emitted in the middle of a large room such that the source can be treated as a point source, for example, gas released from a container nozzle. Define the release position as the origin (0,0,0).

At t = 0, M_0 = 10,000 mg of contaminant is emitted. Assume that spherical turbulent diffusion occurs with D_T = 0.50 m^2/min (chosen for purposes of illustration only), and with no advective air flow such that U = 0. Consider two workers who are near the release position. Worker A's breathing zone is at position (x = 1,y = 0,z = 1), which is 1 m to the right of the origin on the x-axis, at the origin on the y-axis, and 1 m up from the origin on the z-axis. Worker B's breathing zone is at position (x = -1,y = 0,z = 1), which is 1 m to the left of the origin on the x-axis, at the origin on the y-axis, and 1 m up from the origin on the z-axis. For this scenario, both workers are at radial distance 1.4 m from the source, or:

$r = \sqrt{(-1)^2+0^2+1^2} = 1.4$ m. We substitute the appropriate values into Equation (7-3) which then reduces to:

$$C(r=1.4, t) = \frac{635}{t^{1.5}} \times \exp\left[-\frac{1}{t}\right]$$

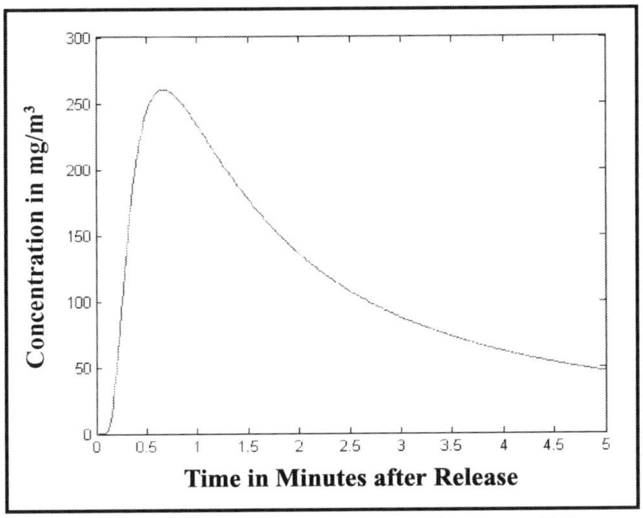

Figure 7.1 — The concentration time series at a radial distance of 1.4 meters following a pulse release of 10,000 mg of pollutant with $D_T = 0.50$ m²/min and without advection.

Figure 7.1 depicts the concentration time series over the first 5 minutes following the pulse release. A "wave" of contaminant moves past each position. The concentration increases to a peak value of 266 mg/m³ at about 0.7 min after the pulse release, and decreases thereafter. If the radial distance were less than 1.4 m, the concentration values would be higher overall, and the peak concentration would occur prior to 0.7 min. On the other hand, if the radial distance were greater than 1.4 m, the concentration values would be lower overall, and the peak concentration would occur later than 0.7 min.

Decreasing the D_T value would slow down the dispersion process such that the contaminant wave would pass each position more slowly, although the peak concentration would be similar. Increasing the D_T value would speed up the process such that the contaminant wave would pass each position more quickly, but the peak concentration would again be similar.

7.3 Spherical Turbulent Diffusion with Advection following a Pulse Release

Consider that contaminant mass M_0 (mg) is released as a pulse at time zero, with the release position at the origin. Mass disperses away from the origin in a spherically symmetrical pattern via turbulent diffusion with coefficient D_T (m²/min), and there is an advective air flow parallel to the x-axis from left to right. The advective air speed is a constant value U (m/min). Let C(x,y,z,t) denote the contaminant concentration in mg/m³ at position (x, y, z) relative to the pulse release position. The concentration equation is as follows:

$$C(x, y, z, t),\ \text{mg/m}^3 = \frac{M_0}{8(\pi\, D_T\, t)^{1.5}} \times \exp\left[-\frac{(x - U\,t)^2 + y^2 + z^2}{4\, D_T\, t}\right] \qquad (7\text{-}5)$$

Due to advection, concentrations at positions *downstream* from the emission point (values of $x > 0$) are greater than at positions *upstream* of the emission point (values of $x < 0$). Because there is no radial symmetry, there is no simple expression for concentration as a function of radial distance from the emission point. Note that if the advective air flow were parallel to the *y*-axis or the z-axis instead of the *x*-axis, the equation could be easily rewritten. For example, if the advective air flow were parallel to the *y*-axis, the numerator of the exponent argument would appear as: $x^2 + (y - U\,t)^2 + z^2$.

Example 7.2

Consider the previous pulse release scenario, except with an advective air flow parallel to the x-axis from left to right such that U = 1.5 m/min (5 fpm). Again, M_0 = 10,000 mg and D_T = 0.50 m²/min. As before, Worker A's breathing zone is at position (x = 1,y = 0,z = 1), which is 1 m *downstream* from the release point on the x-axis, at the origin on the y-axis, and 1 m up from the origin on the z-axis. Worker B's breathing zone is at position (x = -1,y = 0,z = 1), which is 1 m *upstream* from the release point on the x-axis, at the origin on the y-axis, and 1 m up from the origin on the z-axis. To find the expression for the contaminant concentration at these two positions, we substitute the appropriate values into Equation (7-5):

Worker A:

$$C(x = 1, y = 0, z = 1, t) = \frac{635}{t^{1.5}} \times \exp\left[-\frac{2 - 3t + 2.25\,t^2}{2t}\right]$$

Worker B:

$$C(x = -1, y = 0, z = 1, t) = \frac{635}{t^{1.5}} \times \exp\left[-\frac{2 + 3t + 2.25\,t^2}{2t}\right]$$

Figure 7.2 depicts the concentration time series over the first 5 minutes following the pulse release. A wave of contaminant moves past each position, but the

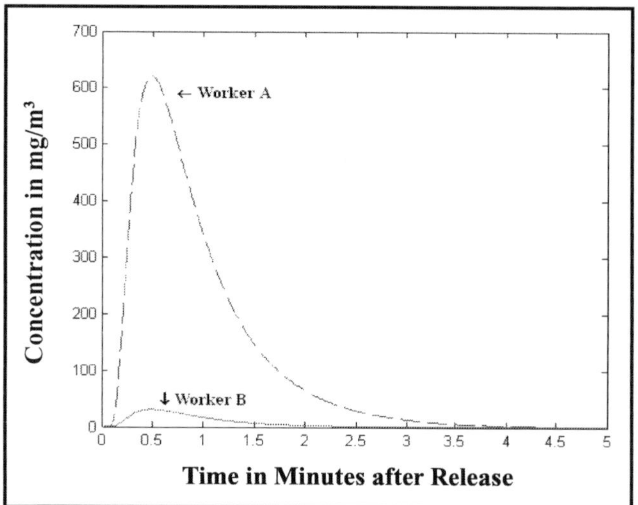

Figure 7.2 — The concentration time series at a position downstream (Worker A) and a position upstream (Worker B) from a pulse release of $M_0 = 10{,}000$ mg of pollutant at the origin (0,0,0). There is advective air flow along the x-axis from left to right with U = 1.5 m/min. The turbulent diffusion coefficient $D_T = 0.50\ m^2/min$. Worker A's breathing zone is at position (x = 1,y = 0,z = 1), which is 1 m *downstream* from the release point on the x-axis. Worker B's breathing zone is at position (x = -1, y = 0,z = 1), which is 1 m *upstream* from the release point on the x-axis.

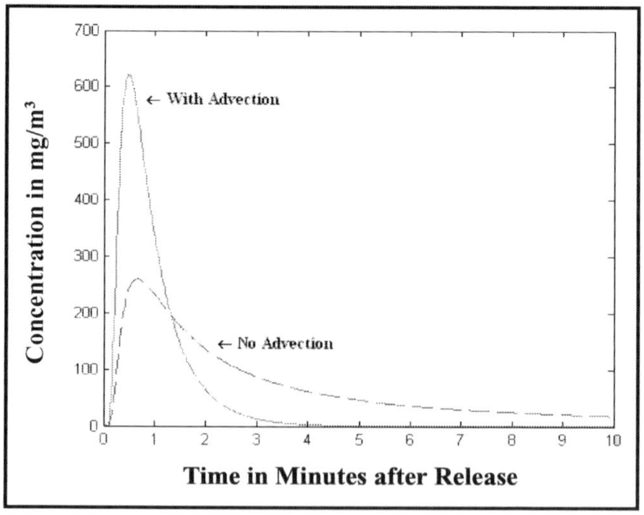

Figure 7.3 — The concentration time series with and without advective air flow following a pulse release of $M_0 = 10{,}000$ mg of pollutant at the origin (0,0,0). The turbulent diffusion coefficient $D_T = 0.50\ m^2/min$. The advective air flow is along the x-axis from left to right with U = 1.5 m/min. The worker's breathing zone is at (x = 1,y = 0,z = 1), which is 1 m to the right of the release point on the x-axis.

concentration values are much greater at worker A's downstream position than at worker B's upstream position relative to the release point. For worker A, a peak concentration of 626 mg/m^3 occurs at t = 0.5 min after the pulse release, and decreases thereafter. For worker B, a peak concentration of 30 mg/m^3 occurs at t = 0.5 min after the pulse release, and decreases thereafter. Although some contaminant moves upstream (toward the x = -1 position) via turbulent diffusion, the advective flow pushes most of the contaminant downstream (toward the x = 1 position).

Advection has two effects which initially seem inconsistent. A worker downstream from the release point will experience a higher peak exposure with advection than without advection, yet the short-term TWA concentration value (over several minutes or more) will be lower with advection. Both effects are seen by comparing the concentration time series at worker A's position (x = 1,y = 0,z = 1) with and without advection, as depicted in Figure 7.3. Advection causes the contaminant wave to reach the downstream position more quickly than if there were no advection. In turn, there is less time for contaminant to diffuse away from the x-axis in a lateral direction (along the y-axis) and in a vertical direction (along the z-axis) before the wave reaches x = 1 m, so the contaminant cloud is more concentrated which leads to a higher peak concentration. On the other hand, because the wave passes through the x = 1 m position more slowly without an advective air flow, the contaminant cloud lingers. For example, at t = 10 min after the pulse release, the concentration at the x = 1 m position is 18 mg/m^3 without advection, but zero with advection. This slower dispersion results in a higher 10-min TWA without advection (74 mg/m^3) than with advection (60 mg/m^3).

An outcome of the higher TWA value without advection (over intervals exceeding several minutes) is the following. If emission occurs continuously, or as a series of pulse releases, the contaminant concentration at a fixed position near the source will be higher without an advective air flow than with an advective air flow.

7.4 Accounting for Non-spherical Diffusion

Spherical diffusion strictly applies to dispersion away from a point source in free space, but few emission sources satisfy that description. Many sources are on flat horizontal surfaces, such as a work table or floor located well away from the room walls. In these situations, diffusion near the source can be viewed as occurring through half a sphere (a hemisphere), because the horizontal surface "reflects" back contaminant that tries to move down through the hemisphere base. Or an emission source might be on a horizontal surface next to a room wall. In this case, diffusion near the source occurs through one-fourth of a sphere, because the horizontal surface and the adjoining vertical wall reflect contaminant that tries to move through these two boundaries. Or an emission source might be on a horizontal surface in the corner of the room. In this situation, diffusion near the source occurs through one-eighth of a sphere, because the horizontal surface and the two adjoining vertical walls reflect contaminant that tries to move through these three boundaries.

For diffusion without advection (which produces an assumed radial symmetry in concentration values), dispersion through different portions of a sphere is accommodated by multiplying the spherical diffusion equation by an integer value as follows:

- Diffusion through a Hemisphere: C (r,t) = 2 × Spherical C (r,t)
- Diffusion through 1/4 of a Sphere: C (r,t) = 4 × Spherical C (r,t)
- Diffusion through 1/8 of a Sphere: C (r,t) = 8 × Spherical C (r,t)

A similar adjustment can be made for diffusion with advection through a hemisphere and through a quarter of a sphere, as long as the advective air flow is not perpendicular to one of the reflecting surfaces:

- Diffusion with Advection through a Hemisphere: C(x, y, z, t) = 2 × C(x, y, z, t) for a Sphere
- Diffusion with Advection through 1/4 of a Sphere: C(x, y, z, t) = 4 × C(x, y, z, t) for a Sphere

7.5 Hemispherical Turbulent Diffusion without Advection and with a Constant Contaminant Emission Rate

Consider that contaminant is emitted at a constant rate G (mg/min). The release position is taken as the origin (0,0,0). The source sits on a horizontal surface with the origin point on the surface, or z = 0, such that z can take on positive values only. Assume that contaminant disperses away from the origin in a symmetrical pattern via turbulent diffusion with coefficient D_T (m^2/min). Assume there is no advective air flow near the source that affects the pattern of dispersion, or U = 0. Let C (x, y, z, t) denote the contaminant concentration in mg/m^3 at position (x, y, z) relative to the source at the origin at time t (min) after the start of emission. All distances are in m. In the alternative, let C (r,t) denote the contaminant concentration at radial distance r (m) from the source, where $r = \sqrt{x^2 + y^2 + z^2}$. The concentration equations are as follows:

$$C(x,y,z,t),\ mg/m^3 = \frac{G}{2\pi D_T \sqrt{x^2+y^2+z^2}} \times \left[1 - \text{erf}\left(\frac{\sqrt{x^2+y^2+z^2}}{\sqrt{4 D_T t}}\right)\right] \qquad (7\text{-}6)$$

$$C(r,t),\ mg/m^3 = \frac{G}{2\pi D_T r} \times \left[1 - \text{erf}\left[\frac{r}{\sqrt{4 D_T t}}\right]\right] \qquad (7\text{-}7)$$

The term erf( ) in the brackets denotes the error function, which is an integral expression. If *a* denotes a real number greater than or equal to zero:

$$\text{erf}(a) = \frac{2}{\sqrt{\pi}} \int_0^a \exp(-w^2)dw \qquad (7\text{-}8)$$

where *w* is a dummy variable of integration. The error function ranges in value from zero (for *a* = 0) to one (for *a* → ∞). An erf(*a*) ≈ 1 corresponds to a value of time close to zero, and a concentration close to zero. An erf(*a*) ≈ 0 corresponds to a large value of time and a concentration close to the steady state value. Many mathematical software packages can compute erf(*a*) values. In Microsoft EXCEL®, the function name is ERF and is listed under the "Engineering" function category. For the mathematically inquisitive reader, the derivation of the spherical form of Equation (7-7) is presented in Technical Appendix I at the end of this chapter. The derivation is based on discretizing continuous emission into a series of pulse releases, where the time-dependent concentration due to each pulse adheres to Equation (7-4).

If a computer software package providing the error function is not available, an approximation for the error function can be used. The following concentration equation in terms of radial distance r from the source provides values close to those computed by the exact equation based on the error function:

$$C(r,t) \cong \frac{G}{2\pi D_T r} \times \left[1 - \sqrt{1 - \exp\left[-\frac{r^2}{\pi D_T t}\right]}\right] \qquad (7\text{-}9)$$

If contaminant emission continues at a constant rate G for a long time, the contaminant concentration at position (x, y, z) will reach a steady state value, regardless of whether an advective air flow is present. The respective steady-state equations in terms of (x, y, z) and r are as follows:

$$C\ (x, y, z)_{SS},\ mg/m^3 = \frac{G}{2\pi D_T \sqrt{x^2+y^2+z^2}} \qquad (7\text{-}10)$$

$$C\ (r)_{SS},\ mg/m^3 = \frac{G}{2\pi D_T r} \qquad (7\text{-}11)$$

There are two points to note about the steady-state equations. First, the concentration value decreases inversely with increasing distance from the emission source. However, if r corresponds to the position of an exhaust air outlet, it will seldom hold that $C(r)_{SS} = G/Q$, which signifies that the turbulent diffusion model does not ensure a mass balance in the room. Second, the concentration value becomes infinite as the distance from the source approaches zero. Because an infinite concentration is impossible, these turbulent diffusion models are not appropriate for positions very close to the emission source, say, within 0.1 m. In the more complex method for modeling turbulent diffusion which will be discussed, a mass balance is maintained and there is a finite limit on the contaminant concentration near the source.

Example 7.3

An emission source located on a large work table emits contaminant at a constant rate of G = 1000 mg/min. Assume that hemispherical turbulent diffusion of contaminant occurs with $D_T = 1.0$ m²/min (chosen for purposes of illustration only), and with no advective air flow in the vicinity of the release (U = 0). A worker's breathing zone is located at radial distance r = 0.76 m from the source (at arm's length). We substitute the appropriate values into the Equation (7-7) for C (r,t) without advection:

$$C(r = 0.76, t) = 209 \times \left[1 - \mathrm{erf}\left(\frac{0.381}{\sqrt{t}} \right) \right]$$

Based on Equation (7-11), the steady-state concentration is 209 mg/m³. The solid-line curve in Figure 4 depicts the concentration time series over the first 30 minutes of emission. About 92% of the steady-state concentration is reached at 30 minutes, and the steady-state value is approached only slowly thereafter. However, the initial buildup in concentration is fairly rapid such that 80% of the steady-state value is reached by five minutes after the start of emission.

Comparison to the Near Field/Far Field Model: How does the hemispherical diffusion model without advection compare to the near field/far field model given a constant contaminant emission rate? The two models are comparable over a small range of D_T values. The problem is the large uncertainty in assigning the appropriate D_T value. To explain, we revisit the toluene scenario posed in the chapter on the Near Field/Far Field model.

A person worked at arm's length (0.76 m) from an emission source located on a table, so we posited a hemispherical near field zone centered on the source with radius r = 0.76 m. The emission rate G was 1000 mg/min. The operation was located in a large room with V = 200 m³ and Q = 20 m³/min. The random air speed in the vicinity of the operation was S = 3.7 m/min (12 fpm). The near field concentration equation was:

$$C_N(t),\ \mathrm{mg/m^3} =$$

$$199.8 - 51.4 \exp(-0.10\,t) - 148.4 \exp(-7.23\,t)$$

The predicted steady-state near field concentration was 200 mg/m³. The dashed-line curve in Figure 7.4 depicts $C_N(t)$ over the first 30 minutes of emission. It is evident that the turbulent diffusion model with $D_T = 1.0$ m²/min and the near field/far field model with S = 3.7 m/min show close agreement in this scenario.

However, $D_T = 1.0$ m²/min was not a predicted value, but was chosen for purposes of illustration only. What if we were to use the Equation (7-1) algorithm for D_T? Assume that H = 3 m, and that $U_{IN} = 0.25$ m/sec (about 50 fpm). For V = 200 m³ and Q = 20 m³/min, the number of air changes per second is .00167. According to the algorithm:

$$D_T = 60 \times (0.0167 \times 0.25^2 \times 3^4 \times 0.08)^{0.33} = 5.3\ \mathrm{m^2/min}$$

If we had assigned $D_T = 5.3$ m²/min, the turbulent diffusion model would predict exposure concentrations near the emission source about 5-fold lower than those predicted by the near field/far field model. On the other hand, if we had assigned $D_T = 0.5$ m²/min, the turbulent diffusion model would predict exposure concentrations near the emission source about 2-fold higher than those

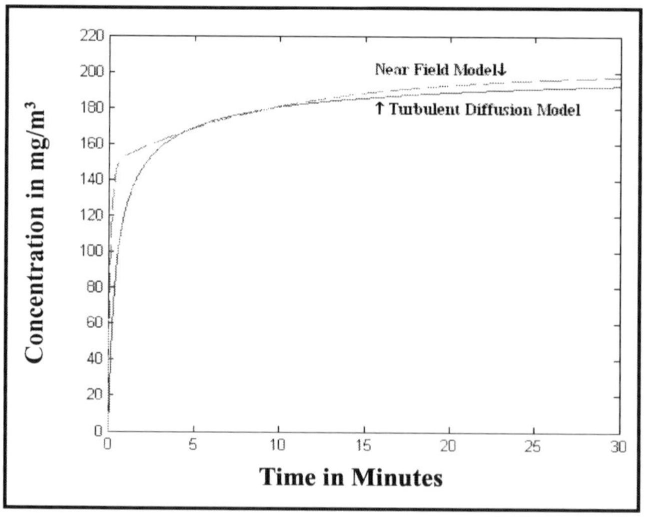

Figure 7.4 — The concentration time series for a contaminant emission scenario as predicted by the near field/far field model (the dashed-line curve) versus the hemispherical turbulent diffusion model (the solid-line curve). A worker's breathing zone is 0.76 m from an emission source located on a table. The pollutant emission rate G = 1,000 mg/min. The operation is in a large room with V = 200 m³ and Q = 20 m³/min. The random air speed in the vicinity of the operation is S = 3.7 m/min. The near field is modeled as a hemisphere with radius r = 0.76 m. Turbulent diffusion is modeled with $D_T = 1.0$ m²/min, and the assumption of zero velocity on average (no advective air flow on average in a given direction).

predicted by the near field/far field model. This outcome illustrates the current problem with applying turbulent diffusion models, namely, the large uncertainty in assigning the D_T value.

7.6 Hemispherical Turbulent Diffusion with Advection and with a Constant Contaminant Emission Rate

Modeling dispersion via advective air flow and turbulent diffusion is more difficult than if there is no advective flow. One result is the lack of a closed-form equation that describes the contaminant concentration as a function of time at position (x, y, z). In contrast, without advective air flow, Equation (7-6) is the closed-form expression for C (x, y, z, t). A numerical method for computing C (x, y, z, t) given advection and turbulent diffusion is available in the literature.(6) However, if all one desires to know is the steady-state concentration value, a relatively simple equation is available. The following expression assumes the emission source is at the origin (0,0,0), and that advective air flow at speed U is parallel to the x-axis from left to right such that x > 0 for downstream positions:

$$C(x,y,z)_{SS},\ mg/m^3 = \frac{G}{2\pi D_T\sqrt{x^2+y^2+z^2}} \times \exp\left[-\frac{U}{2D_T} \times \left(\sqrt{x^2+y^2+z^2} - x\right)\right] \quad (7\text{-}12)$$

7.7 Computing TWA Exposure Levels

The mathematical form of the turbulent diffusion model equations does not lend itself to simple expressions for the TWA concentration. Instead, the recommended method is to compute the concentration time series using small (say, one-second for a 15 minute TWA) time steps over the interval of interest, and compute the average of all the values. For example, assume we use the hemispherical diffusion model without advection given a constant emission rate, and we desire the 15-min TWA from t = 0 to t = 15 min. We compute 900 successive concentration values at 1 sec, 2 sec, and so forth, up to 900 sec after the start of emission. We add all the values and divide by 901 (where the extra one includes the zero concentration value at t = 0). This average is the 15-min TWA concentration value. Clearly, performing all these calculations by hand would not be feasible. However, if the equation is coded into an EXCEL® spreadsheet (or is coded using one of numerous mathematics software programs), computing and averaging 900 successive concentration values poses no problem.

7.8 A Numerical Method

It was previously noted that the traditional turbulent diffusion models do not maintain a contaminant mass balance in the room, and predict a concentration that becomes infinite as one moves close to the emission source. A numerical method based on matrix multiplication overcomes these limitations.(7) The method is outlined below; those readers desiring further details are referred to Reference 7.

Consider the movement of a molecule (or particle) due to turbulent diffusion along the room's x-axis. Divide the x-axis into small cells of length Δx (say, 0.1 m), and divide time into small intervals of duration Δt (say, 0.001 min). Consider a contaminant molecule that is in an interior room cell (not next to a room wall). In one time step Δt, the molecule can do one of three things: (1) it can move one cell to the left, with probability p., (2) it can move one cell to the right, with probability p, or (3) it can hold its position, with probability 1–2p. Due to symmetry, the molecule has the same probability p of moving to the left as it does of moving to the right, in which case the probability that the molecule holds its position is 1–2p. The probability p is related to D_T, Δx and Δt by the following expression:

$$p = \frac{D_T}{(\Delta x)^2}\Delta t \quad (7\text{-}13)$$

A square matrix is formed such that each x-axis cell is represented by a row and by a column. For example, if the length of the x-axis is 5 m and if Δx = 0.1 m, the matrix is 50 × 50. Number the left-hand most cell as #1 and the right-hand most cell as #50. Consider cell #25 near the room center. At matrix position (row = 25, column = 25), insert the value 1–2p; this is the probability that if the particle starts in cell #25, it remains there after one time step. At matrix position (row = 25, column = 24), insert the value p; this is the probability that if the particle starts in cell #25, it moves one step to the left (to cell #24) in one time step. Similarly, at matrix position (row = 25, column = 26), we insert the value p; this is the probability that if the particle starts in cell #25, it moves one step to the right (to cell #26) in one time step. All the other entries in the 25th row are zero, which means that in one time step, a molecule in cell #25 cannot move to cells #1–23 and #27–50.

The analogous procedure is followed for all the other rows in the matrix, except for the boundary rows #1 and #50. If a molecule in cell #1 tries to move to the left, it is reflected back by the wall. Therefore, the matrix position (row=1, column=1) is assigned the value 1–p, and the matrix position (row=1, column=2) is assigned the value p. Similarly, if a molecule in cell #50 tries to move to the right, it is reflected back by the wall. Therefore, the matrix position (row=50, column=50) is assigned the value 1 - p, and the matrix position (row=50, column=49) is assigned the value p. All values in the matrix are now assigned, and we label this matrix **PX**. Note that in every row, the sum of the entries equals one. A square matrix in which the entries in each row sum to one is termed a Markov matrix.

Next, consider that at time zero, a molecule is in cell i, where i = 1,2,..,50. Via matrix multiplication, we compute the probability that after n × Δt time steps have elapsed, this molecule is in cell j, j = 1,2,..,50. That is, we multiply the matrix **PX** by itself n times, and denote the product matrix $\mathbf{PX}^{(n)}$. The entry in the ith row and jth column of the product matrix, which we denote $PX^{(n)}_{i,j}$, is the probability that a molecule starting in cell i is in cell j after n × Δt time steps. If we started with 10^6 molecules in cell i instead of just one molecule, we can interpret $PX^{(n)}_{i,j}$ as the proportion of those molecules starting in cell i that are in cell j after n × Δt time steps.

A similar procedure is used to describe a molecule's movement via turbulent diffusion along the room's lateral y-axis and vertical z-axis. First, we assign the entries in a **PY** matrix and in a **PZ** matrix. The number of rows in each of these latter two matrices is determined by the length of the respective room axis. The entry in the ith row and jth column of **PY**, which we denote $PY^{(n)}_{i,j}$, is the probability that a molecule starting in cell i is in cell j after n × Δt time steps. The entry in the ith row and jth column of **PZ**, which we denote $PZ^{(n)}_{i,j}$, is the probability that a molecule starting in cell i is in cell j after n × Δt time steps.

Consider that a mass M_0 (mg) of contaminant molecules is emitted as a pulse at time zero into a room cell with coordinates (x_0, y_0, z_0). At time t = n × Δt, the proportion of mass in the cell with coordinates (x, y, z) is given by:

Proportion of M_0 in cell (x, y, z) at n × Δt =

$$PX^{(n)}_{x_0,x} \times PY^{(n)}_{y_0,y} \times PZ^{(n)}_{z_0,z}$$

Because D_T is assumed constant over space (and time), the movements along the three axes are considered mutually independent. In turn, the proportion of mass in cell (x, y, z) is the product of the respective proportions (probabilities) along each of the three axes. The contaminant concentration (mg/m³) in cell (x, y, z) at time t = n × Δt after the pulse release is given by the following expression:

$$C(x,y,z,t),\ \text{mg/m}^3 = \frac{M_0}{(\Delta x)^3} \times PX^{(n)}_{x_0,x} \times PY^{(n)}_{y_0,y} \times PZ^{(n)}_{z_0,z} \qquad (7\text{-}14)$$

Because Equation (7-14) accounts for all the contaminant mass emitted, it maintains a mass balance in the room. In addition, the maximum concentration that can exist in room air at the source is $M_0 \div (\Delta x)^3$.

In developing the method to this point, we have not accounted for contaminant removal from room air by exhaust ventilation. To be strict, we should specify where the exhaust outlet(s) is located on each axis, and let contaminant be removed only at the corresponding cells. However, this would invalidate the assumption of mutually independent movement along the three axes and, in turn, would require a more complicated matrix method also described in Reference 7. To permit using the relatively simple approach embodied in Equation (7-14), we adopt an approximation used in an analogous method for modeling turbulent diffusion.(1) In a well-mixed room, the probability that a molecule present in room air at time zero is still present at subsequent time t is equal to exp(− (Q/V) t). Thus, we multiply the right-hand side of Equation (7-14) by this factor to account for removal by exhaust ventilation:

$$C(x,y,z,t),\ \text{mg/m}^3 = \frac{M_0}{(\Delta x)^3} \times PX^{(n)}_{x_0,x} \times PY^{(n)}_{y_0,y} \times PZ^{(n)}_{z_0,z} \times \exp\left[-\frac{Q}{V} \times n \times \Delta t\right] \qquad (7\text{-}15)$$

A final consideration is accounting for ongoing contaminant emission. Consider the simplest case of constant emission at rate G mg/min. In each interval of duration Δt, the mass emitted into cell (x_0,y_0,z_0) is: G × Δt. Consider the mass emitted in the first interval after time zero. At future time t = n × Δt, the contribution of that mass to the contaminant concentration in cell (x, y, z) is:

$$C_1(x,y,z,t),\ \text{mg/m}^3 = \frac{G \times \Delta t}{(\Delta x)^3} \times PX^{(n)}_{x_0,x} \times PY^{(n)}_{y_0,y} \times PZ^{(n)}_{z_0,z} \times \exp\left[-\frac{Q}{V} \times n \times \Delta t\right]$$

The subscript "1" on C_1 (x, y, z, t) denotes the contribution due to emission in the first interval. Next, consider the mass emitted in the second interval after time zero. At future time t = n × Δt, the contribution of that mass to the contaminant concentration in cell (x, y, z) is:

$$C_2(x,y,z,t),\ \text{mg/m}^3 = \frac{G \times \Delta t}{(\Delta x)^3} \times PX^{(n-1)}_{x_0,x} \times PY^{(n-1)}_{y_0,y} \times PZ^{(n-1)}_{z_0,z} \times \exp\left[-\frac{Q}{V} \times (n-1) \times \Delta t\right]$$

The subscript "2" on C_2 (x, y, z, t) denotes the contribution due to emission in the second interval. Because (n–1) × Δt time steps elapse between the second interval and time t, the **PX**, **PY** and **PZ** matrices are multiplied n–1 times, and exhaust ventilation operates over (n–1) × Δt time steps. Next, consider the mass emitted in the third interval after time zero. At future time t = n × Δt, the contribution of that mass to the contaminant concentration in cell (x, y, z) is:

$$C_3(x,y,z,t),\ mg/m^3 = \frac{G \times \Delta t}{(\Delta x)^3} \times PX^{(n-2)}_{x_0,x} \times PY^{(n-2)}_{y_0,y} \times PZ^{(n-2)}_{z_0,z}$$

$$\times \exp\left[-\frac{Q}{V} \times (n-2) \times \Delta t\right]$$

The subscript "3" on C_3 (x, y, z, t) denotes the contribution due to emission in the third interval. Because (n–2) × Δt time steps elapse between the third interval and time t, the **PX**, **PY** and **PZ** matrices are multiplied n–2 times, and exhaust ventilation operates over (n–2) × Δt time steps. We continue in this fashion until we have accounted for the emissions in all the time intervals. At time t = n × Δt, the concentration in cell (x, y, z) is the sum of the individual contributions due to emissions in the n intervals:

$$C(x,y,z,t),\ mg/m^3 = \frac{G \times \Delta t}{(\Delta x)^3} \sum_{i=0}^{n-1} PX^{(n-i)}_{x_0,x} \times PY^{(n-i)}_{y_0,y} \times PZ^{(n-i)}_{z_0,z}$$

$$\times \exp -\left[\frac{Q}{V} \times (n-1-i) \times \Delta t\right] \qquad (7\text{-}16)$$

To compare Equation (7-16) to a traditional spherical turbulent diffusion model, consider the following scenario. Contaminant is emitted into room air at rate G = 1000 mg/min. The room has length × width × height dimensions of 8.1 m × 8.1 m × 3.1 m, such that the room volume V is 203 m^3. The room receives a dilution ventilation air flow corresponding to Q/V = 0.1 min^{-1}. Let Δx = 0.1 m such that: (i) the room's x-axis has 81 cells (#1 through #81 from left to right); (ii) the room's y-axis has 81 cells (#1 through #81 from front to back); and (iii) the room's z-axis has 31 cells (#1 through #31 from floor to ceiling). The contaminant is released at cell coordinate position (x = 41, y = 41, z = 16), which is in the middle of the room at 1.6 m above the floor. Assume D_T = 1 m^2/min. We are interested in estimating the contaminant concentration at 30 minutes after emission begins at cell coordinate positions (x = i, y = 41, z = 16) for i = 1,2,..,81. These positions are along the room's length axis, in the middle of the room from front to back, and at approximate breathing zone height.

For t = 30 min, the traditional spherical diffusion model equation is:

$$C(x,y = 4.1\ m, z = 1.6\ m, t = 30\ min),\ mg/m^3 =$$

$$\frac{79.6}{\sqrt{(x-4.1)^2}} \times \left[1 - \text{erf}\left[\frac{\sqrt{(x-4.1)^2}}{10.95}\right]\right]$$

for 0 ≤ x ≤ 8.1 m. The dashed line curve in Figure 5 depicts the concentrations predicted along the room's length axis by the above equation. In theory, the peak at cell #41 is an infinite concentration, because the value of x corresponds to the emission point. In contrast, the solid line curve in Figure 7.5 depicts the concentrations predicted along the room's length axis based on the method of Equation (7-16) for n = 300,000 and Δt = 0.0001 min. The numerical method predicts concentrations about 40% lower close to the emission source, and predicts concentrations about 60% higher near the room walls.

One drawback of the numerical method is that it requires the ability to write a computer code. Technical Appendix II at the end of this chapter presents a code written for MATLAB® software that illustrates how Equation (7-16) is implemented to generate Figure 7.5. A second drawback is that the method does not account for reflective surfaces due to objects in the room, although it does account for reflection by the walls, floor and ceiling. If the objects are small in scale relative to the room dimensions, and if we are interested in exposures at

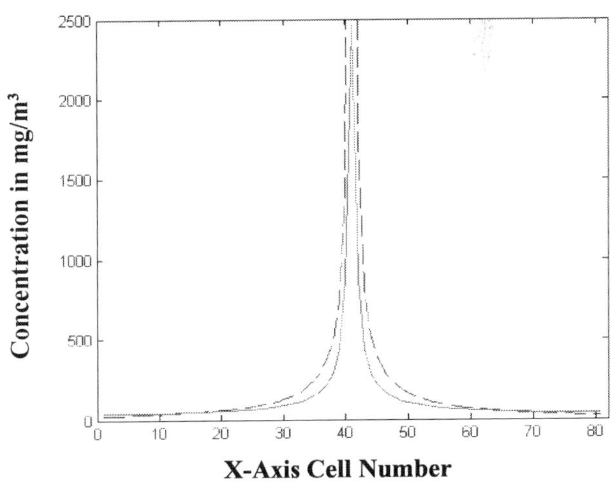

Figure 7.5 — Concentrations at different positions along the room's x-axis, and at height 1.6 m above the floor, at 30 minutes following the start of pollutant emission, as predicted by the traditional spherical turbulent diffusion model (the dashed-line curve) and by the numerical matrix method (the solid-line curve). The room dimensions are 8.1 m × 8.1 m × 3.1 m, and the dilution air exchange rate is 0.1 min^{-1}. The emission source is in the middle of the room at 1.6 m above the floor. Cell positions 1 and 81 are at room walls, and cell position 41 is midway along the x-axis.

positions not close to the source, this drawback is not so important. However, if the emission source is on a table and we are interested in exposures at the table, accounting for reflection by the table is important. As noted, Reference 7 describes a more complicated matrix method that can account for the positions of exhaust air outlets. It can also account for in-room reflective surfaces and for nonventilation loss mechanisms such as adsorption onto surfaces (for molecules) and gravitational settling (for particles).

7.9 Technical Appendix I Derivation of the Spherical Form of Equation

For spherical diffusion (with diffusion coefficient D_T, m^2/min) away from a point source with ongoing constant contaminant emission at rate G (mg/min), the concentration C (mg/m^3) that develops at radial distance r (m) at time T (min) after the start of emission is given by:

(A1) Constant Emission:

$$C(r,T) = \frac{G}{4\pi D_T r} \times \left[1 - \text{erf}\left(\sqrt{\frac{r^2}{4 D_T T}}\right)\right]$$

To derive this expression, consider the equation for spherical diffusion away from a point source following an instantaneous pulse release M (mg):

(A2) Pulse Release:

$$C(r,T) = \frac{M}{(4\pi D_T r)^{1.5}} \times \exp\left(-\frac{r^2}{4 D_T T}\right)$$

where C (r,T) is the concentration at radial distance r at time T after emission. Divide the time interval [0, T] into small subintervals of length Δt and consider that in each subinterval, a mass G × Δt is instantanously released. The continuous emission process is now discretized into a series of many pulse releases. The mass released in the ith subinterval contributes a portion of the concentration C (r,T) according to the expression:

$$\frac{G \times \Delta t}{(4\pi D_T [T - i \times \Delta t])^{1.5}} \times \exp\left(-\frac{r^2}{4\pi D_T [T - i \times \Delta t]}\right)$$

The concentration C(r,T) is the sum of the contributions due to the individual pulse releases:

(A3) Discretized Constant Emission:

$$C(r,T) = \sum_{i=1}^{T/\Delta t} \frac{G \times \Delta t}{(4\pi D_T [T - i \times \Delta t])^{1.5}} \times \exp\left(-\frac{r^2}{4\pi D_T [T - i \times \Delta t]}\right)$$

To be strict, the term T - i × Δt should be written as T – (i – 0.5) × Δt to ensure that there is never a zero quantity in a denominator. This is equivalent to saying that the pulse release occurs at the mid-point of the subinterval Δt.

Rearranging Equation (A3) yields:

(A4)

$$C(r,T) = \frac{G}{(4\pi D_T)^{1.5}} \sum_{i=1}^{T/\Delta t} \frac{1}{([T - i \times \Delta t])^{1.5}} \times \exp\left(-\frac{r^2}{4 D_T [T - i \times \Delta t]}\right) \times \Delta t$$

Note that if we change the order of summation (start from i = T/Δt and proceed to i = 1), the value of the sum does not change. As Δt → 0, the sum becomes a definite integral:

(A5)

$$C(r,T) = \frac{G}{(4\pi D_T)^{1.5}} \int_0^T \frac{1}{t^{1.5}} \exp\left(-\frac{r^2}{4 D_T t}\right) dt$$

Focusing on the definite integral, we make the change of variable:

$$x^2 = \frac{r^2}{4 D_T t} \rightarrow x = \sqrt{\frac{r^2}{4 D_T t}}$$

Several identities follow from this expressions for x as a function of t:

- for $t \rightarrow 0$, $x \rightarrow \infty$
- for $t = T$, $x = \sqrt{\frac{r^2}{4 D_T T}}$
- $x^2 = \frac{r^2}{4 D_T t} \rightarrow t = \frac{r^2}{4 D_T x^2} \rightarrow t^{1.5} = \left[\frac{r^2}{4 D_T}\right]^{1.5} \frac{1}{x^3} \rightarrow \frac{1}{t^{1.5}} = \left[\frac{r^2}{4 D_T}\right]^{-1.5} x^3$
- $dt = \frac{r^2}{4 D_T}(-2x^{-3})\,dx = -\frac{2r^2}{4 D_T}\frac{1}{x^3}\,dx$

Based on the change of variable, the original definite integral becomes:

$$\int_0^T \frac{1}{t^{1.5}} \exp\left(-\frac{r^2}{4 D_T t}\right) dt =$$

$$\int_{\infty}^{\sqrt{r^2/4D_TT}} -2\left[\frac{r^2}{4\,D_T}\right]^{-1.5} x^3 \left[\frac{r^2}{4\,D_T}\right]\frac{1}{x^3}\exp(-x^2)\,dx$$

$$= \int_{\infty}^{\sqrt{r^2/4D_TT}} -2\left[\frac{r^2}{4\,D_T}\right]^{-0.5}\exp(-x^2)\,dx$$

Reversing the range of integration eliminates the minus sign:

$$= \int_{\sqrt{r^2/4D_TT}}^{\infty} 2\left[\frac{r^2}{4\,D_T}\right]^{-0.5}\exp(-x^2)\,dx$$

Bringing a coefficient outside the integral yields:

$$\left[\frac{r^2}{4\,D_T}\right]^{-0.5}\int_{\sqrt{r^2/4D_TT}}^{\infty} 2\exp(-x^2)\,dx$$

Multiplying the integral by the term $\sqrt{\pi}/\sqrt{\pi}$ yields:

$$\left[\frac{r^2}{4\,D_T}\right]^{-0.5}\sqrt{\pi}\int \frac{2}{\sqrt{\pi}}\exp(-x^2)\,dx =$$

$$\left[\frac{r^2}{4\,\pi\,D_T}\right]^{-0.5}\int_{\sqrt{r^2/4D_TT}}^{\infty}\frac{2}{\sqrt{\pi}}\exp(-x^2)\,dx$$

The new integral is recognized as the complement of the error function (erf) such that:

$$\int_{\sqrt{r^2/4D_TT}}^{\infty}\frac{2}{\sqrt{\pi}}\exp(-x^2)\,dx =$$

$$1-\int_{0}^{\sqrt{r^2/4D_TT}}\frac{2}{\sqrt{\pi}}\exp(-x^2)\,dx =$$

$$1-\mathrm{erf}\left[\sqrt{\frac{r^2}{4\,D_T\,T}}\right]$$

It follows that:

$$\int_0^T \frac{1}{t^{1.5}}\exp\left[-\frac{r^2}{4\,D_T\,t}\right]dt =$$

$$\left[\frac{r^2}{4\,\pi\,D_T}\right]^{-0.5}\int_{\sqrt{r^2/4D_TT}}^{\infty}\frac{2}{\sqrt{\pi}}\exp(-x^2)\,dx$$

$$=\left[\frac{r^2}{4\,\pi\,D_T}\right]^{-0.5}\left[1-\mathrm{erf}\left[\sqrt{\frac{r^2}{4\,D_T\,T}}\right]\right]$$

$$=\frac{(4\,\pi\,D_T)^{0.5}}{r}\left[1-\mathrm{erf}\left[\sqrt{\frac{r^2}{4\,D_T\,T}}\right]\right]$$

This identity for the original integral is substituted into Equation (A5):

$$C(r,T)=\frac{G}{(4\,\pi\,D_T)^{1.5}}\int_0^T\frac{1}{t^{1.5}}\exp\left[-\frac{r^2}{4\,D_T\,t}\right]dt$$

$$=\frac{G}{(4\,\pi\,D_T)^{1.5}}\,\frac{(4\,\pi\,D_T)^{0.5}}{r}\left[1-\mathrm{erf}\left[\sqrt{\frac{r^2}{4\,D_T\,T}}\right]\right]$$

$$=\frac{G}{4\,\pi\,D_T r}\left[1-\mathrm{erf}\left[\sqrt{\frac{r^2}{4\,D_T\,T}}\right]\right]$$

Thus, we have derived Equation (A1).

7.10 Technical Appendix II — MatLab Code Used to Generate Figure 7.5.

```
% figure5.m
G = 1000; % emission rate of 1000 mg/min
DT = 1.0; % 1 m2/min
deltaX = 0.1; % interval length in meterw
deltaT = .0001; % one ten thousandth of a minute
p = DT*deltaT/(deltaX^2);

L = 81; % number of cells on x axis
W = 81; % number of cells on y axis
H = 31;% number of cells on z axis

RPX = 41; % Release cell on the length axis, 4.1 m from the left wall
RPY = 41; % Release cell on the width axis, 4.1 m from the front wall
RPZ = 16; % Release cell on the vertical axis, 1.5 m above the floor

N = 300000; % number of time stepsN x deltaT = 30 min
T = N*deltaT; % N x deltaT = 30 min

% The traditional spherical diffusion model
CONCENTRATIONS1 = zeros(L,1);
for x = 1:L
 xx = x/10; % in meters
 a = (xx - 4.1 - 0.01)^2;
 r = sqrt(a);
 CONCENTRATIONS1(x,1) = (G/(2*pi*DT*r))*(1 - erf(r/sqrt(4*DT*T)));
end
PX = zeros(L,L);
for i = 2:(L-1)
 PX(i,i) = 1 - (2*p);
 PX(i,i-1) = p;
 PX(i,i+1) = p;
end
PX(1,1) = 1-p; PX(1,2) = p;
PX(L,L) = 1-p; PX(L,L-1) = p;

PY = zeros(W,W);
for i = 2:(W-1)
 PY(i,i) = 1 - (2*p);
 PY(i,i-1) = p;
 PY(i,i+1) = p;
end
PY(1,1) = 1-p; PY(1,2) = p;
PY(W,W) = 1-p; PY(W,W-1) = p;

PZ = zeros(H,H);
for i = 2:(H-1)
 PZ(i,i) = 1 - (2*p);
 PZ(i,i-1) = p;
 PZ(i,i+1) = p;
end
PZ(1,1) = 1-p; PZ(1,2) = p;
PZ(H,H) = 1-p; PZ(H,H-1) = p;

GperdeltaT = G/(1/deltaT); % 0.1 mg emitted per deltaT
ACM = 0.1; % 0.1 air change per minute
```

```
CellVolume = (deltaX)^3;

PXTEMP = PX;
PYTEMP = PY;
PZTEMP = PZ;
OUTPUTX = zeros(N,L);
OUTPUTY = zeros(N,1);
OUTPUTZ = zeros(N,1);
for n = 1:N
 for x = 1:L
 OUTPUTX(n,x) = PXTEMP(RPX,x);
 end
 OUTPUTY(n,1) = PYTEMP(RPY,41);
 OUTPUTZ(n,1) = PZTEMP(RPZ,16);
 PXTEMP = PXTEMP*PX;
 PYTEMP = PYTEMP*PY;
 PZTEMP = PZTEMP*PZ;
end

% COMPONENTS contains the contributions to the final concentration at
% time N x delatT due to the emissions in the N subintervals.
COMPONENTS = zeros(N,L);
for n = 1:N
 for x = 1:L
 term1 = GperdeltaT/CellVolume;
 term2 = OUTPUTX(n,x)*OUTPUTY(n,1)*OUTPUTZ(n,1);
 term3 = exp(-ACM*n*deltaT);
 COMPONENTS(n,x) = term1*term2*term3;
 end
end

CONCENTRATIONS2 = zeros(L,1);
for x = 1:L
 CONCENTRATIONS2(x,1) = sum(COMPONENTS(:,x));
end
CELLSXAXIS = (1:81)';
plot(CELLSXAXIS,CONCENTRATIONS1,'—',CELLSXAXIS,CONCENTRATIONS2,'-')
xlabel('\bf \fontname{times} X-Axis Cell Number')
ylabel('\bf \fontname{times} Concentration in mg/m^{3}')
axis([0 82 0 2500])
```

7.11 References

1. **Drivas, P.J, P.A. Valberg, B.L. Murphy and R. Wilson:** Modeling Indoor Air Exposure from Short-Term Point Source Releases. *Indoor Air 6*:271–277 (1996).
2. **Franke, J.E., and R.A. Wadden:** "Some Observation of Eddy Diffusivity in Industrial Settings." Presentation # 307, American Industrial Hygiene Conference & Exposition, St. Louis Mo., 1989.
3. **Wadden, R.A., P.A. Scheff, and J.E. Franke:** Emission Factors for Trichloroethylene Vapor Degreasers. *Am. Ind. Hyg. Assoc. J. 50*:496–500 (1989).
4. **Scheff, P.A., R.L. Friedman, J.E. Franke, L.M. Conroy, and R.A. Wadden:** Source Activity Modeling of Freon Emissions from Open-Top Vapor Degreasers. *Appl. Occup. Environ. Hyg. 7*:127–134 (1992).
5. **Conroy, L.M., R.A. Wadden, P.A. Scheff, J.E. Franke, and C.B. Keil:** Workplace Emission Factors for Hexavalent Chromium Plating. *Appl. Occup. Environ. Hyg. 10*:620– 627 (1995).
6. **Nicas, M. and T.W. Armstrong:** Computer Implementation of Mathematical Exposure Modeling. *Appl. Occup. Environ. Hyg. 18*:566–571 (2003)
7. **Nicas, M.:** Modeling Turbulent Diffusion and Advection of Indoor Air Contaminants by Markov Chains. *Am. Ind. Hyg. Assoc. J. 62*:149–158 (2001).

Knowledge Based Models

By
Thomas W. Armstong, PhD, CIH
Bradley W. Jessup, CIH, CSP, CHMM
Susan F. Arnold, CIH

8.0 Overview

Knowledge based models include those that use rules in an "expert system" approach to deliver decisions about exposure. The concept is to mirror how an expert in industrial hygiene exposure assessment assembles details about the exposure scenario and uses judgment to interpret the relative influence of the many variables on the exposure. The expert judgment process applied in the art and science of industrial hygiene exposure assessment uses many comparative rules and logic that branches in an often complex and branching pattern. The branches followed depend very much on the scenario and the strength of information available. The two segments of this chapter cover two somewhat different approaches in knowledge based (or expert system) models. The first discusses WORKSPERT and WORKBOOK.[1–3] WORKSPERT arguably better illustrates the conceptual framework of expert systems than the publications on WORKBOOK do. The EASE 2.0 model covered in the second segment is a hybrid model that incorporates aspects of expert systems with a database. EASE uses branching decision rules to select and report applicable data from an underlying structured database.

8.1 WORKSPERT and WORKBOOK

There are few published examples of knowledge based (or expert system) models implemented for industrial hygiene exposure assessment processes. The two identified that illustrate the concepts and implement certain aspects are the WORKSPERT and WORKBOOK models.[1–3]

8.1.1 WORKSPERT

WORKSPERT was developed to perform three major functions:

1. Define Homogeneous Exposure Groups (also known as Similar Exposure Groups)
2. Assist in qualitatively determining the relative risks of different worker groups' exposures to workplace hazards and defining exposure monitoring program needs
3. Provide recommendations on the need for additional workplace exposure controls.

WORKSPERT accomplishes these objectives by using inference rules in three broad categories of variables: substance (V_s), workplace (V_w), and exposure (V_E). These variables also have associated weighting factors that influence the magnitude of the factor in the program algorithm. The WORKSPERT structure described[1] includes 27 questions, each with multiple choice responses. The Substance category has 8 questions, the Workplace category has 9, and the Exposure category has the remaining 10. Each of the 27 questions has a list of rank scores and a weighting factor for the possible "multiple choice" answers. The results for each category are standardized to a score between 0 and 1. The responses to each may lead to a different branch in the subsequent questions, or truncate the process directly to the conclusion. The answers and weighting variables placed on them by the program designers determine the subsequent questions and the recommendations when the applicable logic branch is completed. Other experts could use different rules or different weightings and end up with similar or divergent recommendations.

Example WORKSPERT Application

The Substance category has 8 questions. For example, a question in the Substance category "Is the substance regulated in the workplace" has a five part answer, and each part has a rank score and weighting factor:

Rank	*Weighting*	*Description*
5	1	OSHA health standard with exposure limits
4	1	OSHA PEL

3	1	Proprietary substance with exposure limit(s)
2	1	Nuisance contaminant with exposure limit(s)
1	0	Proprietary substance without exposure limit)s
0	—	Substance not regulated

The first response "OSHA health standard with exposure limit" would contribute a rank of 5 and weighting coefficient of 1 to the algorithm. This would lead to supporting a recommendation to conduct air monitoring, depending on the responses to subsequent questions. A response "Substance not regulated" contributes a rank score of 0 and no weighting coefficient. This answer would lead (depending on subsequent questions) to a "do not monitor exposures" conclusion.

The other 26 questions are structured similarly, and address different considerations about the substance, workplace and exposure. For further details, see appendix 1 of the original publication.(1)

The WORKSPERT algorithm then uses the scores to the questions to provide the recommendations:

1. Monitor the workplace or do not monitor
2. Survey strategy to employ (initial baseline, periodic update, workplace areas. employee exposures)
3. Sample duration (full shift, part shift, short term)
4. Multiple substance monitoring (or single substance)
5. Recommended number of samples (continuous monitoring, more than 10 samples, 6 to 10 samples, at least 5 samples, at least 2 samples, at least 1 sample or not required)

8.1.2 WORKBOOK

WORKBOOK(2) uses a series of questions to assess whether or not the exposures are acceptable (no further action), require further evaluation, or require additional controls. WORKBOOK's decision logic as described provides a concise illustration of an expert system relevant to industrial hygiene.

The major blocks in WORKBOOK's logic include:

1. Initial characterization; the assembly of information about the workplace environment, processes, hazards, implemented hazard controls and workers
2. Formation of initial Homogeneous Exposure Groups (also known as Similar Exposure Groups): the analysis of work patterns and tasks to stratify the total workforce into group in which each group's members work in a similar way (environment, time, work methods) with similar hazards (composition, quantities)
3. Workplace exposure assessment (qualitative tools, quantitative tools, monitoring/measurements)
4. Decisions on the workers exposures: acceptable, unacceptable, or uncertain
5. Implement the hierarchy of exposure controls as appropriate
6. Evaluate the effectiveness of current exposure controls
7. Re-evaluation (exposure review) frequency (not needed, annual, other)

These stages are undoubtedly familiar to experienced industrial hygienists and reflect approaches commonly used in industrial hygiene investigations. In this sense, the logic reflects an expert's approach and is an expert system model. It is important to note that WORKBOOK has a fixed set of rules implemented to meet the particular design goals of its development team. The rules were chosen to be no more complex and no less complex than needed to accomplish the program's objectives. Thus, it has limitations, and if used outside its design domain, its logic and rules likely fail. By studying the rules given in WORKBOOK's description(2) one can better understand the expert systems strengths and limitations. WORKBOOK has a series of questions following the flow of the 7 steps outlined above that implement the program's function. For example, WORKBOOK question 3a is "What health hazards are posed by the substance?" The user must then choose the most appropriate answer from those listed:

5. Life-threatening or permanently disabling injury/illness
4. Irreversible adverse health effects of serious concern
3. Severe reversible adverse effects of serious concern
2. Reversible effects of moderate concern
1. Reversible effects of limited concern
0. No known of suspected adverse health effects

Each of these choices has a relative weighting factor that the program uses in deriving the recommendations from the assessment. The questions themselves require a degree of expert interpretation and value judgments for appropriate application. For example, what does "high level noise exposure that causes a permanent partial hearing loss" represent? Is it "5" or "4" in the rating scheme outlined above? An organization using WORKBOOK or developing their own expert system may encounter the need for explicit definitions of the values and decisions behind such rules. How would a suspected human carcinogen with equivocal evidence and perhaps moderate to low potency be rated? This example is given mainly to point out that often when one digs deeper into an expert system, the value judgments behind the rules can become somewhat murky.

WORKBOOK was developed in the software package TOOLBOOK. Since WORKBOOK's development (prior to the 1993 publication) expert system development software has evolved somewhat. However, further industrial hygiene expert system development does not seem to have followed at this time. Descriptions of fuzzy sets(4), fuzzy logic and expert system development applications using those approaches also may have utility in mirroring the industrial hygiene decision logic where more often than not the answers "maybe" or "it depends" are more likely than a firm "yes" or "no". Some interesting research to incorporate these judgment decisions in a robust and transparent Bayesian framework has been conducted.(5) This research examines and evaluates the use of Bayesian statistics in combining information from varied sources including models, measured data and 'other' information to estimate worker exposure.

8.2 Estimation and Assessment of Substance Exposure, EASE Modeling Software

A software program entitled "Estimation and Assessment of Substance Exposure" or "EASE" was developed in the 1990s by the United Kingdom's Health and Safety Executive (HSE) for assistance in modeling worker exposures in the workplace. A main application of the EASE software is estimating exposure prior to manufacturing and use of a new chemical. The program is broad-based and user friendly but has several aspects which limit the appropriate uses of the 2.0 Version of the software. The software is in the "expert system" category, and thus uses a set of standard definitions and rules to define the scenario and substance to be estimated. Once the scenario is defined according to the EASE vocabulary, the software links to results from the HSE inspectorate database of factory air monitoring, and reports the range of exposures associated with the most similar substances and scenarios. This hinges on the structure of the database, which is both one of the strengths (based on real data) and restrictions (limited to the database contents) of the software. Once the user knows the EASE system vocabulary and rules structure, assessment may be completed remarkably quickly.

The EASE 2.0 package provides an overview in its help system, extracts of which follow:

> *" ... The system is able to provide estimates for inhalation exposure to dust, inhalation exposure to vapour, and dermal exposure. EASE for Windows starts by requesting basic administrative information: the name of the user, the name of the substance, and the name of a file which will be used to hold a record of the session.*
>
> *After requesting the temperature of the process, the system asks for the physical state of the substance at that temperature. The system's next few questions are dependent on the physical state which was specified, but they are all directed towards determining which types of exposure might be relevant for this substance. For example, inhalation exposure to vapour will only be relevant to a solid which has a significant vapour pressure. At the end of this phase, the system will ask which exposure type should be investigated.*
>
> *The system now moves into its second phase of questions. The actual questions which are asked will depend very much on the exposure type chosen; there are also some questions which are only triggered by particular answers to other questions. ... the estimation of vapour pressure based on chemical information about a substance uses a derivative of the Antoine equation. (Further details of the methods used are given in System Calculations.) The questioning continues until an exposure range has been derived. At any time during the session, the user is able to change values given previously, causing the system to re-run from the point at which the value of the chosen property was specified. The user can also obtain an explanation of the system's conclusion, view an explanation of how to convert exposure ranges, view the current contents of the log file, or exit from the system. After exiting from the system, the log file can be printed or stored, as appropriate."*

HSE published a 127 page research report in 2003 which is entitled: *Evaluation and further development of the EASE model 2.0.*(6) A summary is also available.(7) The following discussion provides a short summary of the findings and recommendations of those reports.

Individuals who are interested in pursuing an in depth analysis of EASE should obtain a copy of the HSE Research Report (available online http://www.hse.gov.uk/research/rrhtm/rr136.htm). In that report, one will find the history of the development of the EASE model, its evaluation by stakeholders, a review of some alternative models and a discussion of the way forward. The last chapter of that HSE report states: "For almost ten years, EASE has provided a way for regulators to estimate occupational chemical exposure. It has served well and there is no evidence that it has resulted in unsafe regulatory decisions. However, there is both documentary evidence and the strong opinion amongst many stakeholders that the predictions from EASE are not sufficiently reliable for future needs. The information used to define the exposure ranges in EASE is likely to be more than 20 years old and there have been many changes in European workplaces during this time. There is also increasing concern about the highly conservative nature of the EASE predictions and the potential impact of these on the competitiveness of European businesses. However, one of the strengths of EASE is that it maps the situation for which an exposure estimate is required onto a situation for which you have some information. The weakness is that this is not always done well."

8.3 Validation Studies

When one is evaluating a model for use, a prime question is, has the model been validated? Although the dictionary definition of validation is clear, the use of the term with respect to exposure models is not so well defined. To some, it means demonstrating that the computer code correctly and reliably executes. To others, validation means do the results of the model compare well (perhaps by some statistical criteria) to actual exposures, possibly as evaluated by reliable monitoring data. A report[(8)] summarizes validation study of an expert system model.

A validation of the EASE model[(9)] reported somewhat mixed results. Two scenarios evaluated with real data (handling of volatile substances in closed systems, handling 1 to 200 tonne of dry powdered substances) compared well. EASE scenarios involving direct handling with natural ventilation or direct handling with local exhaust ventilation did not compare as well.

Both model systems strive to provide for and incorporate expert knowledge into a logical framework. The models' utility are limited by the need for value based judgments that are inherently scenario and often organization specific. Future developments of fuzzy logic and Bayesian statistic based tools may yield more broadly applied systems.

8.4 References

1. **Tait, K.:** The Workplace Exposure Assessment Expert System (WORKSPERT). *Am. Ind. Hyg. Assoc. J. 53(2)*:84–98 (1992).
2. **Tait, K.:** The Workplace Exposure Assessment Workbook (WORKBOOK). *Appl. Occup. Environ. Hyg. 8(1)*:55–68 (1993).
3. **Tait, K. and M. Mehta:** Validation of the Workplace exposure Assessment Expert System (WORKBOOK). *Am. Ind. Hyg. Assoc. J. 58(8)*:592–602 (1997).
4. **Zahed, L.A.:** The Concept of a Linguistic Variable and its Application to Approximate Reasoning, Parts 1 & 2. *Info. Sci. 8*:199–249 (1975).
5. **Sottas, P.E., J. Lavoué, R. Bruzzi, D. Vernez, N. Charrière, and P.O. Droz:** An empirical hierarchical Bayesian unification of occupational exposure assessment methods. *Stat. Med. 28(1)*:75–93 (2009).
6. **Cherrie, J. et al.:** *Evaluation and Further Development of EASE Model 2.0.* Research Report 136. Norwich, U.K.: HSE Books, 2003.
7. **Creely, K.S. et al.:** Evaluation and Further Development of EASE Model 2.0. *Ann. Occup. Hyg. 49(2)*:135–145 (2005).
8. **Elliott, L.J. and R.K. Oestenstad:** Evaluation of the predecitive abilities of a qualitative exposure assessment model. *J. Occup. Environ. Hyg. 4(6)*: 440–447 (2007).
9. **Brendendiek-Kämper, S.:** Do EASE Scenarios Fit Workplace Reality? A Validation Study of the EASE Model. *Appl. Occup. Health Env. 16(2)*:182–187 (2001).

Statistical Models of Exposure Determinants

By
M. Abbas Virji, ScD, CIH
Thomas W. Armstrong PhD, CIH

9.0 Introduction

Exposure monitoring has traditionally focused on collecting full-shift time-weighted average (TWA) exposure measurements with the primary objective of demonstrating compliance with regulatory limits. Sampling strategies have relied on identifying and monitoring the highest exposed workers rather than on random monitoring of workers or groups of workers. Additionally, very little contextual information has been collected during monitoring on characteristics of the processes, work environment, control measures, environmental conditions, job tasks and worker activities. For example, the National Institute for Occupational Safety and Health (NIOSH) Occupational Exposure Sampling Strategies Manual identifies specific information on sampling and analysis to be recorded, but provides only minimal guidance on collecting information on work environment characteristics.[1] Resulting databases, such as the Occupational Safety and Health and Administration Integrated Management Information System (OSHA IMIS), consist of little descriptive information beyond job, industry, year of inspection, company size, and other higher order information.[2] As a consequence, such databases serve limited objectives beyond their intended purposes, despite the potential and need for multiple uses. Nevertheless, the IMIS data have been used successfully to illustrate trends in exposure over time or among industry sectors.[3] In practice, even when some contextual information is gathered during monitoring, it is generally in the form of notes and is not systematically collected, hence limiting its utility. The contemporary practice of industrial hygiene is moving away from compliance-based monitoring toward comprehensive exposure assessment.[4–6] There is also the recognition that exposure modeling is an integral part of exposure assessment[7]; both monitoring and modeling are required to improve the efficiency of a sampling strategy and to achieve a more complete exposure assessment.[8]

This chapter focuses on statistical models applied to measured exposure data to evaluate the underlying nature of exposures and causes of variability. It describes the mathematical basis for these models and provides some applied examples. Citations within the text do not represent a comprehensive review of the literature. The reader is referred to additional useful references for a thorough and complete treatment of the topic.

9.1 Statistical Models

In industrial hygiene, models are used to describe the relationships between exposure levels and factors affecting/determining exposures via deterministic or statistical model designs. Deterministic models are based on principles of physics and chemical properties of the exposure substance and can range from simple vapor pressure models to more complex models characterizing the generation and transport of exposure.[9] Many deterministic models, based on first principles such as the conservation of mass, energy, or momentum, are discussed in other chapters of this book. Deterministic models are focused on underlying mechanisms involved in the exposure paradigm, including the initial process of contaminant generation to the ultimate step of estimating breathing zone exposures.

Other models arise from using observational and experimental data. In exposure assessment, these empirical models are developed to examine and estimate the relationship between a dependent exposure variable and a set of independent predictor variables related to characteristics of the work environment. These models rely on statistical techniques, using exposure measurements combined with either quantitative (continuous) or qualitative (categorical) data on the work environment collected simultaneously with the monitoring data. Therefore, statistical modeling of exposure determinants requires the collection of both exposure measurements and detailed contextual information on the work environment. One of the impediments of statistical modeling using existing or historical exposure data are the lack of systematically gathered contextual information on the work environment.

In exposure assessment models, the dependent variable is usually an exposure metric such as the full-shift TWA exposure or short-duration task exposure. Independent variables represent one or more factors affecting exposure or surrogates for such factors and may or may not be directly related to the physical and chemical factors used in deterministic models. More often, independent variables include a range of factors that describe exposure-generating processes, the transport of exposure and worker interface with exposure.(10) Hence, the model is often a mathematical expression translating observable or measurable information likely to affect exposures and quantifying the impact of that information on exposure levels. A diversity of determinants-of-exposure models have been published in the literature, making it difficult to summarize and understand exposure factors, thereby underscoring the need for a structured approach to studying exposure determinants. Tielemans et al.,(11) have proposed a conceptual model that describes the processes leading to exposure and suggest a set of broadly defined exposure modifying factors within which specific determinants can be identified and used in either deterministic or statistical models.

A number of statistical techniques have been used for modeling exposure determinants including: 1) linear regression; 2) random- and mixed-effects analysis of variance (ANOVA); 3) logistic regression; and 4) other regression techniques (e.g., Tobit regression). These models are briefly described in the next sections.

9.1.1 Linear Regression Models

Linear regression is typically used to study the determinants of exposure when modeling continuous exposure measurements that are combined with information on characteristics of the work environment. By identifying factors of the work environment that contribute most to exposures, these models can assist in prioritizing interventions for controlling workplace exposures. Multiple linear regression modeling is the most commonly used technique to investigate determinants of exposure, and is of the general form:

$$\text{Ln}\,[C] = \beta_0 + \beta_1\,(\text{Var}_1) + \beta_2\,(\text{Var}_2) + \ldots + \beta_n\,(\text{Var}_n) + E \qquad (9\text{-}1)$$

where

- Ln [C] = natural logarithm of the exposure concentration, since most occupational exposure data have a more lognormal than normal distribution,
- β_0 = intercept,
- β_1–β_n = regression coefficients of the independent predictor variables 1 to n,
- Var_1–Var_n = possible independent exposure predictors or modifiers, and
- E = error term.(12)

The intercept represents the overall mean exposure level when all independent variables are equal to zero; it is also often considered an estimate of the background exposure level. Regression coefficients represent the change in exposure levels associated with each one-unit change in the independent variables. Dichotomous independent variables are modeled as (0/1), while polychotomous variables are transformed into dummy dichotomous variables (0/1), selecting one level as the reference level. The error term in the statistical model represents the difference between the predicted and observed exposure measurements, but is not used in the mathematical expression to predict exposure levels. However, the variance associated with the error term is used to construct confidence intervals for predictions. Kleinbaum et al.(12) describe in detail the assumptions associated with using multiple linear regression models, and include:

1) Ln [C] is a random variable dependent on Var_i values,
2) Values of Ln [C] are statistically independent,
3) The mean value of Ln [C] is a linear function of the Var_i values, though product terms can be included that allow for interactions,
4) The variance of Ln [C] is the same for any Var_i, and
5) For any value of Var_i, Ln [C] is normally distributed.

It is noteworthy that the linear regression model is not appropriate for repeat measurements on the same worker or when the data are highly correlated within a job group or worksite. A mixed model (described in the next section) is required for such data. In addition, measurement error in the explanatory variables (Var_i) can lead to biased estimates of the regression coefficients and thus it is important to attempt to understand the sources of measurement error in the explanatory variables. A summary of linear regression models used for studying the determinants of exposure, including examples, is contained in texts by Boleij et al.(13) and Burdorf.(14)

9.1.2 Mixed Models

A large number of published studies continue to show considerable variation in exposure between workers performing the same job, as well as differences in exposure levels from day-to-day.(15) Random effects ANOVA models have been used to model exposure data with repeated measurements on workers to obtain estimates of within- and between-worker components of exposure variability.(13) Understanding of variance components can have important implications for developing strategies for sampling, exposure control, and epidemiological grouping. The more recent use of mixed models permits the identification of important factors affecting exposures, while providing additional information on within- and between-worker variability.[6] The natural logarithm of exposure is modeled as the dependent variable, with person effects declared as random and workplace characteristics as fixed effects. The model is of the general form:

$$Y_{ij} = Ln(X_{ij}) = \mu + \beta_{1\text{-}k} + X_i + \varepsilon_{ij} \quad (9\text{-}2)$$

where

Y_{ij} = natural logarithm of the exposure of the ith person on the jth day,
μ = overall mean of log-transformed exposure,
β_1–β_k = fixed effects of exposure determinants,
X_i = random effect of the ith worker, and
ε_{ij} = random error.

Parameter estimates of the significant fixed effects ($\beta_{1\text{-}k}$) provide estimates of the magnitude of the change in average exposure associated with each factor (exposure determinants). From the model, estimates of the within-worker ($_{ww}S^2$) and between-worker ($_{bw}S^2$) variances are obtained. Assumptions of the mixed models are described in detail by Kleinbaum et al.[12] and Rappaport and Kupper.[6] Mixed models are now commonly used to understand and determine factors affecting exposure and can include any of the explanatory variables included in the regression model (9-1). In addition, mixed models can account for correlation between repeated measurements on the same worker and among samples collected from the same work area of facility. Moreover, mixed models can accommodate a grouping variable to evaluate whether workers in different groups have different exposures and different variance components. Information on variance components can be used to identify types of control measures that are most beneficial, as well as to determine exceedance probabilities and average exposure for individual workers.[6] A summary of mixed models, including examples, is also provided by Burdorf.[14]

9.1.3 Logistic Regression Models

Modeling using dichotomous exposure data (i.e., yes/no) obtained from survey questionnaires, exposure measurements above/below the limit of detection (LOD), or exposure measurements above/below some identified threshold value (e.g., PEL) requires special consideration. Logistic regression models have been used to identify factors affecting exposure from severely censored (>50% LOD) datasets[16] and from data exceeding regulatory limits.[17] Logistic regression models can be considered a special case of linear regression for dichotomous dependent exposure variables. The exposure measurements or scores are dichotomized (1/0) to mark an exposure event. The unconditional logistic regression model is of the general form[12]:

$$\text{Logit}\,(\text{pr}\,Y{=}1) = \alpha + \beta_1\,(Var_1) + \beta_2\,(Var_2) + \ldots + \beta_n\,(Var_n) + E \quad (9\text{-}3)$$

The Logit (pr Y=1) is Ln (Pr Y=1/Pr Y=0). The logit model above describes the ratio of the probabilities for and against an exposure event (the outcome variable) as a function of:

a = constant term,
β_1–β_n = parameter estimates for the exposure factors (Var_1–Var_n),
Var_1–Var_n = possible independent exposure predictors or modifiers, and
E = error term.

From the model described above, odds ratios and confidence intervals for exposure determinants are available. If the probability of an exposure event is high (not rare), then the odds ratios from the logistic model provides biased estimates of the relative risk (in this case of exposure event).[18] Methods have been developed to convert the odds ratio to prevalence ratio[18] or to directly estimate prevalence ratio from a Cox regression model.[19] For repeated measurements on the binary outcome variable, general estimating equations can be used to model exposure determinants.

9.1.4 Other Regression Models

Modeling highly censored exposure data requires special consideration to adjust for truncation of the exposure distribution. Modeling truncated continuous exposure measurements can be achieved through Tobit regression, which uses the maximum likelihood method to estimate parameters (i.e., exposure determinants). The model uses log-transformed exposure measurements as the outcome variable with the assumption that the data below the LOD follow the same distribution as the observed data. The log-likelihood function has two components, one for observed data and the other for data below the LOD.[20] Maximum likelihood estimates of β_1 through β_n are then obtained by maximizing the log likelihood function. Woskie et al.[16] used this approach to model determinants of isocyanate exposures in automobile body repair shops. Methods have also been developed to model highly censored data with repeated measurements using mixed models.[21]

9.2 Model Development Strategies

Many studies have applied the principles of multiplicative models, including the following reviews and examples.[22–25] A review by Burstyn and Teschke[10] illustrates the range of approaches and methods used to develop exposure determinants models. The authors also show that a large number of predictor variables have been used in these models that are often specific to particular workplace circumstances. Hence, a universal model of exposure determinants is not feasible, and would not yield useful results. An important initial step before modeling is to develop a conceptual framework of how various factors may impact exposure levels and the inter-relationships among factors.

The following steps are provided as a framework for readers to develop an exposure model based on specific workplace determinants. Statistical model development is well described by Kleinbaum et al.[12]

9.2.1 Develop a Conceptual Model

Starting with a conceptual model, such as the source-receptor model, allows for a more consistent approach to collecting information on potential classes of exposure determinants. A source-receptor model describes the physical pathway of exposure from its source through different compartments and mechanisms to the route of entry into the body. Figure 9.1 illustrates the three components of a source-receptor model and provides some examples of general factors that may affect exposure (adapted from Armstrong et al.[23]); note that specific operations factors may include additional sub-factors to consider. Additionally, there are higher order factors related to workplace, industry, and calendar time that may also directly or indirectly impact determinants of exposure, thus affecting exposure levels.

Developing a conceptual model requires an understanding and knowledge of: processes, raw materials, by-products and end-products; mechanisms and factors affecting exposure generation; mechanisms and factors affecting transport of exposure; individual activities and opportunities for direct or indirect contact with exposure; and routes of entry and associated factors. A joint ACGIH®-AIHA task group on occupational exposure databases issued a report that identified over 100 data elements to be collected with exposure data.[26] These data elements are exposure determinants that can serve as a starting point toward building the conceptual model. Tielemans et al.[11] have recently proposed a conceptual model for estimating inhalation exposures based on the source-receptor model that describes the source, various transport compartments and mechanisms, and the receptor. A set of nine broadly defined, mutually independent modifying factors are outlined, chosen according to first principles. Each of the modifying factors can include a series of additional factors that could be considered as exposure determinants, and that can be applied to either deterministic or statistical models. Such a structured approach permits for a consistent consideration of compartments and mechanisms of exposures, and at the same time allows for more specific exposure determinants to be developed. It also provides a structured framework for characterizing the workplace environment and collecting contextual information on the work environment characteristics.

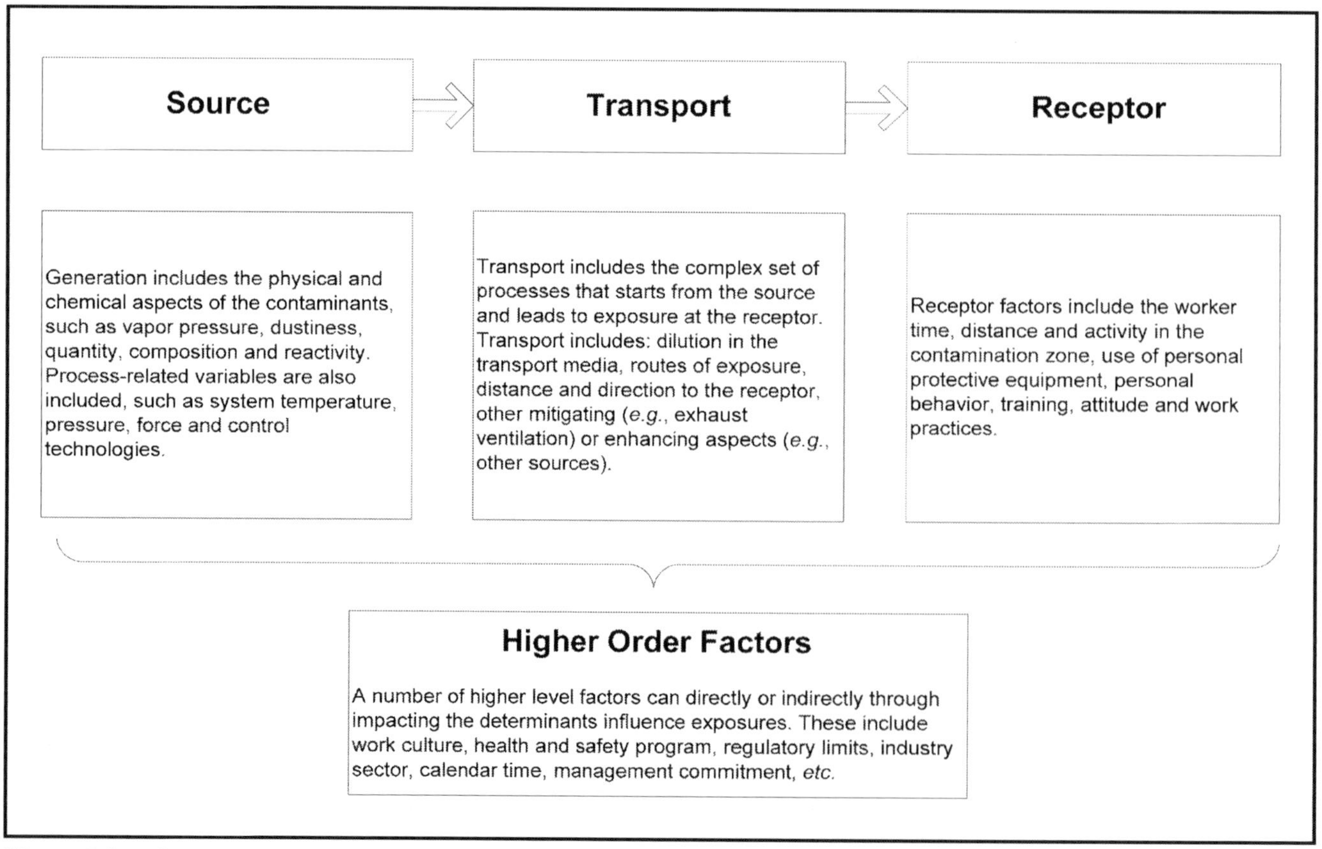

Figure 9.1 — Source-receptor model components with examples of possible exposure determinants.

9.2.2 Collect Information on Exposure Determinants

A prerequisite of statistical modeling is the availability of information on exposure determinants describing different aspects of the work environment. This information is generally collected via questionnaires, worker diaries or direct observation by exposure assessors during exposure monitoring. A conceptual model and initial list of exposure determinants are generally constructed after an initial assessment that includes a walk-through of the facility and a review of the processes, materials, etc., as described in the AIHA strategy. Much has been covered on individual versus group measurements, with well-formed and evaluated Similar Exposure Groups (SEGs) widely used for efficient workplace evaluations. Exposure monitoring records for SEGs may be particularly informative if the underlying exposure determinants are documented or can be reconstructed. Details on exposure assessment strategies and construction or validation of SEGs is available elsewhere.[4,5] Additional information on the integration of professional judgment of industrial hygienists with the available exposure data are provide elsewhere in this book. Further information on the assembly of sources for and methods of collecting exposure determinant for exposure reconstruction is available from Viet et al.[27]

9.2.3 Evaluate Exposure Data

Exposure monitoring is conducted for a variety reasons such as compliance assessment or exposure evaluation using different strategies, which may result in variable data quality, potential biases and limitations to data interpretation. For example, Lavoué et al.[28] investigated the determinants of formaldehyde exposures in the reconstituted wood panel industry in Québec by combining two exposure datasets collected using different sampling strategies. Results indicated significant differences in the estimated yearly geometric means for jobs and areas between exposure data collected by the government as compared to data collected by researchers. Likewise, in another study, Lavoué et al.[3] reported some differences in formaldehyde exposures in the OSHA IMIS dataset between data collected for programmed versus non-programmed inspections. Aspects of an exposure dataset that may impact its utility in exposure assessment include but are not limited to the purpose of sampling; the sampling strategy including the manner in which samples were collected, whether specific jobs or work sites were intentionally included or excluded, and details of sampling and analysis; and the extent of contextual information recorded or extractable from notes or other information sources. Depending on the specific purpose for modeling exposure determinants, these factors can impact or limit the utility of exposure datasets in achieving the desired objectives. Tielemans et al.[29] proposed guidelines for evaluating the quality of exposure data to assess the relative value of data for risk assessment. Such a procedure ensures systematic evaluation of exposure datasets and identifies biases or limitations. Furthermore, such a procedure ensures consistency among assessors as well as in pooling or aggregating exposure datasets from different sources. Several researches have provided details of the evaluation of data quality and the development of datasets used in their studies.[30–33] Before performing exposure modeling, it is critical to understand data quality as well as the representativeness of the data to enhance interpretation of modeling results.

9.2.4 Evaluate Exposure Determinants

After evaluating the exposure distribution and identifying the appropriate models to use, exposure determinants are investigated in univariate models to identify factors that significantly affect exposure. Factors that are significant in explaining the variability in exposure are further evaluated to investigate their correlations and to develop a strategy to handle correlated variables. Several approaches can be used to handle correlated variables including selecting only a single variable that is the easiest to interpret or that has the greatest impact on exposure; developing separate models for each of the correlated variables; categorizing continuous variables; and creating new variables by combining the correlated ones.[10,34] Principle components analysis or latent variable analysis can also be used to generate factors that can then be used as independent variables in multiple regression models.

9.2.5 Build Multiple Regression Models

A number of techniques have been used to build the multiple regression models, including forward selection, backward elimination, or step-wise process with different criteria for model entry significance level, and significance level to remain in the model. The statistic C_p is also a useful means of comparing regression models with many explanatory variables. Details of model building and variable selection criteria are described by Kleinbaum et al.[12] The following studies serve as examples of model building strategies.[34,35] Generally, automatic model building is not recommended. Some of these techniques have application to mixed models as well.

9.2.6 Use Standard Techniques to Evaluate Models

An integral component of model building is model evaluation. Standard regression diagnostic techniques to evaluate the model have been described by Kleinbaum et al.[12] and include assessing goodness of fit of the model, assessing distributional assumptions, examining residuals and standardized residuals, and detecting outliers and points of influence, including sensitivity analysis.

9.3 Model Validation

Several problems can be associated with predictive models that cannot be detected using standard regression diagnostics identified above. These problems can include variable selection, over-fitting the model to the data, estimation of the parameters, and prediction error.[36,37] Hence, validation using data not used in model building is necessary. However, such data are often unavailable and alternative model evaluation strategies using bootstrap or jackknife methods can be used. Three options are briefly presented.

9.3.1 Internal Validation by Splitting Data

When large datasets are available, exposure data can be randomly split into model building and model validation datasets. The predictions from the final model in the model building dataset are then compared to the exposure measurements in the validation dataset. Hornung et al.[35] provide an example of data splitting for validation. Several measures can be used to assess the degree of agreement between the model estimations and observed exposure measurements. These measures include bias, precision and accuracy.[35] Additional measures of agreement that provide information on the degree of concordance and sources of disagreement between two measurements include the concordance correlation coefficient and the intra-class correlation coefficient; a summary of these approaches is described in Virji et al.[38]

9.3.2 External Validation using Different Data Source

The model-generated exposure estimates can be compared to external data sources such as comparison to additional exposure data from the same or different workplaces within the same industry; comparison to data reported in the literature; comparison to exposures predicted by deterministic models; and/or comparison to expert assessment of exposure. An example from the published literature includes a study of formaldehyde exposures among embalmers that used measurements from other surveys for validation.[39]

9.3.3 Evaluation Using Monte Carlo Methods

The traditional internal and external validation approaches require large amounts of data and are not immune to errors associated with over-fitting the model to the data. However, computer-intensive methods, referred to as re-sampling methods, are available to address these issues. These methods are well described in a text entitled "Randomization and Monte Carlo Methods in Biology."[40] The bootstrapping technique has consistently yielded optimum results and is one of several computer-intensive approaches used to develop and evaluate models; it is ideal for small sample sizes.[41] This approach is based on sampling with replacement from the original dataset of measurements. The original sample is treated as a population from which typically 1000 datasets are drawn with replacement, each the same sample size as the original data. The 1000 datasets can then be used to develop predictive models and identify the important factors affecting exposures as well as to estimate parameters.[41,42] The results from each of the 1000 datasets are combined to generate a distribution of parameter estimates and their confidence intervals. Recently, Hein et al.[33] used a combination of the split data validation and bootstrap methods to validate estimates of benzene, toluene and xylene exposures from statistical models. For a selected number of operations, they drew 1000 datasets each of which was split into model building and validation subsets. The predicted exposures from the model-building subset were compared to the observed data in the validation subset via correlation coefficients, and summarized over the 1000 datasets to obtain their distributions.

9.4 Examples of Application

The objective of modeling in industrial hygiene is to investigate and describe the relationship between exposure levels and workplace factors to identify the sources of exposure variation. In workplace environments, exposures tend to vary significantly from day-to-day, location-to-location, job-to-job, and between workers due to differences in a host of factors including workplace conditions, environmental factors, and personal work practices. Statistical models have been developed to identify factors that are most influential in determining exposures, thereby providing opportunities to identify and prioritize control measures, to estimate current exposures for un-sampled workers or historical exposures of individuals, or to assess time trends in exposures. Such statistical models have previously been applied to a variety of industrial and environmental situations. While an exhaustive literature review is not the focus of this chapter, readers are urged to conduct a literature search using the phrase "exposure determinants" to access more studies. Several examples are discussed in this section.

9.4.1 Identifying Exposure Factors for Control Measures

A study was conducted to examine the determinants of bulk and airborne microbial exposures, and to assess the impact of changes in exposure determinants on bulk and airborne microbial levels.[34] These investigators used modeling to identify factors affecting exposure and to develop effective exposure control strategies. They developed three separate models to evaluate the effects of environmental, process and personal factors on airborne viable microbial levels, and three additional models to evaluate the effects of fluid characteristics, process and environmental factors on bulk material viable microbial levels. The determinants evaluated were based on a conceptual framework of exposure to metal working fluids (Figure 9.2).

The final full model for airborne microbial levels accounted for 24% of the total variability, whereas the final full model for bulk material microbial levels captured 45% of the total variability. To evaluate the impact of each independent factor on exposure levels, the value of one variable was changed at a time from its minimum to maximum value while holding the remainder of the variables at the median value (Figure 9.3). The airborne model showed that (1) decreasing bulk microbial concentrations, machine density and production rate, (2) increasing worker distance, and (3) use of enclosures were all associated with reducing airborne microbial levels (Figure 9.3a). The bulk material microbial model showed that increasing pH and decreasing machine RPM, percent tramp oil, and sump size of the fluid were all associated with decreasing bulk material microbial contamination levels (Figure 9.3b). Results from these models can be used to estimate potential reduction in worker exposure or fluid contamination due to introduction of certain control measures.

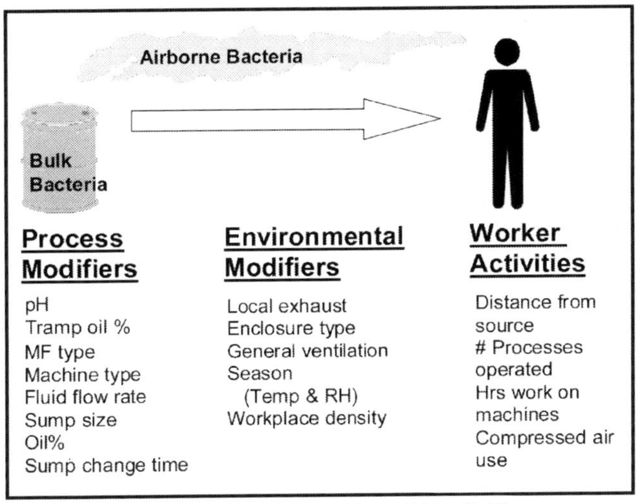

Figure 9.2 — Illustration of metal working fluid exposure model.

9.4.2 Predicting Exposures

A weighted linear regression model was developed as part of an epidemiologic study to estimate historical worker exposure to ethylene oxide (EtO) in the sterilizing industry.[35] The model considered independent variables identified as possible exposure determinants based on walkthrough surveys of the operations by experienced investigators. The individual 8-hr TWA exposure data were grouped by plant, year, and type of sampling media for each job and location, resulting in 205 average measurements. A multiple regression model was developed that included the significant effects of exposure category, product type, size of the sterilization unit, selected engineering controls, days after sterilization and calendar year. The model explained a remarkable 85% of the variability and showed a steady decline in exposures over time. The model predicted EtO exposures were validated based on a fraction of the data not used in model building (validation by splitting data). Comparing the predicted exposures with measured data showed a remarkably small bias of 1.1 ppm. The model was used to predict historical EtO exposures for workers based on the combinations of the independent variables in the final model.

9.4.3 Evaluating Time Trends

Among the assumptions of retrospective exposure assessment is that exposure levels change over time due to

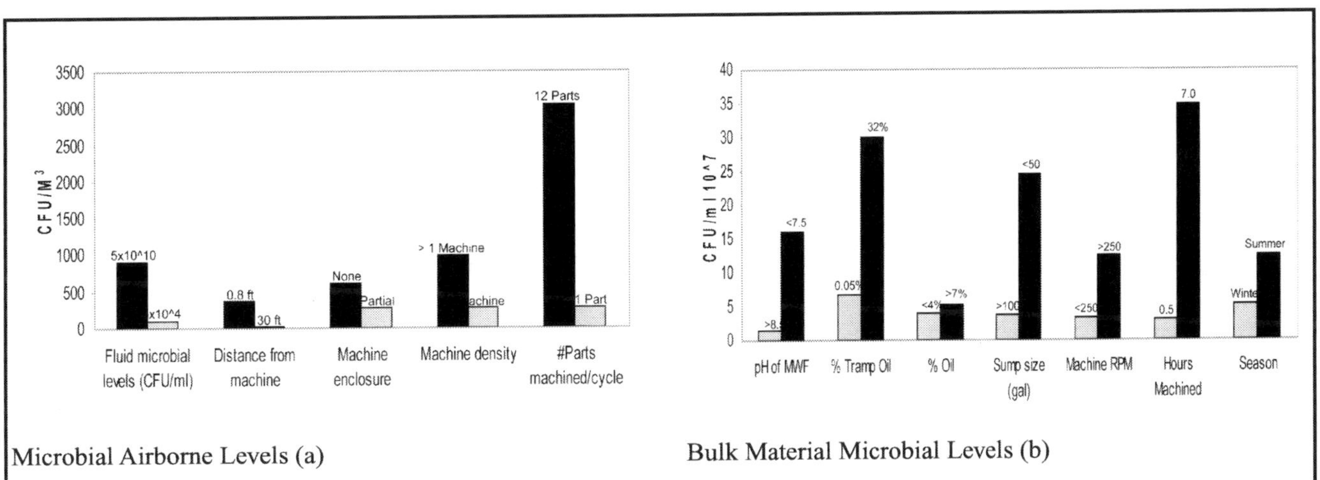

Figure 9.3 — Evaluating the impact of changes in exposure determinants on microbial levels.

factors that include changes in processes, ventilation controls, and regulations among others. Therefore, to more accurately estimate past exposures, current exposure estimates need to be adjusted for such changing factors. A number of studies have quantitatively documented time trends in exposures using statistical models.[43–45] These studies have not only reported trends in the reduction of exposures over time, but have also demonstrated the need to understand factors affecting exposure. For example, Burstyn et al. (2000)[45] constructed a database of bitumen fume and vapor and benzo(a)pyrene exposures among asphalt workers from numerous sources and several countries. Using a mixed-effects model, they showed a steady decline in exposure levels of 6–14% per year for the three exposure agents over a 20-year period between 1970 and 1990. Their statistical models explained 36–43% of the total variability for the three exposure agents and included significant effects of various production factors and asphalt temperature.

Several other studies have reported such declines in exposure levels over time, including decline of 12% per year in toluene exposure among commercial painters.[46] Symanski et al.[44] reported declining total nickel exposures of 7 to 9% per year in nickel mining, smelting and refining sectors, but an increasing trend in exposures (4% per year) within the milling sector. Yu et al.[47] used data on formaldehyde exposures among foundry workers to point out the need for step-change considerations in deterministic exposure modeling, noting that changes in exposure can be sudden and may occur at a given point in time.

9.5 Summary

In this chapter, we have attempted to describe statistical models of exposure determinants and demonstrate their utility in understanding factors affecting exposure. These models generally require large quantities of exposure measurements along with contextual information on characteristics of the work environment that are collected during monitoring. Additionally, some of the exposure determinants models utilizing full-shift exposure measurements capture only a modest fraction of exposure variability and identify few specific activities or control measures.[48] Perhaps developing these models using exposure determinants from a conceptual model based on the first principle of mass conservation may improve the fraction of exposure variation explained by the models. Additionally, consideration of time-varying exposure determinants with real-time exposure measurement can identify specific activities and control measures and can have a significant impact on intervention strategies.[48]

Improvements in exposure assessment methodologies require both exposure monitoring and modeling, both statistical and deterministic. This chapter provides only a brief overview of techniques available for using exposure data to generate models to estimate historical exposures, to identify exposure factors contributing to increased or decreased exposures, and to develop a targeted control strategy. Consultation with statistical texts and software packages are recommended prior to proceeding. However, with this overview information and additional reading of studies in journals, the reader is encouraged to evaluate exposure databases to determine if exposure determinants can be evaluated. While this approach requires significantly more data than may exist in small worksites, multi-site industries or multi-industries with similar processes may aggregate exposure data and develop models to determine key workplace environmental factors that affect exposures. While a mixture of factors and branches of logic may be used in different situations, a conceptual framework can help to organize the many considerations. Efforts by regulators, researchers, and industry groups are encouraged to pursue data aggregation, with clearly and uniformly defined datasets including information on workplaces, materials, work methods, and environmental data that may contribute to task- and industry-specific workplace exposure determinant models. Using data to identify key contributors to exposure is critical to understanding and controlling exposure risk factors, and to providing evidence-based methods for prioritizing intervention strategies.

9.6 Acknowledgement

The authors would like to acknowledge Gregory Day, Susan Woskie, and Stanley Shulman for their review of the chapter and their thoughtful comments and suggestions which improved the quality of the chapter.

9.7 References

1. **Leidel, N. A., K.A. Busch, and J.R. Lynch:** *Occupational Exposure Sampling Strategy Manual.* U. S. Department of Health, Education, and Welfare. Public Health Service. Center for Disease Control. National Institute for Occupational Safety and Health. DHEW (NIOSH) Publication No 77-173. Cincinnati, OH: NIOSH, 1977.
2. **Gómez, M.R.:** Factors associated with exposure in Occupational Safety and Health Administration data. *Am. Ind. Hyg. Assoc. J. 58(3)*:186–195 (1997).
3. **Lavoué, J., R. Vincent, and M. Gérin:** Formaldehyde exposure in U.S. industries from OSHA air sampling data. *J. Occup. Environ. Hyg. 5(9)*:575–587 (2008).
4. **Mulhausen, J.R. and J. Damiano:** *A Strategy for Assessing and Managing Occupational Exposures*, 2nd edition. Fairfax, VA.: American Industrial Hygiene Association, 1998.

5. **Ignacio, J.S. and W.H. Bullock (eds.):** *A Strategy for Assessing and Managing Occupational Exposures*, 3rd edition. Fairfax, VA.: American Industrial Hygiene Association, 2006.
6. **Rappaport, S.M. and L.L. Kupper:** *Quantitative Exposure Assessment*. El Cerrito, CA: Stephen Rappaport, 2008.
7. **Jayjock, M.A., C.F. Chaisson, S. Arnold, and E.J. Dederick:** Modeling framework for human exposure assessment. *J. Expo. Sci. Environ. Epidemiol. 17(1)*:S81–S89 (2007).
8. **Schneider, T.:** Improving exposure assessment requires measurements and modeling. *Scand. J. Work Environ. Health. 28(6)*:367–369 (2002).
9. **Tan, Y., M.R. Flynn:** Experimental evaluation of a mathematical model for predicting transfer efficiency of a high volume/low pressure air spray gun. *Appl. Occup. Environ. Hyg. 15(10)*:785–793 (2000).
10. **Burstyn, I. and K. Teschke:** Studying the determinants of exposure: A review of methods. *Am. Ind. Hyg. Assoc. J. 60(1)*:57–72 (1999).
11. **Teilemans, E., et al.:** Conceptual model for assessment of inhalation exposure: Defining modifying factors. *Ann. Occup. Hyg. 52(7)*: 577–568 (2008).
12. **Kleinbaum, D.G., L.L. Kupper, K.E. Muller, and A. Nizam:** *Applied regression analysis and other multivariable methods*, Third Edition, Pacific Grove, CA: Duxbury Press, 1998.
13. **Boleij, J., E. Buringh, D. Heederik, and H. Kromhout:** *Occupational Hygiene of Chemical and Biological Agents*. Amsterdam, The Netherlands: Elsevier Sciences, 1995.
14. **Burdorf, A.:** Analysis and modeling of personal exposure. In *Exposure Assessment in Occupational and Environmental Epidemiology*. Nieuwenhuijsen, M.J. (ed.). New York, NY: Oxford University Press, Inc. (2003).
15. **Symanski, E., S. Maberti, and W. Chan:** A meta-analytic approach for characterizing the within-worker and between-worker sources of variation in occupational exposure. *Ann. Occup. Hyg. 50(4)*:343–357 (2006).
16. **Woskie, S.R., et al.:** Determinants of isocyanate exposures in auto body repair and refinishing shops. *Ann. Occup. Hyg. 48(5)*:393–403 (2004).
17. **LaMontagne, A.D. and K.T. Kelsey:** Evaluating OSHA's ethylene oxide standard: Exposure determinants in Massachusetts hospitals. *Am. J. Public Health. 91(3)*:412–417 (2001).
18. **Thompson, M.L., J.E. Myers, and D. Kriebel:** Prevalence odds ratio or prevalence ratio in the analysis of cross sectional data: What is to be done. *Occup. Environ. Med. 55*:272–277 (1998).
19. **Lee, J.:** Estimation of prevalence rate ratios from cross-sectional data: A reply. *Internatl. J. Epidemiol. 24(5)*:1066–1067 (1995).
20. **Lubin, J.H., et al.:** Epidemiologic evaluation of measurement data in the presence of detection limits. *Environ. Health. Perspect. 112(17)*: 1691–1696 (2004).
21. **Thiébaut, R. and H. Jacqmin-Gadda:** Mixed models for longitudinal left-censored repeated measures. *Comput. Methods Programs Biomed. 74(3)*:255–260 (2004).
22. **Dodgson, J., J. Cherrie, S. Groat:** Estimates of past exposure to respirable man-made mineral fibres in the European insulation wool industry. *Ann. Occup. Hyg. 31(4B)*:567–582 (1987).
23. **Armstrong, T.W., et al.:** Retrospective benzene and total hydrocarbon exposure assessment for a petroleum marketing and distribution worker epidemiology study. *Am. Ind. Hyg. Assoc. J. 57(4)*:333–343 (1996).
24. **Lewis, S.J., et al.:** Retrospective estimation of exposure to benzene in a leukaemia case-control study of petroleum marketing and distribution workers in the United Kingdom. *Occup. Environmental Med. 54(3)*:167–175 (1997).
25. **Glass, D.C., G.G. Adams, R.W. Manuell, and J.A. Bisby:** Retrospective exposure assessment for benzene in the Australian petroleum industry. *Ann. Occup. Hyg. 44(4)*:301–320 (2000).
26. **American Conference of Governmental Industrial Hygienists® (ACGIH®):** Data elements for occupational exposure databases: Guidelines and recommendations for airborne hazards and noise. *Appl. Occup. Environ. Hyg. 11(11)*:1294–1311 (1996).
27. **Viet, S. M., M. R. Stenzel, C.P. Rennix, T.W. Armstrong, and J.R. Couch (eds.):** *Guideline on Occupational Exposure Reconstruction*. Fairfax, VA: American Industrial Hygiene Association, 2008.
28. **Lavoué, J., C. Beaudry, N. Goyer, G. Perrault, M. Gérin:** Investigation of determinants of past and current exposures to formaldehyde in the reconstituted wool panel industry in Quebec. *Ann. Occup. Hyg. 49(7)*:587–602 (2005).
29. **Tielemans, E., H. Marquart, J. De Cock, M. Groenewold, J. Van Hemmen:** A proposal for evaluation of exposure data. *Ann. Occup. Hyg. 46(3)*:287–297 (2002).
30. **Vincent, J.H. and M.A. Werner:** Critical evaluation of historical occupational aerosol exposure records: Applications to Nickel and Lead. *Ann. Occup. Hyg. 47(1)*:49–59 (2003).
31. **Burstyn, I., H. Kromhout, P.J. Cruise, and P. Brennan:** Designing an international industrial hygiene database of exposure among workers in the asphalt industry. *Ann. Occup. Hyg. 44(1)*:57–66 (2000).
32. **De Vocht, F., et al.:** A database of exposures in the rubber manufacturing industry: design and quality control. *Ann. Occup. Hyg. 49(8)*:691–701 (2005).
33. **Hein, M. J., M.A. Waters, E. van Wijngaarden, J.A. Deddens, and P.A. Stewart:** Issues when modeling benzene, toluene, and xylene exposures using a literature database. *J. Occup. Environ. Hyg. 5(1)*:36–47 (2008).

34. **Virji, M. A., S.R. Woskie, S. R. Sama, D. Kriebel, and D. Eberiel:** Identifying the determinants of viable microorganisms in the air and bulk metalworking fluids. *Am. Ind. Hyg. Assoc. J. 61(6)*:788–797 (2000).
35. **Hornung, R.W., et al.:** Statistical model for prediction of retrospective exposure to ethylene oxide in an occupational mortality study. *Am. J. Ind. Med. 25(6)*:825–836 (1994).
36. **Schumacher, M., N. Holländer, and W. Sauerbrei:** Resampling and cross-validation techniques: a tool to reduce bias caused by model building? *Stat. Med. 16(24)*:2813–2827 (1997).
37. **Steyerberg, E.W., et al.:** Internal and external validation of predictive models: a simulation study of bias and precision in small samples. *J. Clin. Epidemiol. 56(5)*:441–447 (2003).
38. **Virji, M.A., et al.:** Agreement between task-based estimates of the full-shift noise exposure and the full-shift noise dosimetry. *Ann. Occup. Hygiene.* [In press] (2009).
39. **Hornung, R.W., et al.:** An experimental design approach to retrospective exposure assessment. *Am. Ind. Hyg. Assoc. J. 57(3)*:251–56 (1996).
40. **Manly, B.F.J.:** *Randomization and Monte Carlo Methods in Biology*. New York, NY: Chapman and Hall, 1991.
41. **Jensen, R. L. and G.M. Kline:** The resampling cross-validation technique in exercise science: modelling rowing power. *Med. Sci. Sports. Exerc. 26(7)*: 929–933 (1994).
42. **Molinaro, A.M., R. Simon, and R.M. Pfeiffer:** Prediction error estimation: a comparison of resampling methods. *Bioinformatics. 21(15)*: 3301–3307 (2005).
43. **Vermeulen, R., J. de Hartog, P. Swuste, and H. Kromhout:** Trends in exposure to inhalable particulate and dermal contamination in the rubber manufacturing industry: effectiveness of control measures implemented over a nine-year period. *Ann. Occup. Hyg. 44(5)*:343–354 (2000).
44. **Symanski, E., C.C. Chang, W. Chan:** Long-term trends in exposures to nickel aerosols. *Am. Ind. Hyg. Assoc. J. 61(3)*:324–333 (2000).
45. **Burstyn, I., H. Kromhout, T. Kauppinen, P. Heikkilä, and P. Boffetta:** Statistical modelling of the determinants of historical exposure to bitumen and polycyclic aromatic hydrocarbons among paving workers. *Ann. Occup. Hyg. 44(1)*:43–56 (2000).
46. **Burstyn, I. and H. Kromhout:** Trends in inhalation exposure to hydrocarbons among commercial painters in The Netherlands. *Scand. J. Work Environ. Health 28(6)*:429–438 (2002).
47. **Yu, R. C., W.Y. Tan, R.M. Mathew, D.A. Andjelkovich, and R.J. Levine:** A deterministic mathematical model for quantitative estimation of historical exposure. *Am. Ind. Hyg. Assoc. J. 51(4)*:194–201 (1990).
48. **Meijster, T., E. Tielemans, J. Schinkel, and D. Heederik:** Evaluation of peak exposures in the dutch flour processing industry: implications for intervention strategies. *Ann. Occup. Hyg. 52(7)*:587–596 (2008).

Uncertainty 10

By
Michael Jayjock, PhD, CIH
Gurumurthy Ramachandran, PhD
Susan F. Arnold, CIH

10.0 Overview

When selecting models and values to enter into them, modelers may fret about the precision of their input data. For example as hygienists know, velocity measurements taken to quantify airflow in a potential spill area may change significantly or even dramatically over the course of a shift, let alone over seasons. Uncertainties in actual vapor pressure of a specific hazardous component in a mixture are also a common concern. This chapter provides both guidance to understand the role of uncertainty in modeling computations and tools to evaluate the impact of uncertainty in concentration estimations and exposure assessments.

10.1 Uncertainty: Where Does It Come From?

Uncertainty about exposures comes from lack of knowledge, which stems from one of two sources. The first source of uncertainty is natural variability. In the risk assessment world, this is frequently referred to as 'category' or 'Type I'. Some examples of natural variability include differences in body weights within a population of workers or differences in the amount of air a worker breathes in workday. Another example of natural variability is the daily or hourly airborne concentration of a contaminant in the workroom. Even in a relatively well controlled environment, this variability results in a log-normal distribution with geometric standard deviations (GSD) typically in the range of 1.5 to 4. With sufficient monitoring data, we can characterize and quantify this kind of variability using statistics.

The second kind of variability, referred to as Type II uncertainty, represents a fundamental ignorance. Typically, Type II uncertainty is the dominant kind of uncertainty. Limited availability of empirical information, imperfections in measurement instruments, models or techniques used to develop our representation of reality all contribute to Type II uncertainty.(1) An example is provided to better ground the reader in the fundamental differences of these two types of uncertainties.

10.2 LIBBY: An Example of Natural Variability Versus Lack of Knowledge

One of the authors (Jayjock) has been using his faithful if somewhat neurotic dog (Libby) to help students understand the difference between the critical components of overall uncertainty in risk assessment, namely, the distinction between "natural variability" and ignorance or "lack of knowledge."

This lesson starts by informing students that he left Libby in his residential garage before coming to class because she suffers from separation anxiety and would do damage if he left her in the house. His sense is that the students, and probably the reader, wonder why he would share such information. The value or utility of at least some of this information will become apparent latter.

The situation is posited to the class that Libby's weight is a critical factor or variable being fed into a particular risk assessment. The details of this connection are (conveniently) not important to the example or the student. We simply need to know for the purpose of this exercise that if we overestimate Libby's weight we will overestimate the ultimate risk and, conversely, if we underestimate her weight we will underestimate the risk.

10.2.1 Initial Condition: Lack of knowledge is high but not too high

Jayjock tells the students that the only information he will give them is that Libby is a full grown female dog and of pure breed. He tells them that they all have some experience with dogs and are thus somewhat "expert" in this area. Jayjock asks if Libby can weigh less than two pounds, and the usual response from the class is a resounding "No!"

NB: He had one student advise the class that the smallest living adult dog in the Guinness Book of World Records is 1.4 lbs ("Ducky," the Chihuahua, in 2007). This shows the possible existence of unusual and remarkable "outliers."

Jayjock asks if it's likely that Libby weighs more than 200 lbs and the response again is "No." At that point the discussion within the class usually comes up with a range for Libby's weight. Keeping in mind the admonition that overestimation of weight is tantamount to overestimate of risk, the students often, and correctly, purposely overestimate the range. A typical range, after discussion and sometimes after voting, would be 50–200 lbs. This could be represented in a Monte Carlo simulation as a Uniform Distribution (min, max) = UNI(50, 200). Thus, the ascribed value for Libby's weight would be a random number from 50 to 200 lbs with all values having an equal probability of being sampled.

A graphical representation of this uniform distribution is presented by the green bar in Figure 10.1, where there is a constant probability assigned to all weights between 50 and 200 pounds and zero probability outside of this range.

10.2.2 Second Condition: Some information provided — Lack of knowledge lower but still high

Jayjock tells the students at this point that Libby is an English Springer Spaniel. This typically sparks some discussion about the size of the breed and even allows anyone with an Internet connection to quickly lookup the characteristics of the breed including weight. The following is a cut and paste from a Google hit for "English Springer Spaniel weight": "Dog Weight: about 50 lb. (22.7 kg). Bitch Weight: about 40 lb. (18.2 kg)."

These facts spark a discussion that perhaps the students should use 40 lbs as the estimated value. Often, some of the more prudent students remind everyone that these are *averages* and Libby may not be average. A typical conclusion by the class might be that the range of Libby weight is now narrower from 40 lbs on the low end

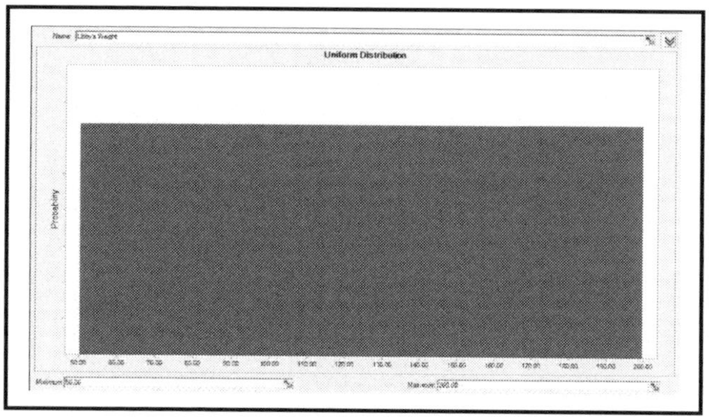

Figure 10.1 — Uniform distribution assigned to Libby's weight

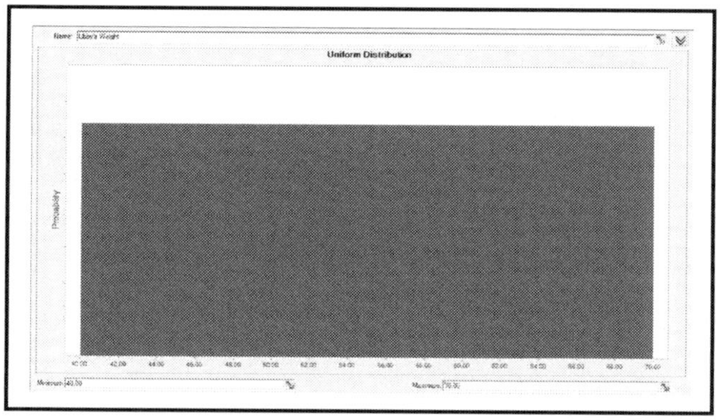

Figure 10.2 — Revised uniform distribution of Libby's weight, with increased knowledge

to 70 lbs on the high end. This approach is consistent with a precautionary approach which purposely biases input toward overestimation of Libby's weight (and subsequent risk) in the face of imperfect information. Thus, given this new information our new estimate is UNI(40,70), illustrated in Figure 10.2. Note that this new information narrowed the confidence range significantly.

10.2.3 Third Condition: Actual Data are Gathered — Lack of knowledge much lower

At this point, Jayjock reminds the students that his home address is no secret and that a suitably motivated person could drive to his house and look into the garage to see Libby. If the person had a good eye for estimating weight he or she would conclude that Libby was well above the average of 40 lbs and indeed might weigh as much as 70 lbs! Feeding this information to the class "panel" might result in a new, more refined estimate. Indeed, instead of a Uniform distribution in which all values in the range have an equal probability of being chosen, a Triangular Distribution might be chosen. In this distribution there is a min and max but there is also a central tendency or most likely value.

The class might conclude that Libby is between 60 and 70 lbs with a central tendency at 65 lbs. Triangular (min, central tendency, max) = TRI (60, 65, 70). A graphic representation of this distribution is shown in Figure 10.3.

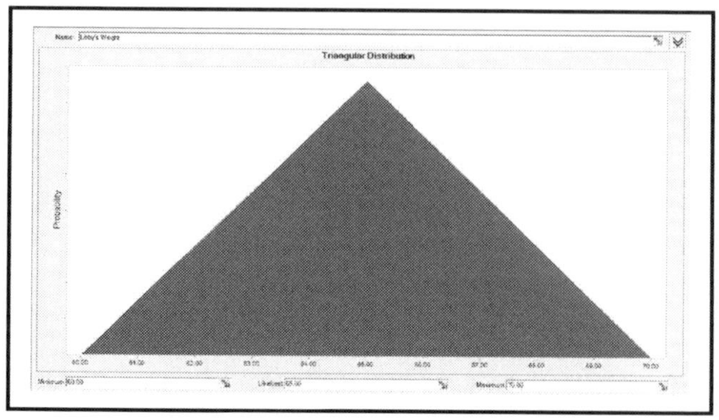

Figure 10.3 — Revised triangular distribution of Libby's weight, with increased knowledge

10.2.4 Fourth Condition: True Variability is Determined — Lack of knowledge is nil

If an assessor could gain access to the interior of my garage they could proceed to actually weigh Libby daily over an extended period of time, say one month. Assume this was performed, and the results follow, including a revised distribution in Figure 10.4:

Average:	63.4 lb
Standard Deviation:	1.76 lbs
N:	30

Our lack of knowledge has been reduced to essentially nil and this distribution represents our confident knowledge; that is, the true variability of this input to the hypothetical risk assessment.

Look at the difference between the estimates from our initial condition (40–200 lbs) and the measured values! Note, also that it took considerable resources (i.e., time and effort) to get there.

There is one final lesson that this example can teach us; namely, we must have at least some threshold of information or knowledge to render a useful answer. The initial condition was highly uncertain but still useful because we all have knowledge or "data" on the weight of dogs and the universe was limited to just full grown, pure breed dogs.

The assessment would be much different if it were much more open ended; e.g., "There is an animal in the garage of unknown species — how much does it weigh?" In this case the uncertainty would be so great as to render essentially any estimate to be meaningless. Thus, as exposure/risk assessors we need to understand that there is a threshold of information that we *must* have before even attempting to render an analysis.

10.3 Uncertainty in the Practice of Industrial Hygiene

The science and art of industrial hygiene is a human endeavor; we can't possibly know everything about a particular exposure scenario, and therefore uncertainty is an inextricable component. Type I uncertainty can be addressed for those aspects of exposure that can be quantified and characterized by taking measurements and conducting statistical analyses. This can be done by taking exposure measurements, area and personal samples, and analyzing the results using statistics. Type II uncertainty is reduced by gathering more information about the exposure scenario: both qualitative and quantitative data can help close the knowledge gap. Most exposure assessments regarding the workplace are completed with little or no empirical exposure data. The industrial hygienist makes a qualitative judgment about the acceptability of the exposure. To make that judgment, the industrial hygienist is using a model — consciously or subconsciously to calculate the exposure. The standard of excellence today requires the hygienist to be as transparent and explicit as possible in documenting the model and its assumptions — and the associated uncertainty. This is sometimes quite painful and perhaps even a little embarrassing, but it is critical to the integrity of the process and credibility of the exercise.

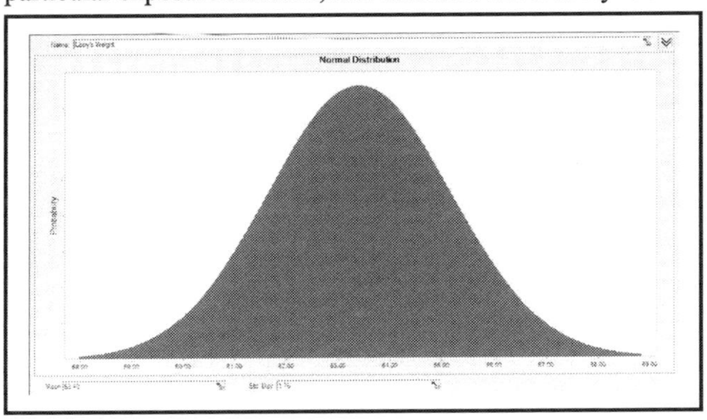

Figure 10.4 — Final revised distribution of Libby's weight, with no more lack of knowledge

10.4 Modeling and Uncertainty

A model is a hypothesis or belief about how a system works or responds to changes in its inputs. The model may be based on statistical principles, regression tools, physical-chemical laws or some other framework, as discussed previously in this book. A model is a simplified version of the real system — the purpose of the model is to represent the system *as accurately as necessary*.[1] In doing so, we accept some degree of uncertainty because we purposefully exclude part of the system by using a simpler representation. By employing a precautionary approach, data are traded for conservatism.

10.5 Deterministic Modeling Techniques and Uncertainty

Most of the models in this book are presented as deterministic models that assume the input parameters are known *precisely*. Thus, if a given set of input parameters,

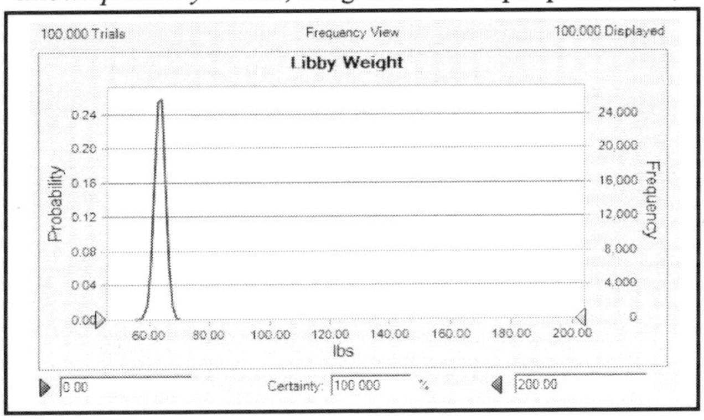

Figure 10.5 — Final distribution of Libby's weight, over the range of all values considered in this exercise

then there is one value of the output for the model. Variability in input (and therefore output) parameters is essentially ignored. So how are we to interpret these results; do they present a best-case scenario? A worst-case scenario? Do they represent the average?

Consider the simple indoor air model from Chapter 2:

$$C = \frac{G}{Q} \tag{10-1}$$

Where

C: equilibrium airborne concentration of a toxicant (mg/m^3)
G: the steady state generation rate (mg/hour)
Q: the steady state ventilation rate (m^3/hour)

One method used to address uncertainty with these models is to present *reasonable* best- and worst-case estimates, whereby an upper and lower bound estimate are calculated using the lowest plausible generation rate and highest plausible removal rate for the best-case and the highest generation rate in combination with the lowest plausible removal rate to calculate the worst-case exposure. One can also examine the impact or *sensitivity* of each parameter by calculating the results of a unit change in the predictors from maximum to minimum individually, for each. In examining this equation, we know that the largest generation-rate estimate combined with the smallest ventilation rate will provide us with the largest concentration estimate, or our "worst-case." Conversely, the smallest generation rate and the largest ventilation rate will yield the smallest or "best-case" concentration estimate.

To illustrate this technique, let us assume we have some emission data to suggest a generation rate that is normally distributed with a mean of 50 mg/hour and a standard deviation of 5 mg/hour. Note this information reflects Type I uncertainty or natural variability.

Let us assume a worst-case generation rate equal to the mean generation rate plus 3 standard deviations. The worst-case generation rate then is = 50 +15 mg/hour = 65 mg/hour. Similarly, let us assume the best-case generation rate is equal to the mean G minus 3 standard deviations. This gives us a best-case G = 50 –15 mg/hour = 35 mg/hour. So we have:

Reasonable worst-case G = 65 mg/hr
Average G = 50 mg/hr
Reasonable best-case G = 35 mg/hr

Let us also assume there is no measured data available for Q. However, we learn from the published literature that local exhaust ventilation is seldom used in this industrial setting. We also learn from the literature that it is unlikely the airflow in this setting would be less than 0.2 air changes per hour (ACH) or greater than 30. From this information, we can compute a range of values for airflow using: Q = air changes per hour or ACH x room volume. In a real world setting where the room dimensions were not implicitly known, we would use a range of dimensions to simulate room volume but for the sake of this example, we will keep it simple and use dimensions for a relatively small room: 3m x 3m x 2m. Since we do not have information on the average case or estimate of central tendency we cannot estimate $Q_{average}$.

Q is calculated for worst- and best-case accordingly:

Q worst-case = 0.2/hr x 18 m^3 = 3.6 m^3/hr ← smallest estimated Q

Q best-case = 30/hr x 18 m^3 = 540 m^3/hr ← largest estimated Q

Applying a reasonable worst-case/best-case approach, we can calculate the airborne concentration, C by using the reasonable worst- and best-case values for G and Q:

$$C_{\text{reasonable worst case}} = \frac{65\text{mg}}{\text{hr}} \; \frac{\text{hr}}{3.6\text{ m}^3} = 18.1 \text{ mg/m}^3$$

$$C_{\text{reasonable best case}} = \frac{35\text{mg}}{\text{hr}} \; \frac{\text{hr}}{540\text{ m}^3} = 0.065 \text{ mg/m}^3$$

Our prediction of C varies 278 fold from reasonable worst-case to best-case. We can also gain insight into the relative impact or sensitivity of the parameters by examining the spread across each parameter from worst- to best-case, individually. Since Q varies by a factor of 150 and G less than 2, Q is obviously the most sensitive. Given limited resources to reduce uncertainty in this scenario, gathering data on Q would be the most cost effective.[(2)]

10.6 Probabilistic Modeling Techniques: Monte Carlo Analysis

As discussed in the previous sections, complete and precise knowledge of the values of the model input parameters (or anything else, for that matter) is impossible. The exposure determinants (the input parameters) are known with some amount of uncertainty. Therefore, selecting only one value for each of the input parameters is not advisable. It is better to find a way to represent the uncertainty in the input parameters and consider how this uncertainty leads to uncertainty in the model output. Monte Carlo sampling analysis is a powerful technique for propagating the uncertainties in the model inputs to determine the uncertainties in the model outputs. This technique can provide insights about the relative importance of the various assumptions and uncertainties in the model inputs to the model's conclusions. Incorporating the parameter variability into the model can help decide whether it is worthwhile gathering more information to reduce uncertainty. This technique also helps avoid disputes over the best-value chosen for point estimates, since the full range of possible results is considered.

The conceptual basis for error propagation can be illustrated using a simple example of a model where the input and output parameters are discrete variables. Let y be a function of x_1 and x_2.

$$y = x_1 + x_2 \tag{10-2}$$

Assume also that both x_1 and x_2 can take only one of three possible values. However, these values have different probabilities associated with them. Thus x_1 can only take the values of 0.3, 1.0, and 1.7, and the probability of x_1 taking each value is 0.2, 0.6, and 0.2, respectively. Note that these probabilities add up to 1.0. Similarly, x_2 can only take the values of 0.5, 1.0, and 1.5 with probabilities of 0.33, 0.34, and 0.33, respectively, again adding up to 1.0. To investigate the possible interactions between the effects of all the inputs at various levels, one has to look at all possible combinations of input values. This is performed by constructing a *probability tree* as shown in Figure 10.6. The uncertainty in the inputs can be expressed as a discrete probability distribution. There will be probabilities attached to each branch of the probability tree. Thus, y can take nine possible values. The probability of each value is obtained by the product of the probabilities of the two input variables. Thus, y can take a value of 0.8 only if $x_1 = 0.3$ ($p = 0.2$) and $x_2 = 0.5$ ($p = 0.33$). Thus the probability associated $y = 0.8$ is $P = 0.2 \times 0.33 = 0.066$. The probabilities of all the possible values that y can take also add up to 1.0. Figure 10.6 shows the values of the input variables and their associated probabilities.

Thus, Figure 10.6 shows how the uncertainty in the two input parameters is reflected in the uncertainty in the

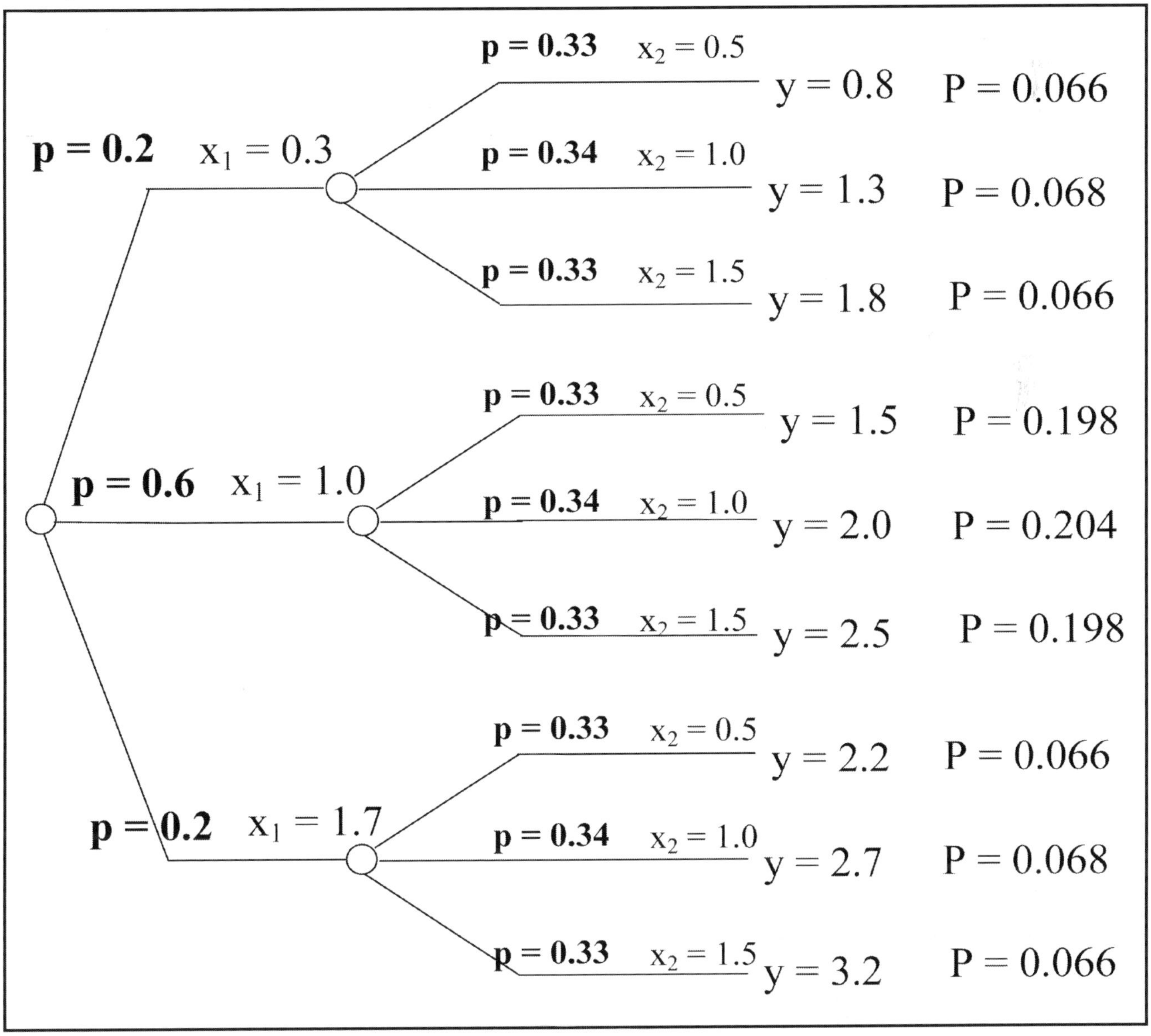

Figure 10.6 — Propagation of uncertainty through a model with two discrete variables as input parameters.

model output. While the example is illustrative, it does show the limitation in using such an approach: consider what the tree would look like when the input variables have more than two terms restricted to a limited number of variables. Also, in many cases, the input variables are *continuous*, such as the values allowed for Libby's weight. With continuous variables, the probability tree will have an infinite number of nodes and branches and would be impossible to calculate by hand.

One way out of this quandary is to select a moderately-sized, random sample of all possible combinations of allowable variables for evaluation. This is called *Monte Carlo sampling*. Each scenario is generated by selecting each branch at a node according to its assigned probability. The branch values are generated from the underlying continuous probability distribution. The accuracy of Monte Carlo sampling can be improved by increasing the sample size.

The uncertainty in occupational exposure measurements is typically expressed using normal or lognormal distributions. However, there is often a more fundamental ignorance about the single predictor variables or parameters used in exposure modeling, and this usually dominates the overall uncertainty. For these parameters, a variety of other distributional shapes can be used to represent this uncertainty. Uniform distributions (where every value between an upper and lower limit is equiprobable) and triangular distributions (with a minimum, maximum, and modal value) are examples of distributions used to represent such uncertainties. Remember that these were used in the above discussion of Libby's weight. Figure 10.7 illustrates the shapes of these commonly used distributions.

If there are available measurements for an input parameter, e.g., the ventilation flow rate through a work area, then one can fit these data to estimate the mean and standard deviation of these measurements and represent the distribution of the parameter using a normal or lognormal distribution. The measurements obtained are assumed to be from the estimated underlying distribution of air flow rates. If, however, there is very little or no information regarding a parameter, then the probability distribution is a reflection of the subjective state of uncertainty or lack of knowledge on the part of the assessor. For example, if one were asked to estimate the air flow rate (air changes per hour) through a room in the absence of any measurements, then an 'expert' might estimate it to be between 5 and 9 air changes/hr, with each value being equally likely (Uniform distribution 5,9). But, this distribution is simply an estimate from an expert since there is no population of air flow measurements. The estimate is subjective and based on his or her experience and thus it will differ from one 'expert' person to the next.

When a distribution is put forward as the best representation of the parameter, it follows that the value of each input parameter is not just a single number but instead is sampled as a number from its own previously determined probability distribution. In Monte Carlo simulations, a single value of each input parameter is randomly selected from each of their probability distributions — one distribution for each parameter. In the

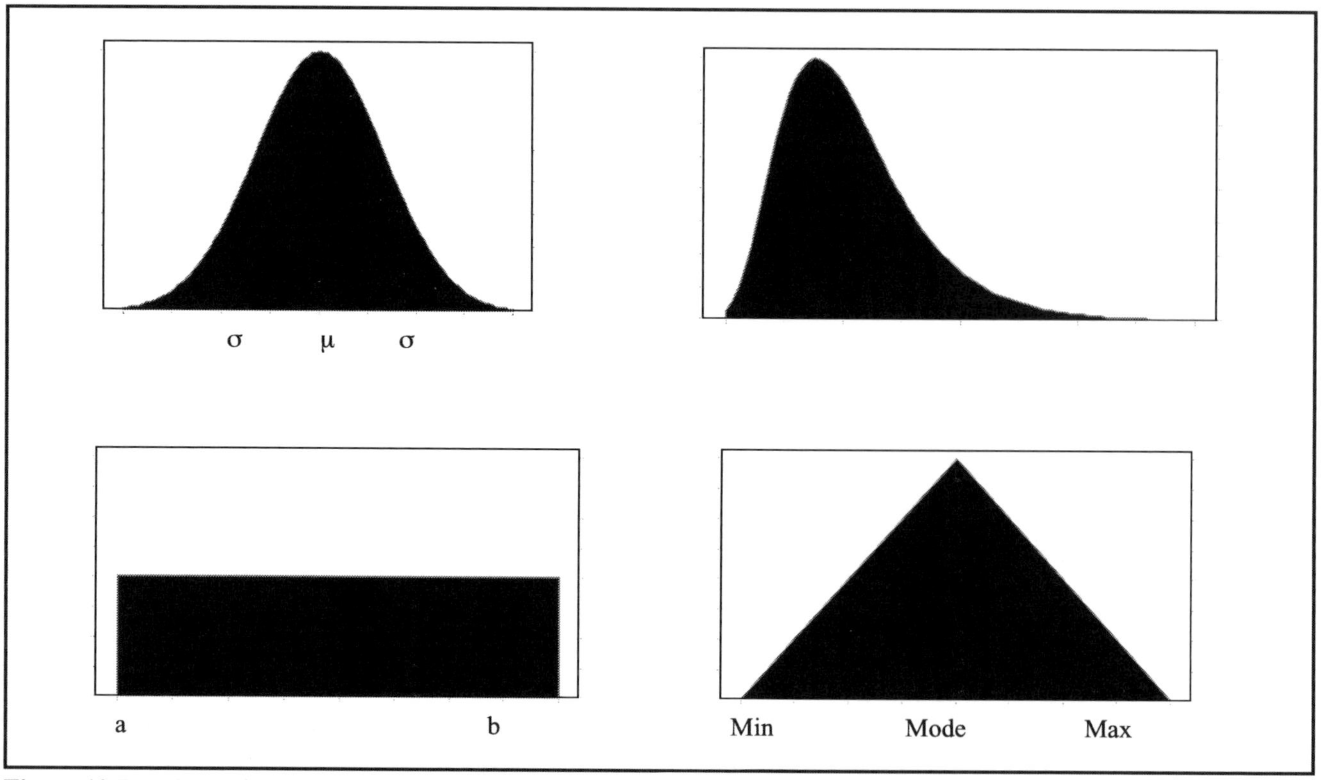

Figure 10.7 — Examples of probability distributions commonly used in exposure modeling. Clockwise from top left: normal, lognormal, triangular, and uniform distributions.

first step of this process, a single estimate of the "concentration" is computed by plugging these selected parameter values into the appropriate equation, such as those expressed earlier in this book. In the second step, a new random selection of input variables are selected from their respective distributions and the same model is used to compute a second estimate of "concentration." This is repeated again and again to obtain many independent values of the modeled "concentration." Once a pre-determined number of these values is obtained, say 100,000, a set of estimated "concentrations" are now available. These 100,000 values of the output are plotted as a probability distribution (e.g., a histogram). From this, an expected value and uncertainty information are obtained.

The following example illustrates the use of Monte Carlo sampling in exposure modeling. The work comes from Ramachandran (2001), which focused on the reconstruction of historical exposures to airborne nickel aerosol in a smelter.(3) Little data exist that characterize exposures that occurred several decades ago (1960–1978). However, modeling using process information was used to 'reconstruct' air concentrations and exposures. Such process information is sometimes available even though air monitoring data may not be available. This includes data on production rates, ventilation, and emission factors. While these data were not available on an annual basis, the few measurements that were available from some years were used to make reasonable assumptions about the values of these parameters for the other years. For the year of interest, 1965, the nominal production of Bessemer matte, an intermediate product in the manufacture of nickel, was 330,800 tons. This is a measurement that is, in fact, available on an annual basis from company records. While no ventilation rate information was available for 1965, there was one ventilation survey carried out in 1970 that estimated a net flow rate of 160,000 m³/min. Therefore, it was assumed that in 1965 the ventilation flow rate had a uniform

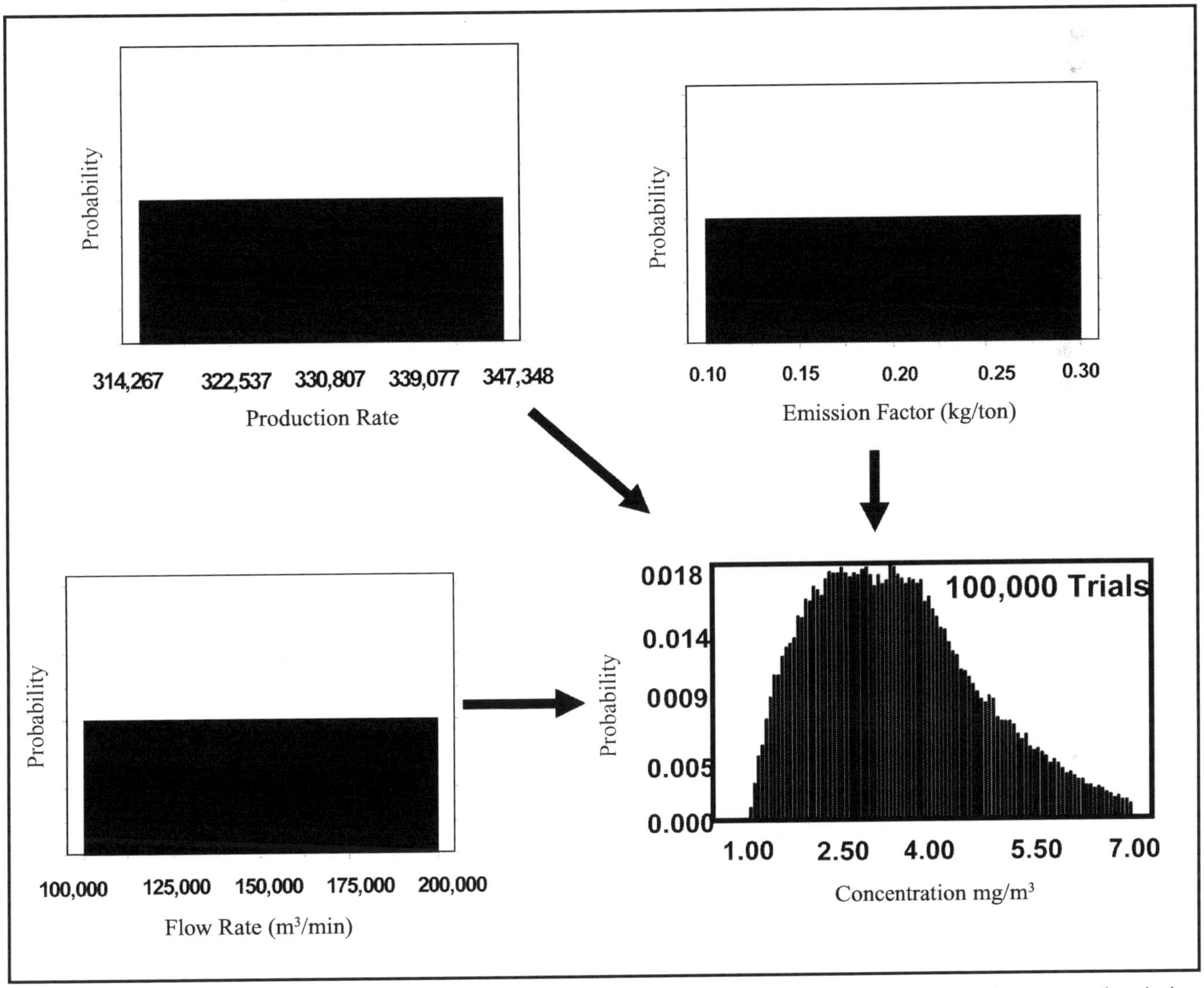

Figure 10.8 — Monte Carlo sampling from the probability distributions for the ventilation rate, production rate, and emission factor to estimate the probability distribution of the steady state aerosol concentration in a nickel smelter (from Ramachandran, 2001). Note that proper unit conversions were done in calculating the concentration.

distribution between 100,000 m^3/min and 200,000 m^3/min. The emission factor (kilograms of aerosol emitted per ton of Bessemer matte processed) was measured once in 1979 in a rather crude way, and this parameter was estimated to be 0.19 kg/ton. On this basis, it was assumed that the emission factor in 1965 had a uniform distribution between 0.1 and 0.3 kg/ton.

The simplest form of the well-mixed room model was used for predicting steady state concentrations of aerosol in the workplace.

$$C_{\text{steady state}} = \frac{G}{Q} = \frac{PR \bullet EF}{Q} \qquad (10\text{-}3)$$

Where G is the aerosol mass generation rate and Q is the volumetric flow rate of air through the workplace. The generation rate is modeled as the product of the production rate (*PR*) of Bessemer matte (tons/year) and the emission factor (*EF*) (kg of aerosol/ton of matte produced). Other factors such as the contribution of the settling velocity of the aerosol particles and removal by other mechanisms were considered negligible.

Using this model and the probability distributions described above, a Monte Carlo simulation was carried out. One hundred thousand sets of the three input parameters (PR, EF, and Q) were obtained by randomly sampling from their probability distributions. For each set of input parameters, the steady state concentration was calculated using the model. The histogram of these 100,000 output values represented a distribution of, including the variability or uncertainty in, the estimate of the steady state concentration of aerosol in the nickel smelter in 1965. Figure 10.8 illustrates this process.

To conduct Monte Carlo simulations, a simple spreadsheet method was illustrated in Chapter 2 of this book. However, software packages are available. For this chapter, versions of Crystal Ball and Simulación 4.0 were used to analyze data and generate the figures incorporated into this chapter.(4,5)

10.7 Final Thoughts

Uncertainty has been argued to be the bane of the exposure and risk assessment process.(6) Given estimates of risk that are highly uncertain, the reasonably conservative assessor, defaulting to worst-case will provide an appropriate and necessary overestimation of the exposure and risk. The degree of overestimation will be directly related to the level of uncertainty. These overestimates, ultimately borne of basic ignorance, could ultimately lead to a lack of credibility for both the process and those providing the analysis. On the other hand no assessor wants to underestimate an exposure or a risk. We believe that the answer lies in providing the most intellectually honest and complete analysis of the exposure and risk using the information at hand and tools outlined in this book. If the results are too uncertain for those responsible for making decision then the only logical answer is to get more data or develop better models which invariably also means getting more data.

10.8 References

1. **Cullen, A.C. and H.C. Frey:** *Probability Techniques in Exposure Assessment, A Handbook for Dealing with Variability and Uncertainty in Models and Inputs.* New York: Springer Publishing Co., 1999.
2. **Jayjock, M.A.:** Uncertainty Analysis in the Evaluation of Exposure. *Am. Ind. Hyg. Assoc. J. 58(5)*:380–382 (1997).
3. **Ramachandran, G.:** Retrospective Exposure Assessment using Bayesian Methods. *Ann. Occup. Hyg. 45(8)*:651–667 (2001).
4. **Oracle Corporation:** Crystal Ball Software. Redwood Shores, CA: Oracle Corp. http://www.crystalballservices.com/Software/OracleCrystalBall/CrystalBalliEditionsi/CrystalBall.aspx (2009).
5. Simulación 4.0, Monte Carlo simulator freeware, developed by José Ricardo Varela, http://www.cema.edu.ar/~jvarela/index_eng.htm (2009).
6. **Jayjock, M.A.:** Uncertainty as the Bane of the Risk Assessment Process. Testimony before the President's Commission on Risk Assessment and Risk Management, Washington, D.C., September 14, 1995.

Exposure Modeling and Bayesian Decision Analysis

By Monika Vadali
Gurumurthy Ramachandran, PhD, CIH
Perry W. Logan, CIH
Paul Hewett, PhD, CIH
John Mulhausen, PhD, CIH, CSP

11.0 Overview

The American Industrial Hygiene Association (AIHA) exposure assessment strategy relies upon the classification of workers into similarly exposed groups (SEGs).[1,2] The occupational hygienist uses a combination of professional judgment, personal experience with a given type of operation, review of exposures from similar operations, and/or exposure predictions developed using physical/chemical exposure modeling techniques to assign a subjective initial "exposure rating" and prioritize their SEGs. SEGs that merit a high initial exposure rating or involve high toxicity substances are typically placed at the top of the priority list for quantitative studies. Next, a baseline monitoring campaign is carried out and the measurement data collected are used to refine the initial rating and determine if the distribution of exposures for each SEG is well characterized and if the exposure distribution is acceptable. Acceptability is commonly evaluated by comparing an upper percentile such as the true group 95th percentile to the occupational exposure limit (OEL). In the AIHA strategy the 95th percentile of the exposure profile is estimated along with its upper confidence limit (UCL). Based on the magnitude of the group 95th percentile and its UCL relative to the OEL, the exposure is classified into one of four categories: "highly-controlled", "well-controlled", (nominally) "controlled", or "poorly controlled" (see Table 11.1). Bayesian Decision Analysis provides a tool for better decision making in selecting the exposure category.

11.1 Exposure Assessment and Bayesian Decision Making

Hewett et al. proposed a decision-making framework using Bayesian statistical analysis, where a key conceptual advance was to determine the probability of the 95th percentile of the exposure distribution being located in each of the four AIHA exposure categories.[3] This Bayesian Decision Analysis (BDA) framework is designed to explicitly take into account both monitoring data and professional judgment and other sources of information.

Table 11.1 — Exposure category rating scheme. A Similar Exposure Group is assigned an exposure rating by comparing the 95th percentile exposure ($X_{0.95}$) of the exposure distribution with the full-shift, TWA occupational exposure limit, OEL.

AIHA Exposure Rating	*Proposed Control Zone Description*	*Qualitative Description*	*AIHA Recommended Statistical Interpretation*
1	Highly-controlled (HC)	Exposures infrequently exceed 10% of the limit.(1)	$X_{0.95} \leq 0.10$ OEL
2	Well-controlled (WC)	Exposures infrequently exceed 50% of the limit and rarely exceed the limit.(1,2,3)	0.10 OEL $< X_{0.95} \leq 0.5$ OEL
3	Controlled (C)	Exposures infrequently exceed the limit.(1,3)	0.5 OEL $< X_{0.95} \leq$ OEL
4	Poorly controlled (PC)	Exposures frequently exceed the limit.(3)	OEL $< X_{0.95}$

(1) 'Infrequently' refers to an event that occurs no more than 5% of the time.
(2) Rarely' refers to an event that occurs no more than 1% of the time.
(3) High concentrations' are defined as concentrations that exceed the TWA limit or STEL.

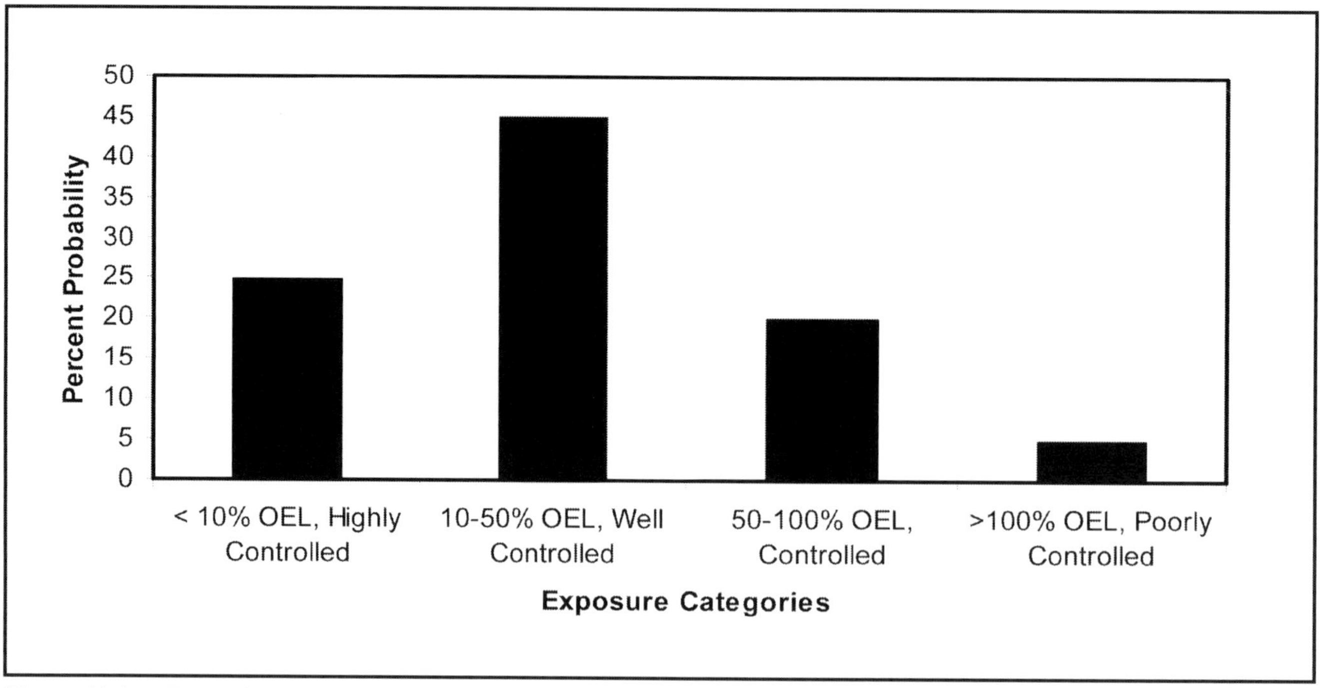

Figure 11.1 — Example exposure judgment for a given task. The bar chart shows the probability that the 95th percentile of the exposure distribution lies in each of the four categories.

The procedure results in a distribution of probabilities that can be graphed as a "decision chart" (Figure 11.1). The interpretation of such a decision chart is straightforward. For example, Figure 11.1 shows that there is a 30% probability that the SEG is highly-controlled, 45% that it is well-controlled, 20% that it is controlled, and a 5% probability that it is poorly-controlled. By inspection of these decision probabilities, one can decide on a course of action that is consistent with the exposure profile most likely being well-controlled (i.e., category 2).

The output of BDA is a set of three Decision Charts: the Prior Decision Chart, the Likelihood Decision Chart, and the Posterior Decision Chart. Figure 11.2 shows such a series of charts. *The Prior Decision Chart* is used to quantitatively capture the *a priori* belief of the industrial hygienist that the true SEG 95th percentile (or other decision statistic) falls within one of the four AIHA exposure control categories. These *a priori* probabilities are determined after considering information from multiple sources, including process knowledge, physical-chemical exposure models, surrogate data, historical and other applicable data, and information related to past exposures for a given SEG. The category probabilities in the *Likelihood Decision Chart* are calculated using the likelihood function described in Hewett et al.[3] These decision probabilities are calculated solely from the exposure data. The *Posterior Decision Chart* represents the probabilities of the exposure categories after combining the information contained in the Prior Decision Chart and the information contained regarding the exposure data in the Likelihood Decision Chart, using the Bayes equation modified to fit the industrial hygiene decision making problem.

The AIHA strategy suggests that 6 to 10 measurements be collected for most SEGs that are to be evaluated using exposure monitoring. In practice, since collecting even 6–10 measurements per SEG can be a challenge, exposure judgments are often made based on fewer measurements. In many situations, exposure assessment may be required for several chemical species simultaneously. Many plants have several thousand process, task, and substance combinations, making a complete quantitative exposure assessment for each all but impossible to accomplish. Anecdotally, it is estimated that greater than 90% of exposure ratings may be based on professional judgments without any monitoring data. Thus, there is a heavy reliance on the accuracy of professional judgments and the ability of occupational hygienists to correctly integrate them with monitoring data to reach an accurate exposure determination.

In the Bayesian context, an informative prior is one where different probabilities are assigned to different exposure categories, while a non-informative prior is one where all four categories are considered equally likely. It has been suggested that given an accurate informative prior, fewer monitoring data would be required to arrive at a decision with a high degree of confidence.[3] This is the promise of Bayesian decision-making, i.e., strategies based on high quality priors will be highly effective and efficient.

However, the converse is also true: inaccurate and overconfident priors will lead to incorrect decisions when monitoring data are sparse. While industrial hygienists can directly provide a prior decision chart based on their professional judgment, the rationale behind it can be unclear and human judgments are subject to cognitive

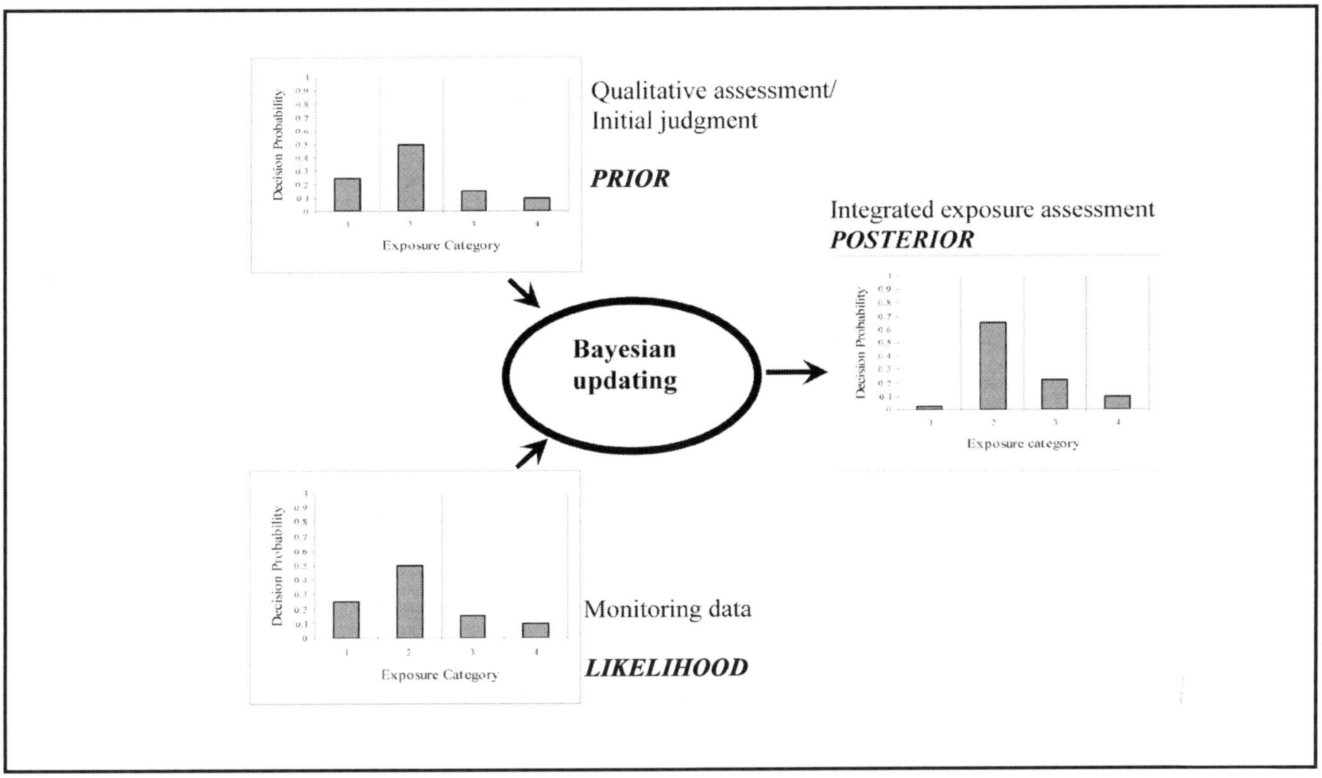

Figure 11.2 — Schematic representation of the Bayesian model. The prior is the subjective judgment provided by the hygienist based on experience and knowledge of task or, as proposed here, on the output of an exposure model, likelihood is the decision based on monitoring data alone and the posterior decision is the synthesis of prior and likelihood within the Bayesian framework.

biases that may affect such judgments in a potentially unquantifiable manner. However, a disaggregation of the problem (e.g., making an exposure judgment directly) into several smaller parts (e.g., making a judgment about an input parameter to an exposure model) may reduce the level of uncertainty.[4] Therefore, the use of mathematical exposure models may offer a step in the direction of reducing cognitive biases.

Mathematical models play an important role in any good comprehensive exposure assessment strategy.[5,6] Exposure modeling makes the rationale behind the judgment transparent and quantifiable. Models provide an explicit description of the mechanism of generation and dispersal of the contaminant and the subsequent exposures of the workers through various routes. The input parameters to the exposure models can be obtained either through measurements or subjective assessments and can be described in the form of a probability distribution. Even if the element of subjectivity is not eliminated, the use of models may reduce uncertainty and bias in exposure estimates.[7]

As shown in Figure 11.2, the decision charts are essentially probability distributions of the likely location of the true 95th percentile of the exposure distribution relative to the OEL. Since there is uncertainty and variability in the inputs of any exposure model, there will be a corresponding uncertainty in the model output (i.e., the exposure distribution). Our goal, therefore, is to arrive at a prior decision chart (similar to those shown in Figures 11.1 and 11.2) based on the output of an exposure model rather than on the direct subjective judgment of an occupational hygienist. In a later section, a method to accomplish this will be described. The priors thus obtained can be coupled with available monitoring data to arrive at a final decision.

11.2. One- and Two-Dimensional Monte Carlo Simulations

A hypothetical example similar to that originally used by Jayjock[7] for a scenario using the steady state solution of the general ventilation model can be used to illustrate different approaches to handling model uncertainty. The steady state concentration 'C', can be calculated as:

$$C(\text{mg/m}^3) = \frac{G(\text{mg/hr})}{Q(\text{m}^3\text{/hr})} \qquad (11\text{-}1)$$

Monte Carlo (MC) simulations allow for distributions to be imposed on each of the model variables. For example, uniform distributions can be imposed on the generation rate, G, and ventilation rate, Q, as described briefly in Chapter 2 and in more depth in Chapter 10. Consider the upper and lower bounds for G and Q to be

25 mg/hr and 75 mg/hr and 3.6 m^3/hr and 540 m^3/hr respectively. MC simulations can be carried out using these distributions. In MC simulation, a single value of G and Q is selected randomly from their respective distributions and the corresponding concentration, C, calculated as C=G/Q. This process is repeated a large number of times (~ 10,000) to yield many (~10,000) values of C which can be arranged as a histogram that approximates the exposure distribution of C. That histogram can then be used to determine a point estimate of the 95th percentile for the exposure distribution (e.g., the 9,500th value). However, it does not provide a probability distribution of the likely 95th percentile value from which a decision chart can be constructed.

The 2-dimensional (2-D) MC method can be considered as a simple nesting of two one-dimensional MC simulations. The inner simulation represents the variability in model parameters and the outer simulation represents the uncertainty (i.e., lack of knowledge) surrounding them. The variability of a model parameter occurs naturally due to the underlying physical and chemical processes. For the general ventilation model, these can be described by distributional parameters for G and Q (e.g., the lower and upper bounds in the case of a uniform distribution or the mean and standard deviation for a normal distribution). Since there will be variability in the generation and ventilation rates, these can be represented in the inner loop 1-D MC simulation. However, the distributional parameters are not known with certainty either. By introducing the second layer of simulation, the uncertainty in the distributional parameters of G and Q can be accounted for. These are represented by the outer loop 1-D MC simulation imposed on the distributional parameters of G and Q.

Using the previous example, uniform distributions can be imposed on G and Q with upper and lower limits. An additional layer of distributions can then be imposed on the lower and upper limits for both G and Q to reflect the uncertainty in those limits. The distributions for the lower and upper limits of G are called G_{min} and G_{max}, the corresponding distributions for the lower and upper limits of Q are called Q_{min} and Q_{max} respectively. G_{min} and G_{max} are represented as uniform distributions with lower and upper limits with G_{min1} = 10 mg/hr and G_{min2} = 40 mg/hr, and G_{max1} = 60 mg/hr and G_{max2}= 90 mg/hr, respectively. Similarly for Q, Q_{min1} = 1 m^3/hr, Q_{min2} = 5 m^3/hr, Q_{max1} = 350 m^3/hr and Q_{max2} = 650 m^3/hr. A single value is picked randomly from the distributions of G_{min} and G_{max} respectively, to define one distribution for G. Similarly, a random pick from Q_{min} and Q_{max} defines one Q distribution (outer loop). Using the G and Q distributions thus obtained, the inner loop of the MC simulation is run to produce a distribution of concentration C (Figure 11.3).This entire process is simulated a large number of times (e.g., 1000 times) to obtain a large number of C distributions. Each concentration distribution yields one point estimate of the 95th percentile and thus a distribution of the 95th percentile is obtained that can be displayed as a histogram. This 95th percentile histogram can be expressed in terms of the AIHA exposure categories

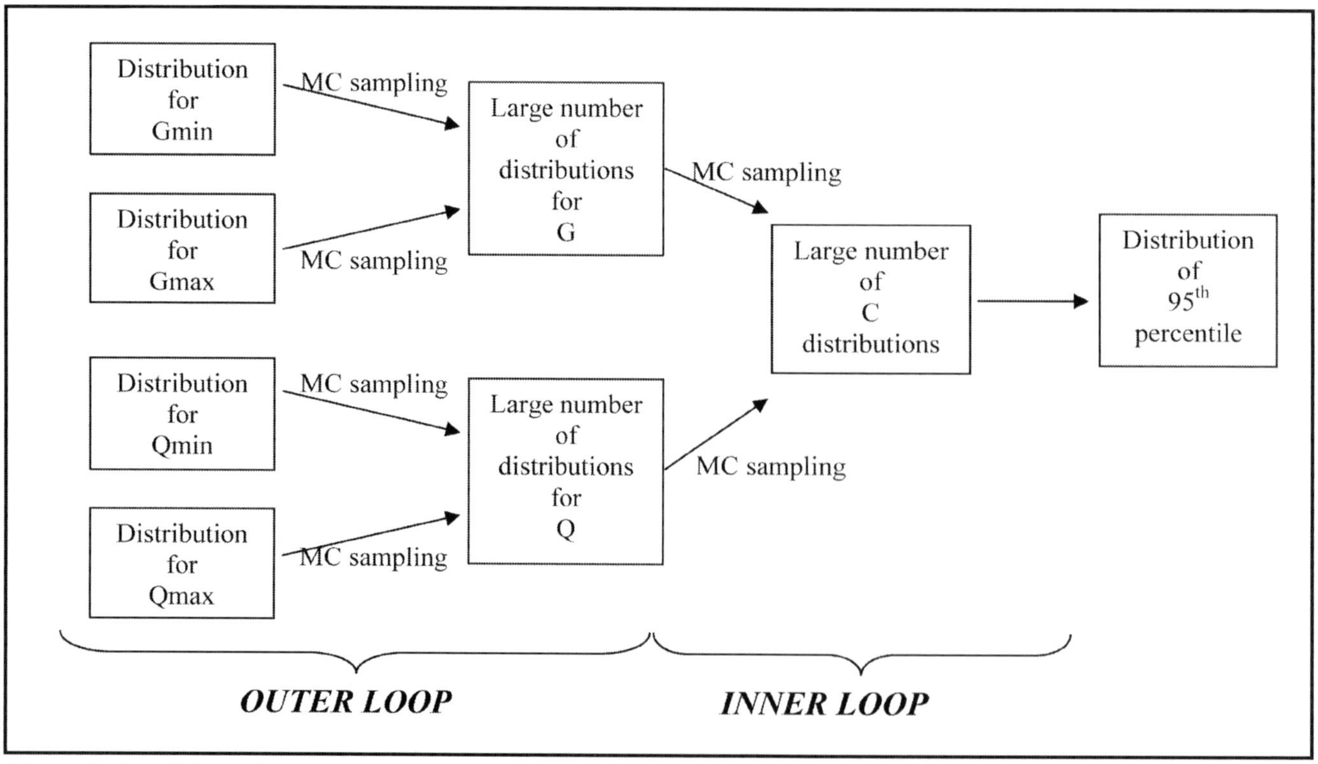

Figure 11.3 — Schematic representation of the process of obtaining the distributions of the 95th percentile using 2 dimensional Monte Carlo simulations. MC pairing from the distribution of G_{min}, G_{max}, Q_{min} and Q_{max} yield a large number of distributions of G and Q. MC sampling from them yields a large number of distributions of C, each of which has a 95th percentile. These, in turn, yield the distribution of the 95th percentile.

as a decision chart, when compared with the OEL (Figure 11.4). As before, the category with the largest probability would most likely drive further actions. Decision categories arrived at using different exposure models can be compared as part of the data analysis.

11.3 Decision-Making with Competing Models

Earlier chapters discuss the various options and the importance of selecting the most appropriate model. There are a wide variety of deterministic models that differ in their level of sophistication. Each level increases the cost of using the model due to the amount of information needed as inputs to the model. For example, the saturation vapor pressure model can be considered to be a rather simple model because it requires only knowledge of the temperature and the saturation vapor pressure of the chemical. The near field-far field model requires knowledge of room ventilation and contaminant generation rates in addition to a parameter known as the inter-zonal ventilation rate — requiring a non-trivial investment in obtaining this information. An even more sophisticated model such as an eddy diffusion model, which takes into account concentration gradients that exist around pollution sources, requires even greater investments.

While costs increase as the level of sophistication increases, more complex models also yield more refined exposure estimates. However, we may run into situations where different models provide significantly different exposure estimates that may lead to different decisions. In this section, we explore this idea with three models with different levels of complexity that lead to different decisions.

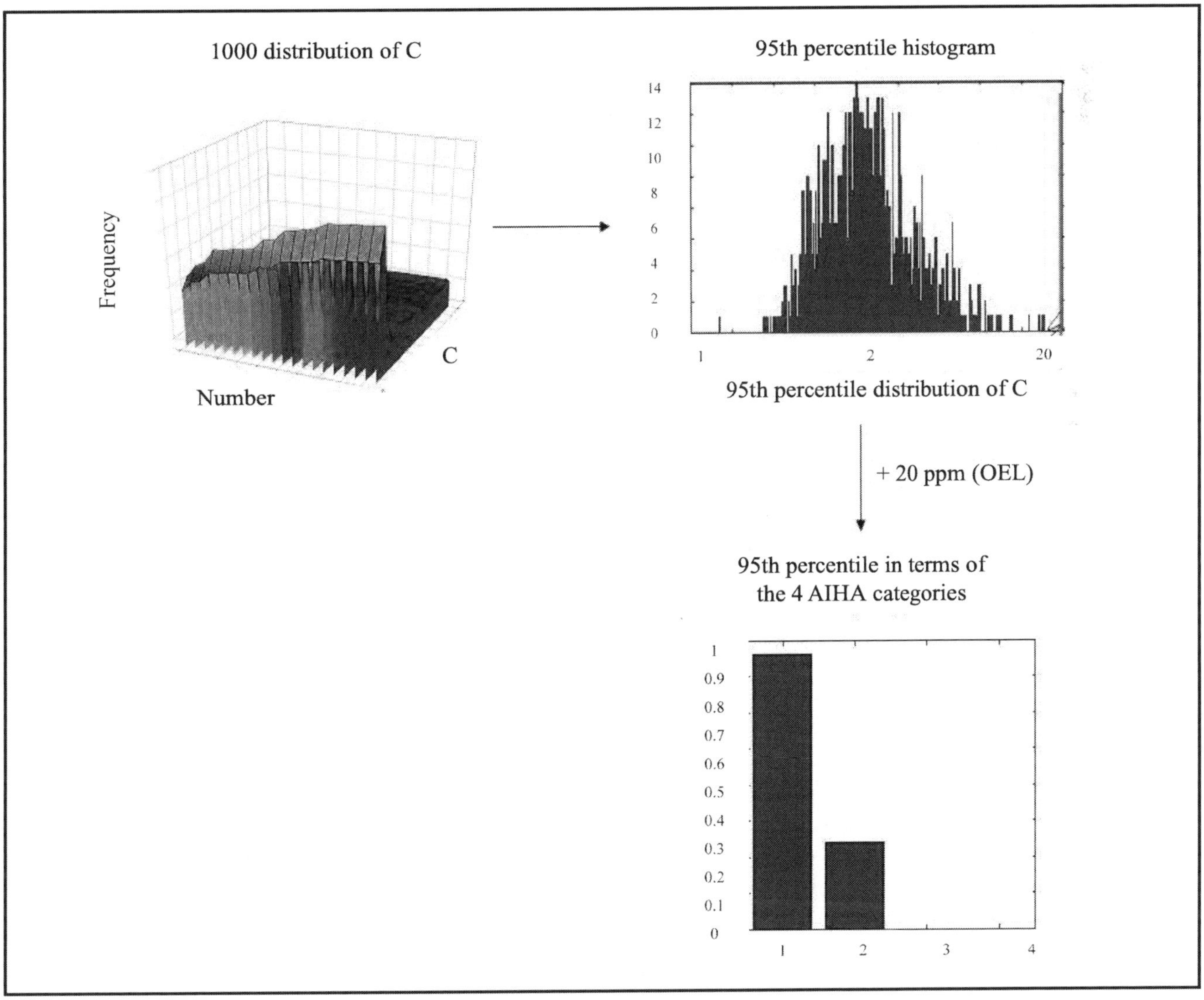

Figure 11.4 — Schematic representation of process for obtaining a decision chart from the output of a two-dimensional Monte Carlo simulation. The large number of C distributions yields a distribution of the 95th percentile. By comparing this distribution with an OEL, a decision chart in terms of the AIHA categories is constructed. For this example, the OEL is assumed to be 20 ppm. The chart indicates that modeling predicts category 1 to be the most likely exposure category.

Model choice has an important effect on the final decision and hence the required action regarding control measures. As the number of variables in the models increase, their behavior becomes increasingly complex, and decisions recommended by the model output may or may not be consistent with our expectations. With several competing models recommending different decisions, the choice of the estimate on which to base one's final decision and action becomes important.

Let us consider a very simple exposure scenario. Suppose there are 500 mL of a volatile chemical in an open container kept in a room at 20°C, constantly evaporating into its immediate environment. The vapor pressure of the chemical at 20°C is 80 mm of Hg. Let us also assume that the ventilation rate for the room is described by a uniform distribution. The lower limit of the distribution is between 1 and 5 m^3/hr and the upper limit of the uniform distribution is between 350 and 650 m^3/hr. Chemical vapors are generated at a rate that is also described by a uniform distribution. The lower limit of the distribution lies between 10 and 40 mg/hr and the upper limit of the distribution lies between 60 and 90 mg/hr. The occupation exposure limit for this chemical is 20 ppm. For this scenario let us consider 3 different models. *Model 1:* The saturation vapor pressure model (Chapter 3), assumes that the chemical is in a room with no ventilation and steady state concentration is calculated when the room reaches saturation. C_{sat} is calculated as:

$$C_{sat}\,(\text{ppm}) = \frac{VP_{sat}}{P_{atm}} \times 10^6$$

where VP_{sat} is the saturation vapor pressure at some temperature and P_{atm} is the atmospheric pressure.

Model 2: The general ventilation model (Chapter 4) assumes that the chemical is placed in a well ventilated room with distributions of ventilation and generation rates as described above and the steady state concentration is given as:

$$C_{steady\ state}\,(\text{mg/m}^3) = \frac{G}{Q}$$

Model 3: Near field-far field model (Chapter 6), where in addition to the conditions in model 2, a near field is assumed to be present around the pollutant generation source. In addition to the parameters G and Q, this model also accounts for the air flow between the two compartments using a parameter for the inter-zonal ventilation rate called β. The *steady state far field concentration* is calculated as:

$$C_{far\ field}\,(\text{mg/m}^3) = \frac{G}{Q}$$

and the *steady state near field concentration* is calculated as:

$$C_{near\ field}\,(\text{mg/m}^3) = \frac{G}{Q} + \frac{G}{\beta}$$

This inner compartment has its own ventilation rate denoted as β, which will be assumed in this example to have a triangular distribution with minimum, mode and maximum values. The minimum, mode and maximum will in turn be assumed to have uniform distributions. The minimum lies between 1 and 5 m^3/hr, the mode lies between 6 and 11 m^3/hr and the maximum has a distribution between 12 and 17 m^3/hr.

Results comparing the three exposure model predictions for the 3 models are presented in Table 11.2. As expected, the highly conservative saturation vapor pressure model yields a category 4 decision. The general ventilation model estimates that the highest exposure category is category 1. The NF-FF model estimates near field and far field exposure concentrations. The far field concentration is consistent with a category 1 exposure. Since this is the same as the general ventilation model, it is entirely expected. The near field concentration, however, is more consistent with a category 3 exposure. The near field exposure is the estimated exposure of the worker in the immediate vicinity of the source. Thus, we have three model-based prior judgments that could potentially lead to three substantially different decisions regarding control.

The choice of the "correct" model is obviously of great importance. There are at least two ways to make this choice. The first is to defer to professional judgment at this stage. Knowledge of the workplace environment would be the deciding factor for model choice. One could argue that the saturation vapor pressure model is excessively conservative and its assumptions do not reflect the exposure conditions; likewise, the general ventilation model is too lenient since it does not account for proximity of the worker to the source. The near-field model best represents the exposure scenario and therefore the most appropriate prior is a category 3 decision.

The second approach to choosing the most appropriate model output is to see which of the model estimates is most consistent with available monitoring data. This model would then be used in future assessments. This method for model selection can also be used as a feedback loop, where monitoring data for a particular task can be used to generate the likelihood, and this can be compared with the prior from each model. The model that best suits the likelihood can be picked as the most appropriate model for future use. For the above example with an OEL of 20 ppm, let us assume that we have five monitoring data points: 9.1 ppm, 7.9 ppm, 8.7 ppm, 10 ppm and 8.6 ppm. These data can be used to construct a likelihood function that can be used to update the prior. The likelihood decisions and the posteriors for each of the 3 ventilation models are shown in Table 11.3. As we can see, the likelihood decision (obtained based only on monitoring data) lies in category 3, and is most consistent with the prior for the near field model. Bayesian model comparison is a sophisticated version of this technique.[8]

Table 11.2 — Comparison of the final exposure category obtained when using 3 different ventilation models.
The Saturation vapor pressure model shows the exposure category in 4; conservative model. The General ventilation model predicts the exposure category as 1 and this is consistent with the prediction form the Near field – Far field model. The Near field – Far field model shows the category as 3.

Model Type	95th Percentile	Exposure Category
Saturation Vapor Pressure	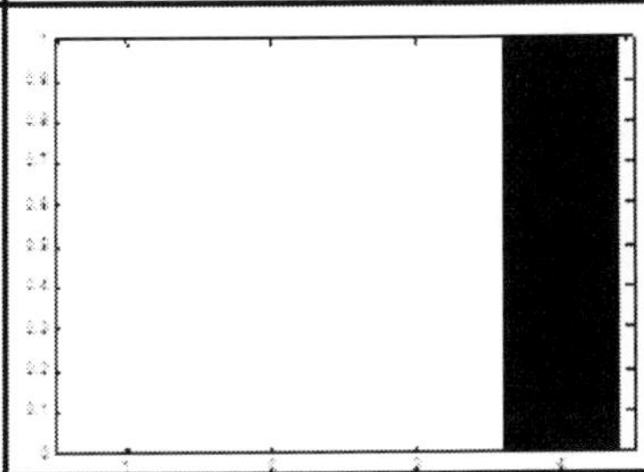	4
General Ventilation	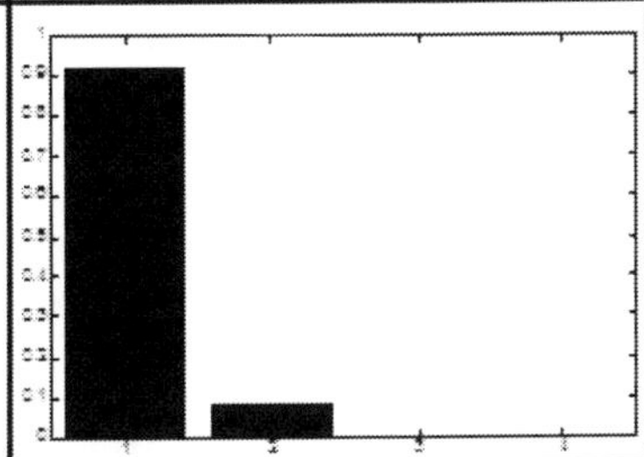	1
Near Field to Far Field	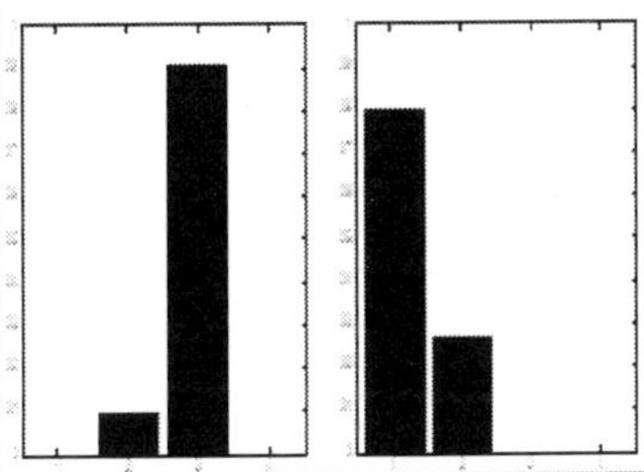	NF – 3 FF - 1

11.4 Concluding Remarks

One of the challenges that exposure modelers have faced is to show how the outputs of models can be used for occupational hygiene decision-making. The Bayesian analysis outputs are in terms of probability that the decision statistic falls into a given exposure control category. These probabilities are usually much easier to communicate and translate into a decision by the practitioner than the standard model outputs with their uncertainty estimates. This chapter demonstrates the applicability of exposure models in conjunction with 2-dimensional Monte Carlo methods for making exposure judgments in the context of Bayesian decision analysis. From the example above, the Posterior Decision Charts in Table 11.3 can be interpreted as "given both the sampling data and the model-based priors there is a greater than 95% probability that the exposure profile is a Category 3."

These model-based exposure judgments may improve upon direct subjective judgments regarding exposure provided by the occupational hygienist since they explicitly incorporate information about determinants of exposure in the form of model input parameters. Model input parameters may be obtained through actual measurements or professional judgments. Thus, subjective professional judgments of exposure are not eliminated but replaced with professional judgments about exposure determinants. Representing parameters as probability distributions allows a level of interaction between the hygienist and the model and also enables a better understanding of the overall system. This allows for a detailed description of the system and in the case of exposure judgments may help provide focus on the actual workplace parameters that influence exposures.

The output in the form of decision charts makes it amenable for use as priors in Bayesian decision making frameworks. The model-based priors can then be used in combination with exposure measurements to arrive at a more refined final decision category. The 2D MC method can be used to develop a multi level screening or tiered model approach for these chemicals.

The Bayesian method gives us a transparent method to utilize the strategy outlined in this manual and incorporate professional judgment or modeling with traditional sampling data to arrive at a final decision. This integration is very helpful for situations where there is minimal or no sampling data but there is a wealth of experience in the

TABLE 11.3 — Likelihood function and posterior obtained from the Bayesian Decision Analysis software [3]

The data predicts that the exposure category is category 3. The prior predicted from the 2D MC simulations predicted exposure category 3 for the near field model.

Model Type	Prior	Likelihood	Posterior	Final category
Saturation vapor pressure	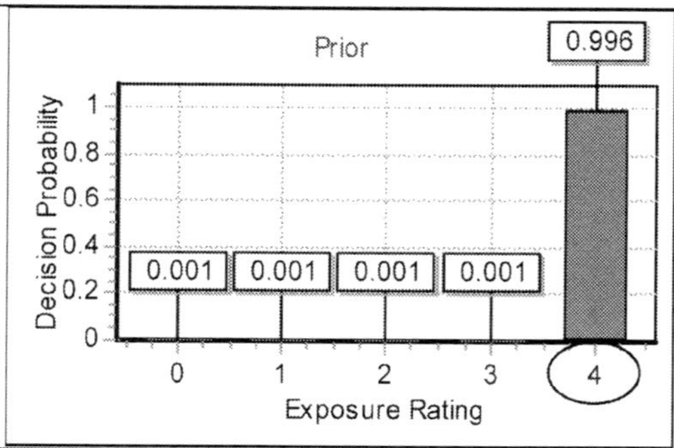	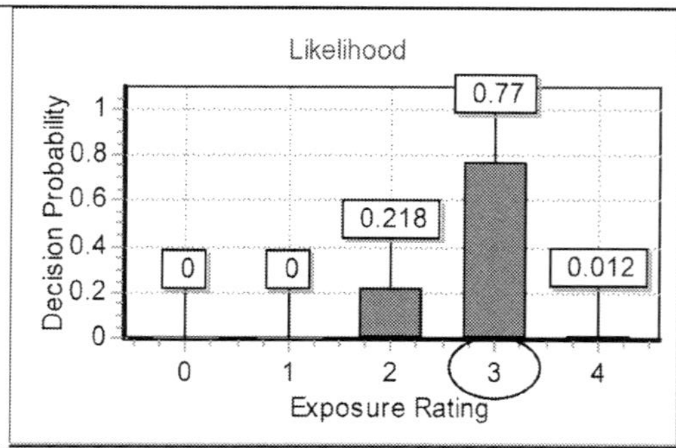	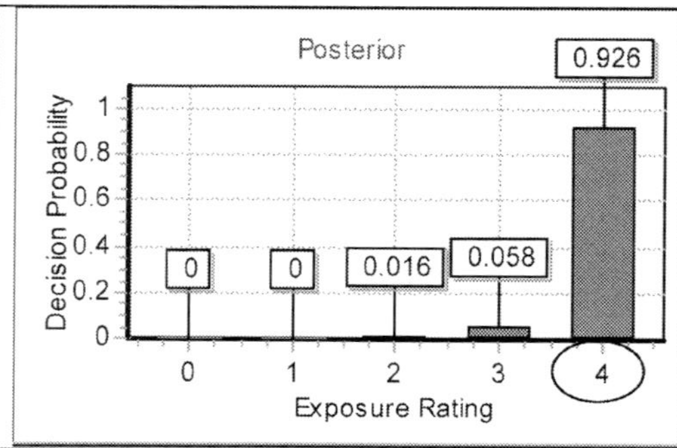	**4**
General ventilation	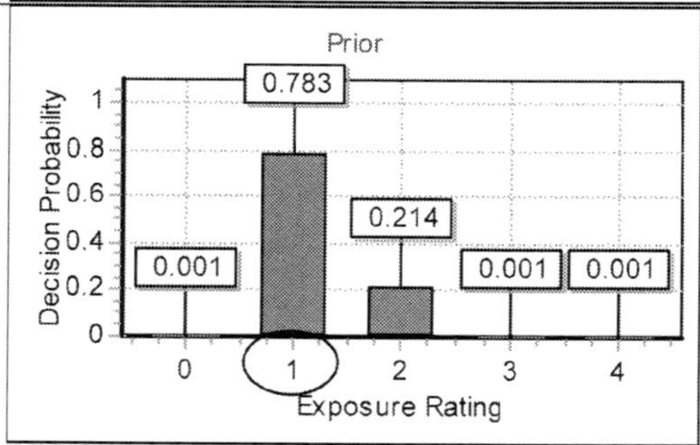	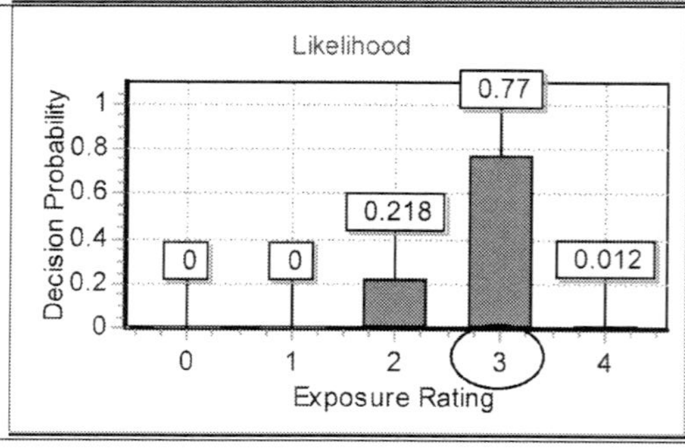	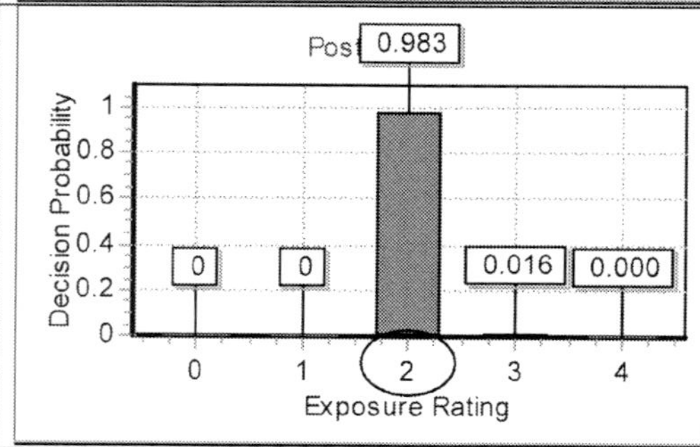	**2**
Far field	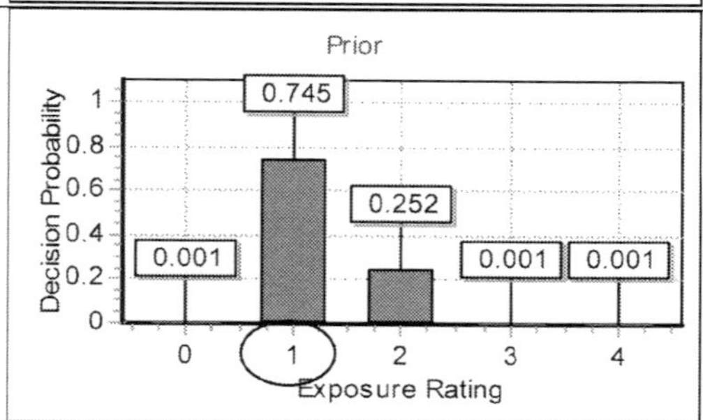	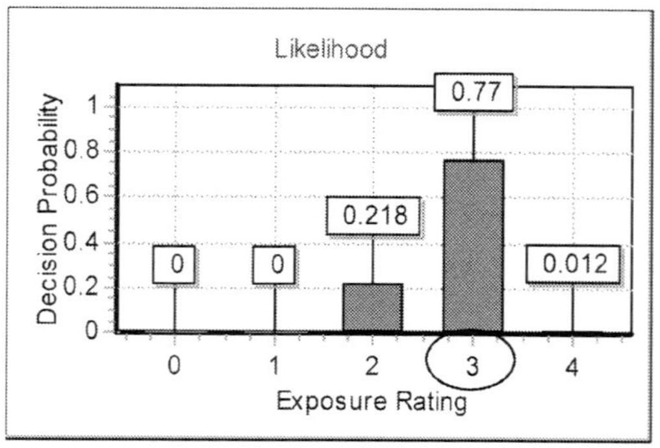	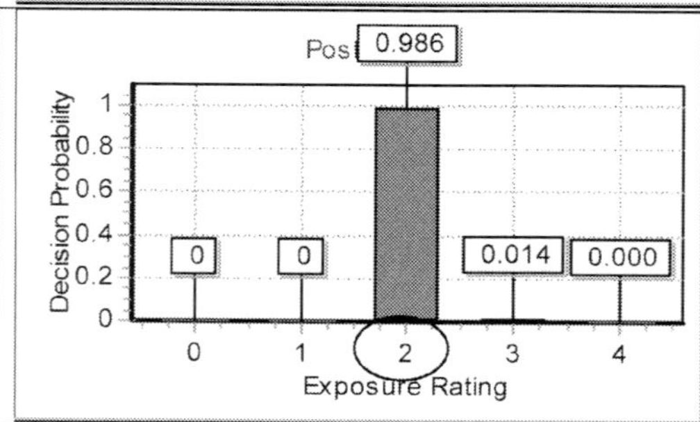	**2**
Near field	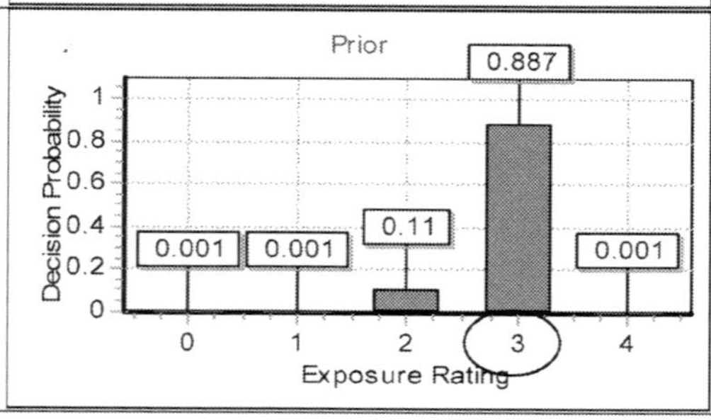		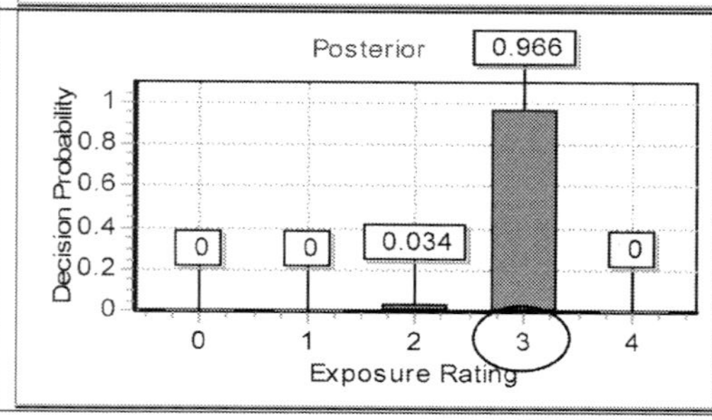	**3**

form of a professional judgment. It is also useful in situations where there may be a wealth of monitoring data for a similar SEG for the same chemical. In this case the Likelihood output could be used as the Prior for the related SEG. This ability to integrate professional judgment and data analysis in a transparent way allows for better use of data across SEGs, a site, or an organization, or other organizations with similar processes and SEGs.

The Bayesian AIHA model also allows the industrial hygienist to easily update judgments and in the process better understand prior professional judgments. This provides a framework and toolset for a more rigorous method for increasing professional judgment efficiency, accuracy, and validation. As monitoring data are collected the accuracy of the initial professional judgment can be systematically evaluated by comparing the professional judgment-based Prior Decision Chart to the Likelihood Decision Chart, which is calculated using only the data. If there a significant discrepancy the industrial hygienist can conclude that the initial professional judgment was faulty or collect additional exposure data. If the industrial hygienist is consistently high or low then perhaps a recalibration of professional judgment is indicated.

11.5 References

1. **Corn, M. and N.E. Esmen:** Workplace exposure zones for classification of employee exposures to physical and chemical agents. *Am. Ind. Hyg. Assoc. J. 40(11)*:47–57 (1979).
2. **Mulhausen, J.R. and J. Damiano (eds.):** *A Strategy for Assessing and Managing Occupational Exposures*, 2nd edition. Fairfax, VA: American Industrial Hygiene Association, 1998.
3. **Hewett, P., P.W. Logan, J. Mulhausen, G. Ramachandran, and S. Banerjee:** Rating exposure control using Bayesian decision analysis. *J. Occup. Environ. Hyg. 3(10)*:568–581. (2006).
4. **Morgan, M.G. and M. Henrion:** *Uncertainty: A Guide to Dealing with Uncertainty in Quantitative Risk and Policy Analysis.* New York: Cambridge University Press, 1990.
5. **Jayjock, M.A. and N.C. Hawkins:** A proposal for improving the role of exposure modeling in risk assessment. *Am. Ind. Hyg. Assoc. J. 54(12)*: 733–741 (1993).
6. **Jayjock, M.A.:** Uncertainty analysis in the estimation of exposures. *Am. Ind. Hyg. Assoc. J. 58*:380–382 (1997).
7. **Nicas, M. and M.A. Jayjock:** Uncertainty in exposure estimates made by modeling versus monitoring. *Am. Ind. Hyg. Assoc. J. 63(3)*:275–83 (2002).
8. **Gelman, A., J.B. Carlin, H.S. Stern and D.B. Rubin:** *Bayesian Data Analysis*, 2nd edition. Boca Raton, FL: Chapman and Hall/CRC, 2004.

Model Selection 12

By
Susan F. Arnold, CIH
Gurumurthy Ramachandran, PhD
Michael Jayjock, PhD, CIH

12.0 Overview

One of the challenges of using a mathematical model to estimate exposures is the initial model selection. Several conventions have been established for choosing a model and these have been incorporated into the framework that is presented here. The framework has been used and tested to a limited degree but not in a rigorous sense. It will surely evolve as we define its limitations and identify better approaches. Nevertheless, it represents the best practice in a young field.

12.1 Which Model Should I Use?

A 'tiered and iterative approach' to modeling has been recommended.[1] Applying a tiered precautionary approach balances the need for data against the available resources for generating data. Simple, conservative models can quickly provide overestimates of exposure that can be used for rapid screening of exposure scenarios. The more advanced models can trade off conservatism as data become available, yielding more realistic estimates. The level of sophistication of the model selected should reflect the objective of the modeling exercise. Examples of different modeling objectives and their relative levels of required precision include:

- Prioritizing an exposure scenario relative to hundreds of others, the model should provide a ballpark estimate of the exposure
- Decision making regarding risk management decisions, e.g. engineering controls or the use of personal protective pquipment (PPE), the model should render an estimate within an order of magnitude or less
- Withstanding legal rigor, the model should provide an estimate of the exposure with a quantifiable level of confidence
- Comparing exposure estimates against a benchmark that is an Occupational Exposure Limit (OEL), a Derived No-Effect Level (DNEL) or a Physiologically Based Pharmacokinetic Benchmark, the model should be as sophisticated as the benchmark.

An important factor in selecting the most appropriate model is cost. Using a more sophisticated model requires more data than the simple models and these data must almost always be generated *de novo*. Thus, the motivation to use a higher tier model must be balanced against the cost. Consider that a relatively simple chamber study to generate an empirically derived generation rate typically costs $25,000–$50,000. These costs may be very justifiable if the output provides credible data on which to base best practice recommendations and stewardship policies for a high volume product, but less justifiable for a low volume, low margin product.

The American Industrial Hygiene Association (AIHA) and the National Institute for Occupational Safety and Health (NIOSH) are collaborating on a data sharing initiative to develop a framework for a repository for high quality occupational exposure and supporting data that is publicly available.[2] When this project is completed exposure determinant data will be more readily available and using these refined models will be easier — and less expensive.

12.2 General Selection Guidance

1. Keep it as simple as possible — but not simpler. Conservative models are simple but come with very conservative (overestimating) assumptions. Screening level models are appropriate as initial assessment tools, especially when many assessments are being ranked and prioritized. While many exposures will be found to be unacceptable or uncertain when these models are used because of the conservative assumptions,

very low exposure scenarios will be identified and can be deemed acceptable. If an inherently over-estimating model produces an exposure estimate that is a small fraction of the occupational exposure limit a decision regarding the acceptability of the exposure can be made. Thus while additional iterations may be required for many exposures, those *de minimis* exposures are quickly and efficiently screened out.

2. If the first iteration does not produce an outcome that meets the defined needs, i.e. if the model estimate is too conservative or too general, consider refining the model parameter estimates or applying a more sophisticated (higher tier) model.
3. Put it together and pull it apart. This facilitates a deepened understanding of the contribution of each parameter. For deterministic models, doubling or halving the value of a given parameter is a simple but effective way to observe the impact on the model output and provides a crude but valuable 'sensitivity analysis.' Monte Carlo tools may be applied facilitating a more rigorous sensitivity analysis as well as quantify the level of uncertainty — and the corresponding confidence when using stochastic models.
4. Step back and ask, "Does it make sense?" and "Is it reasonable?" Exposure assessment practitioners can and should use their professional judgment. A predicted airborne concentration of a chemical that differs significantly from what was anticipated begs further investigation. If the exposure scenario violates any of the model's assumptions (as they often do) then careful consideration of the impact on the estimated airborne concentration is essential. This can be challenging; the published literature is lean, reflecting very limited research to date on this subject.
5. Critically evaluate the model. The application of Bayesian techniques (see Chapter 11) with an integrated approach to exposure assessment using modeling and sampling is a promising and practical evolution in exposure assessment. Financial and practical constraints limit the ability to adequately characterize exposures by sampling alone. But sampling data collected in conjunction with a more rigorous characterization of the workplace (i.e., evaluating airflow conditions and ventilation) serve to 'validate' or indicate the need to refine model estimates. In Bayesian terminology, the sampling data help refine the Bayesian Prior, producing a Bayesian Posterior. Where the data shows agreement with the model output (estimated airborne concentration) the Posterior Prediction is assigned a higher level of confidence. So the integration of the two strategies produces an estimate of the exposure with an associated level of confidence, but also a deeper understanding of the determinants of exposure. These are very valuable products indeed where decision making based upon the output is necessary.

12.3 Using a Tiered Approach

The models discussed in this book are loosely organized into tiers. Models designed for screening level or ballpark estimates form the base tier, Tier I. Tier II models can provide estimates that are within an order of magnitude or two, and Tier III models provide an opportunity to further refine these estimates and quantify the level of certainty in the estimates. It is worth noting at this point that the metrics of exposure and toxicity must be in the same units.

Tier I models are screening level models and are designed to purposefully overestimate exposures. Models in this tier are simple, conservative deterministic estimators. These are point estimates generated from point estimate input parameters. For example, the generation rate would be given as a single value and the corresponding predicted airborne concentration would be a data point. These ballpark estimates are then compared with some Occupational Exposure Limit (OEL) or other appropriate benchmark of toxicity such as a No Observable Adverse Effect Level, possibly divided by a safety factor. Note that these benchmarks used for exposure acceptability decision making are also single values. In actuality we know that these are not safe "lines in the sand" but that health effects are manifested across a range of exposure levels.

Tier I models include:

- 'Rule of 10' model, (ballpark estimate of airborne concentration based upon the nature of the ventilation system)
- Zero ventilation models described in earlier chapters
- Well Mixed Box (or Room) models. These are appropriately used for non-point sources. However an important footnote here is the potential for this model to UNDER estimate exposure where there is a point source allocated proximally to the worker.
- Simple mass balance models ignoring sinks and losses other than from ventilation.
- Estimation and Assessment of Substances Exposure (EASE) (a British Health and Safety Executive developed database driven model (described in Chapter 8).

To apply these models, a description of the pollutant generation must be obtained. For more information on pollutant generation and generation rates, see Chapter 3. For some of the Tier I models, a description of room airflow, denoted by Q are also needed. As a rough guide, a worst case estimate of the ventilation rate for an indoor residence is 0.2 air changes per hour (ACH).[3] In an industrial setting, a worst case guess, based on limited experience is 3 ACH.

To demonstrate the use of a Tier I model, let us consider a company that performs coating of metal substrates using solvent based paints. Two very common solvent carriers used in this type of process are toluene and xylene. The main coater is semi-isolated in a coating room that is supplied with approximately 3–6 ACH of outside fresh air.

The objective of this assessment is to prioritize the coater room operator's exposures against other workplace/worker exposures for further action. A reasonable first tier model in this situation is the 'rule of 10' model. This model suggests that in an environment with limited general ventilation, the airborne concentration will be somewhere between 1/10th and 1/100th of $C_{saturation}$. For toluene, this would suggest a concentration in the range of 280–2800 ppm. For xylene, the concentration range would be 120–1200 ppm. To put this into perspective, the lower explosive limit (LEL) for toluene is 10,100 ppm, and the PEL (TWA) is 200 ppm. There is also a 10 minute ceiling standard of 500 ppm. For xylene, the LEL is 9000 ppm, and the PEL is 100 ppm. While the model suggests concentrations that are less than 25% of the LEL, a conclusion regarding the acceptability of the exposure for health cannot be made. Since the estimated exposure using this model which is known to overestimate exposure exceeds the PEL, the exposure can either be classified as unacceptable or reassessed using a higher tier model to obtain a more realistic estimate.

Tier II models may be appropriate to 'reasonably' estimate the true airborne concentration of a chemical. "Reasonable" can be defined as within in an order of magnitude. Experience with these models has suggested that applied correctly, they provide an estimate much closer than an order of magnitude. Typically, output from these models would be benchmarked against an OEL or a working OEL, a DNEL, or a reference concentration or dose (RfC or RfD).

Tier II models include:

- The two-zone model (Near Field (NF)/Far Field (FF))
- Well mixed room models with more sophisticated generation rate estimators including;
  - The exponentially decreasing emission rate model
  - The drum filling model
  - The back-pressure model
  - Cyclic process emission modeling
- The eddy diffusion model with or without advection

These models require airflow information such as mechanical or natural ventilation rates as well as the dispersion pattern of that air. There are different algorithms for different patterns of air movement. Some example constructs for describing air movement are: air moving between theoretical boxes within a room, air dispersing in a random fashion resulting in spherical or hemispherical diffusion and plume modeling.
How readily the air is dispersed is also important; if air movement away from the source is impeded in some way by process equipment or other physical barriers, models can be modified to account for the geometry of the exposure scenario.

For a non-directional point emission source, such as the example of the toluene and xylene source in the coater room, the two-zone model should provide a reasonable estimate of the true airborne concentration for the coater operator, who is located in the near field between the source and the far field. Additional information is needed for this scenario before the two-zone model can be applied. The movement of air immediately surrounding the operator when he works at the coater must be characterized and quantified. The room volume must also be estimated for the outer box or far field volume. The sizing of the inner box or near field volume can be estimated using the convention of including a volume around the source containing the worker's breathing zone. The estimated steady state NF and FF airborne concentrations for toluene and xylene in this example are calculated below.

The generation rate can be estimated from one of a number of evaporation rate algorithms. In this case the Hummel, Braun and Fehrenbacher[4] algorithm was used. Q is calculated using the room volume to convert ACH (air changes per hour) to units of m^3/min. In this case, since Q was estimated at 3–6 ACH, we select the worst case scenario of 3 ACH. Since little information is provided on Beta, a reasonable worst case value for workplace wind speed (1.5 m/s, with local ventilation) is used, based on professional judgment. Beta is calculated assuming a radius of 0.78 m; approximating 'arm's length.' The parameters are shown in Table 12.1. The results of this model alongside the results of the "Rule of 10" model are shown in Table 12.2.

Table 12.1 — Deterministic Parameters for NFFF Model

	G (mg/m³)	*Q (m³/min)*	*β(m³/min)*	*Vol FF (m³)*
toluene	1000	15	2.9	300
xylene	250	15	2.9	300

Table 12.2 — Rule of 10, NF and FF Predicted Airborne Concentrations (ppm) at the Coater

	Rule of 10	*C NF*	*C FF*	*PEL*
toluene	276–2763	92.3	17.7	TWA 200 ppm; 10 min. ceiling 500 ppm
xylene	118–1184	23.7	16.7	TWA 100 ppm

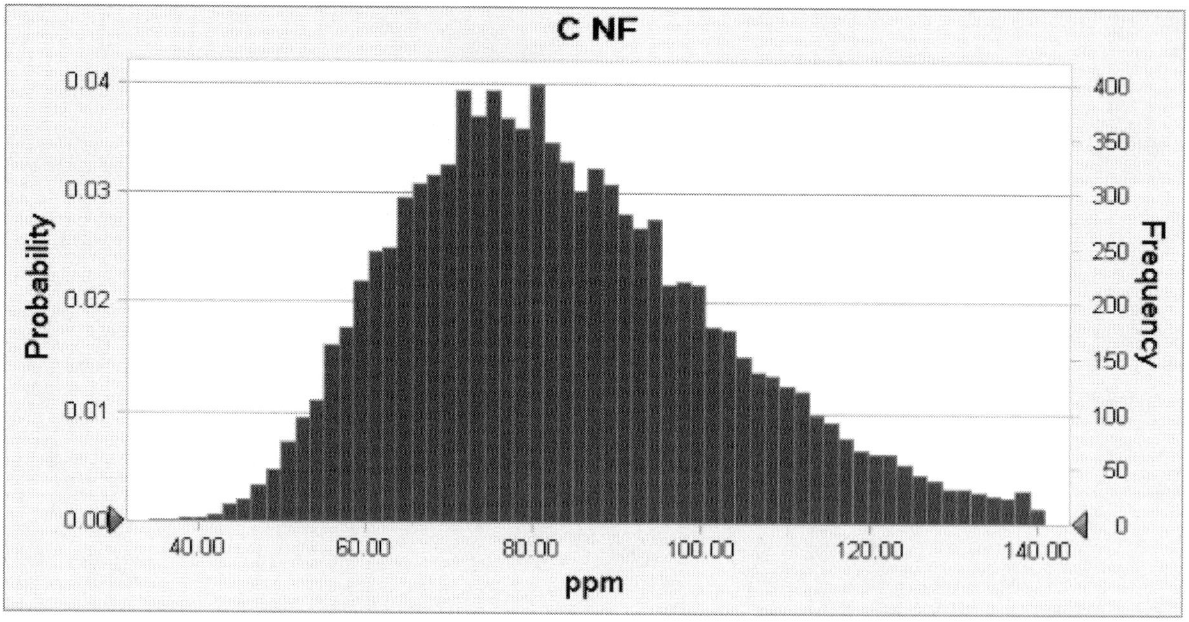

Figure 12.1 — Near Field Toluene Concentration (ppm)

The results of the two-zone model clearly indicate the potential for a significant concentration gradient such that the airborne concentration in the near field is much higher than the far field. The results also estimate NF airborne concentrations that is approximately 50% of the PEL. But how do we put this into context? Does this represent best case? Worst case? Average case? Can we conclude this exposure is acceptable? Deterministic models are limited here. At best, one can take the 50% of the estimate, and 200% of the estimate as crude upper and lower bounds to suggest a best and worst case, respectively.

The use of stochastic or probabilistic techniques with the models represents a third tier of exposure modeling (Chapter 10). Working with ranges or distributional inputs, the models render distributional outputs. The Monte Carlo simulator randomly selects a value from within the defined distribution for each parameter. The outcome (airborne concentration) is calculated, and the result is saved. This process is repeated hundreds, thousands or hundreds of thousands of times, creating a distribution of outcomes. The characteristics of the output distribution provide the user with the quantitative information necessary to quantify the uncertainty in the output.

For example, revisiting the coater operator's exposure scenario, several of the parameters relevant to the two-zone model are defined by using distributions. The spread of the distribution reflects the degree of certainty surrounding the estimation of the 'true' value for that parameter and the distributional characteristics reflect the analyst's understanding of the distributional pattern. The distributional characteristics of Toluene for the Coater Operator Scenario are presented in Table 12.3. The remaining parameters (Near Field Volume and Far Field Volume) are defined as being fixed values. The results of the Monte Carlo Simulation are presented in Table 12.4 and illustrated in Figure 12.1.

Table 12.3 — Distributional Characteristics for Toluene in the Example Scenario

Parameter	*G (mg/m³)*	*Q (m³/min)*	*β(m³/min)*	*Vol FF (m³)*
Distribution	Lognormal	Uniform	Lognormal	Fixed
Values	GM = 1000	3–6 ACH	GM = 4	300
	GSD = 1.13	(15.00–30.00)	GSD = 1.10	

Table 12.4 — NF and FF Concentrations for Toluene using the two zone model.

Location	*Mean (ppm)*	*95th Percentile (ppm)*
Near Field	76.7	162.9
Far Field	11.8	16.9

The stochastic model provides the data needed to quantify our uncertainty in the results. The estimated mean toluene airborne near field concentration is 76.7 ppm. The upper bound (95th percentile) estimate is 162.9 ppm. This upper bound estimate of the NF concentration helps us place the exposure estimate into context, giving the assessor a greater degree of confidence in classifying the exposure and making recommendations for follow up.

The Monte Carlo approach also allows us to investigate which of the parameters has the greatest sensitivity or impact on the outcome. For more information on uncertainty and sensitivity analysis, see Chapter 10.

Regardless of the specific model selected, the step that should always follow application of the model is to 'pull it apart.' Revisit the assumptions made and question whether they make sense. Was the model chosen for the scenario, or was the scenario made to fit the model? Do the assumptions of the model hold true for the scenario that was evaluated and if not, what is the impact on the output of these violations? Are the parameter values used to model the exposure reasonable? Are they based on robust data sets, real but limited data or pure professional judgment? Robust data sets should return realistic estimates with tight distributions while limited data and professional judgment using default assumptions will be reflected by more conservative estimates with broader boundaries.

Finally, confidence in model selection and models in general increases with the critical evaluation of the model selected to evaluate the particular scenario, using quantitative data. This includes limited air sampling data, other emission data, airflow data, and supporting details surrounding their collection.

12.4 References

1. **Keil, C.B.:** A Tiered Approach to Deterministic Models for Indoor Air Exposures. *Appl. Occup. Environ. Hyg. 15*:145–151 (2000).
2. **Middendorf, P.J., T. Havics, M. Jayjock, P.W. Logan, and K.A. Murray-del Aguila:** NOEDB: Making the Case for Sharing Exposure Related Data. Roundtable 224 AIHce 2008. Minneapolis, MN, 2008.
3. **Murray, D. and D. Burmaster:** Residential Air Exchange Rates in the United States: Empirical and Estimated Parametric Distributions by Season and Climatic Region. *Risk Anal. 15(4)*:459–465 (1995).
4. **Hummel, A., K. Braun, and M. Fehrenbacher:** Evaporation of a liquid in a flowing airstream. *Am. Ind. Hyg. Assoc. J. 57(6)*:519–525 (1996).

Dermal Exposure Modeling

By
Jennifer Sahmel, CIH, CSP
Mark Boeniger, CIH
Jeffrey Knutsen, PhD
Wil ten Berge, PhD
M. Cathy Fehrenbacher, CIH

13.0 Overview

A number of modeling methods have been developed to estimate exposure and absorption of substances via the dermal route. The modeling techniques presented in this chapter may assist the industrial hygienist or risk assessor in estimating dermal loading on the skin, dermal absorption into the stratum corneum, or dermal penetration into the viable epidermis and bloodstream. Although only detailed sampling or biological monitoring measurements of individual exposure scenarios provide the most accurate estimates of dermal exposures, for situations where it is impossible or impractical to collect any sampling data, modeling may be a useful surrogate for exposure estimation or risk prioritization. The methods presented in this chapter include basic models to characterize chemical loading on the skin surface (defined by mass per unit surface area) as well as more involved models that estimate the mass of absorption through the stratum corneum for both organic and inorganic chemicals in an aqueous solution. Dermal exposure to chemicals in a gas or vapor form is also briefly discussed. The U.S. EPA has also addressed certain non-aqueous exposure scenarios such as exposure to soil, but these are less well developed than the models for aqueous solutions and are therefore not discussed in this chapter. Finally, examples are presented to the user of several of the modeling techniques presented in the chapter.

13.1 Background

The skin is an extremely complex physical, metabolic, and immunologically active organ which, described in the simplest terms, consists of the outer stratum corneum, the viable epidermal and the vascular dermis.[1] The stratum corneum is an extremely thin layer of flattened, dead keratinized skin cells in a lipid matrix. In healthy skin, the stratum corneum provides an effective barrier to some foreign agents, such as bacteria, and helps to maintain the body's water balance and temperature. In certain cases, the stratum corneum may also reduce or delay the absorption of chemicals through the skin. It should be noted that skin that has been damaged or compromised by previous chemical exposure or environmental factors such as wind, water, or humidity will provide a very different level of protection from chemical exposures than healthy skin.[2] Substances that contact the skin may evaporate or be physically transferred away from the surface of the skin, may be absorbed into the stratum corneum and be reversibly or irreversibly retained, or may penetrate into the viable epidermis and subsequently the vascular dermis and bloodstream where they may be metabolized. The stratum corneum has a lipid structure that holds the dead skin cells in place, and this structure is believed to provide the predominant barrier to chemical skin absorption through its densely packed arrangement of both lipid and protein components.[1] The viable epidermis is composed of living cells and is more aqueous and hydroscopic. This bi-layer configuration of the lipid-rich stratum corneum and the aqueous epidermis further contributes to the barrier properties of the skin. The skin is a complex system, with several potential pathways for transport and loss of substances contacting its surface. Many factors affect sorption into the skin, including physicochemical properties of the contaminant, skin age, skin condition, anatomical location of the contact, skin temperature, and hydration.[3,4]

This chapter provides a brief overview of methods available for estimating dermal exposure and skin permeation and discusses data sources for select model input values. Where available, default input values are identified, although these should be evaluated carefully for applicability to specific exposure scenarios. Biological monitoring is the ultimate tool for assessing total body exposure, although the number of reference standards and methods currently available may limit the applicability of this approach.

13.2 Modeling Dermal Exposure Potential

The process of judging dermal exposure potential is subject to the risk assessor's level of expertise and familiarity with the work tasks being performed, as well as the amount of detailed information collected during the assessment.[5] The U.S. EPA has noted that, in general, exposure duration, exposure frequency, and contact rate are likely to be the most sensitive parameters in an exposure assessment.[6] Several studies have noted that the processes of transferring chemicals to the clothing and skin can be quite variable and complex. Clothing and skin can become contaminated via impaction of aerosols, splashes, immersion, and contact with surfaces containing the chemical. The clothing may completely protect the skin, or it may become a depot for long-term transfer.[7–9]

There have been several relatively detailed approaches developed in order to standardize the approach for assessing dermal exposure. Most of these methodologies are based on a detailed conceptual model presented by Schneider et al.[10] This conceptual model describes the processes of mass transport of contaminants between the environment and the skin and is an excellent reference on the general mechanisms of skin exposure. Exposure assessment tools that have been developed based on this model typically include a series of standard questions that may be presented using a computer questionnaire format. The exposure assessment tool may then automatically generate ranking or default exposure values based on the responses provided, compared against an empirical database of exposure measurements.[11–13] An example of this conceptual model using default exposure parameter factors has been described in detail for spray painters' exposure to solvents.[14,15] Such exposure assessment tools may help provide a more consistent and detailed perspective regarding the workplace variables that could influence exposure magnitude than the industrial hygienist or risk assessor might otherwise develop.

One dermal exposure assessment tool using a computer questionnaire survey instrument is called DREAM (DeRmal Exposure Assessment Method). DREAM uses a structured approach to dermal exposure assessment of chemical or biological agents and provides semi-quantitative estimates of exposure.[16] It guides the development of estimates for exposure on the outside clothing layer (potential exposure) as well as on skin (actual exposure), and it provides insight into the distribution of dermal exposure over the body. Exposure route estimates obtained from the tool incorporate the frequency, probability and intensity of exposure.

The REACH guidance recommends a conservative estimate of 100% dermal absorption for a substance with a molecular weight of smaller than 500 and 10% with a MW greater than 500 and a log K_{ow} (octanol/water partition coefficient) of less than -1 or greater than 4. However, compound specific differences in the dermal absorption process and the kinetic aspects of dermal absorption in this guidance are essentially ignored.[17]

The above approaches and references can be useful in refining an individual's ability to better understand and identify the magnitude of dermal exposures. Similar efforts to further refine and validate dermal exposure assessment models will hopefully continue into the future using advanced statistical approaches and other methods.[18] The ability of these models to reflect actual exposures, and the extent of their practical value, depends on the industrial hygienist's level of expertise and the accuracy of the model inputs.

13.3 Modeling Dermal Risk Potential

The quantification of exposure is only meaningful in the context of our knowledge and understanding of the health risks associated with particular levels of exposure. The RiskofDerm toolkit is one example of a method available to assist in dermal exposure risk assessment and management. This free toolkit was developed within the European Union (EU) and was primarily designed for use by small to medium-sized businesses.[19,20] This tool guides the user through a decision tree based on questions about chemicals used and the exposure situations encountered. It then translates the responses into broad data categories of hazard and exposure that lead to a rough estimate of health risks related to dermal exposure. RiskofDerm can be downloaded free over the internet.[21]

When an existing dermal risk assessment or estimation tool is not available, an occupational exposure limit (OEL) can be used in some instances to help differentiate acceptable from unacceptable exposures. However, nearly all OELs for chemical agents are presented in units of airborne concentrations in the breathing zones of workers. The need for the development of dermal exposure limits, along with possible approaches to setting these limits, have been discussed by many others.[22–25] Limits related to dermal exposure could be based on: (1) workplace surface contamination levels; (2) skin or clothing contamination levels; (3) estimations or biological measures of exposure or internal dose; or (4) an allowable duration of unprotected exposure. Although OELs directly addressing quantitative dermal exposures have not yet been systematically developed, there are approaches available that can be very useful, even in the absence of dermal-specific OELs.

Five different types of models are presented below to assist in the evaluation of dermal exposures. The first provides a methodology to calculate a dermal or total body dose equivalent based on an existing airborne OEL (§13.3.1). The second model compares this calculated OEL equivalent to a simple exposure estimation based on skin loading and a fractional or complete skin absorption value for the agent of interest (§13.3.2). The third model uses Fick's second law of diffusion and applies a permeability coefficient to estimate the absorbed dose of aqueous chemical solutions or neat chemicals through the stratum corneum (§13.3.3). The fourth model uses an OEL equivalent value and a permeability coefficient to determine an allowable skin exposure duration time for an agent of interest based on a calculated estimate of skin absorption (§13.3.4). The fifth model can be used to estimate hand-to-mouth contact related to dermal exposure (§13.3.5). The sixth model is used to estimate risk-based surface wipe screening levels (§13.3.6).

13.3.1 Estimating a Dermal OEL Dose Equivalent Using an Existing Airborne OEL

When dermal exposure limits are unavailable and the toxic effect of a chemical is systemic, an airborne OEL, expressed in units of airborne concentration, may be converted to a metric of dermal exposure. Although a worker's respiration rate varies with activity, workers may breathe about 11–19 m^3 of air in an 8-hour workday when working at moderate physical activity. The U.S. EPA's Exposure Factors Handbook (Table 5-25) reports measured inhalation rates for adults at a variety of exertion levels.(26)

Thus, an estimated total body dose can be created by converting an inhalation exposure limit expressed in mg/m^3 to a dose in units of mg/day using the following relationship

$$\text{Exposure dose (mg/day)} = \text{OEL (mg/m}^3\text{)} \times \text{inhaled air volume (m}^3\text{/day)} \quad (13\text{-}1)$$

One can then estimate the potential total body dose in mg/day and compare it to an estimated dermal loading. A few publications have proposed related strategies for addressing the contribution of dermal exposure.(30,31) Note that this approach is valid only for compounds that exert their toxicity as a systemic effect (e.g., carcinogens affecting target organs other than the skin, liver toxicants, kidney toxicants). It is not appropriate to use this methodology for agents for which the primary health effect is local tissue response (e.g., corrosives, eye irritation, or contact allergens). Additionally, it is important to review the documentation for the OEL and to apply this technique only in cases where the studies used to derive the inhalation OEL were based on systemic toxicity via multiple routes of exposure rather than inhalation toxicity alone. When dealing with local skin effects, toxicological data specific to that effect is required for assessing risk. It should be noted that the level of physical activity used for this calculation should be selected carefully, as this parameter can affect the allowable exposure dose significantly. And finally, a careful review of the literature supporting the inhalation OEL must be conducted before using this approach because factors such as the percent of the inhaled dose actually absorbed at the OEL can be critical considerations.

Table 13.1 — Summary of Adult Inhalation Rates for Short-Term Exposure Studies

Arithmetic Mean (m^3/hr) Activity Level					
Rest	*Sedentary*	*Light*	*Moderate*	*High*	*Reference*
0.5	0.5	1.4	2.4	3.3	Adams, 1993 (Lab protocols)[27]
—	0.6	1.2	1.8	—	Adams, 1993 (Field protocols)
0.4	0.4	0.7	1.4	3.6	Layton, 1993 (Short-term exposure)[28]
0.4	—	0.6	1.5	3.0	Layton, 1993 (3rd approach)
—	—	1.0	1.6	3.0	Linn et al., 1992[29]

13.3.2 Dermal Exposure Estimation Assuming Complete or Fractional Absorption

A simple, but potentially imprecise, model for estimating dermal exposure dose is to assume that a finite mass of a contaminant is deposited on the skin and a complete or fractional amount is subsequently absorbed. Variations of this approach can be used to estimate skin uptake of chemicals in soil or uptake associated with a thin film of liquid deposited on the skin.(32) This approach assumes that deposition over the exposed skin surface area is uniform and that the process of absorption is sufficiently slow so that one can separate the deposition from the absorption process. The method also can be used to preliminarily interpret wipe sampling data.(33) A considerable amount of dermal deposition data has been collected to assess occupational exposure to pesticide handlers and operators, but data for most industrial operations are limited.(34) Input values for the quantity deposited on the skin and the absorption fraction will likely contribute substantially to the uncertainty associated with this model.

Some considerations related to parameter selection must be carefully evaluated before using this model. First, although empirical measurements of the fraction of chemical absorbed have been published in the literature, such data must be used with care because of the specificity of these measurements to the particular scenario under evaluation. Care is required because reporting of the fraction of a chemical absorbed makes some assumptions about steady state conditions, where in reality, the dermal absorption rate is variable over time. Specifically, for a single applied dose, the rate of absorption decreases as a consequence of absorption that has already occurred, depleting the surface concentration over time. As a result, it may be difficult or impossible to use data from a study reporting fraction or percent absorbed to estimate the mass of a chemical absorbed for other dermal exposure scenarios. Studies reporting percent absorption may also report the percent absorbed over a 24-hour period rather than an 8-hour work day. Further, many other factors including skin condition, the presence of chemical mixtures, and skin washing or decontamination practices may dramatically affect percent absorbed values.

Second, a major assumption of this model is that skin loading (i.e., mg/cm^2) is proportional to the number of loading events, such that each event is additive and contributes equally to the final loading. Once deposited, it is assumed that the mass of chemical on the skin does not deplete over time. However, intuition and limited studies suggest that this is an unrealistically high loading. Third, the selection of a value for number of events per day can dramatically affect the model results and must be carefully considered in order to appropriately estimate dermal loading.[35]

This simplified absorption model will provide a typically conservative dermal exposure dose that can be used to estimate the potential daily dermal absorbed dose in units of mg/day, assuming deposition on the skin is followed by absorption. Once the estimated dermal absorbed dose has been determined, it can be compared to an OEL expressed in units of mg/day (Equation 13-1). Equation 13-2 presents the methodology for estimating dermal absorbed dose, assuming deposition followed by absorption.

$$DA = S \times Q \times FQ \times ABS \times WF \quad (13\text{-}2)$$

where:

DA = dermal absorbed dose rate, mg/day
S = surface area of skin available for contact, cm^2
Q = quantity of mixture deposited on the skin per event, mg/cm^2 – event
FQ = contact frequency, number of events per day, #/day
ABS = fraction of applied dose absorbed through the skin per event, unitless
WF = weight fraction of the substance in the mixture, unitless

S (Surface Area of Skin)

To calculate the surface area of potentially exposed skin, the industrial hygienist should gather information on activities and tasks performed by each worker, along with the types and number of skin surfaces (e.g., one hand, two hands) potentially exposed during the activities. Worker activities can significantly influence the amount of a substance deposited on the skin, as well as the location of contact. The U.S. EPA's Exposure Factors Handbook (Table 6-4) provides estimates of skin surface areas for both males and females[26]:

It may be necessary to compute dermal absorbed doses for different parts of the human body, as the quantity of deposition may vary by surface area. As a result, this equation should be used to calculate a dose only for limited surface areas of skin where loading is likely to be uniform (i.e., the hands).[36] Although the hands are often identified as a site of higher exposure, other parts of the body also may receive substantial exposures. In a 1992 study, for example, during mixing/loading of pesticides, high exposure at the lower leg occurred because most of the operations involved pouring liquid from one container to another below the waist level.[37] Another researcher found unusually high

Table 13.2 — Surface Area by Body Part for Adults (m^2)

Body part	Men						Women					
	N[a]	Mean	(sd)[b]	Min.	-	Max.	N	Mean	(sd)	Min.	-	Max.
Head	32	0.118	(0.0160)	0.090	-	0.161	57	0.110	(0.00625)	0.0953	-	0.127
Trunk (Incl. Neck)	32	0.569	(0.104)	0.306	-	0.893	57	0.542	(0.0712)	0.437	-	0.867
Upper extremities	48	0.319	(0.0461)	0.169	-	0.429	57	0.276	(0.0241)	0.215	-	0.333
Arms	32	0.228	(0.0374)	0.109	-	0.292	13	0.210	(0.0129)	0.193	-	0.235
Upper arms	6	0.143	(0.0143)	0.122	-	0.156	-	-	-	-	-	-
Forearms	6	0.114	(0.0127)	0.0945	-	0.136	-	-	-	-		-
Hands	32	0.084	(0.0127)	0.0596	-	0.113	12	0.0746	(0.00510)	0.0639		0.0824
Lower extremities	48	0.636	(0.0994)	0.283	-	0.868	57	0.626	(0.0675)	0.492	-	0.809
Legs	32	0.505	(0.0885)	0.221	-	0.656	13	0.488	(0.0515)	0.423	-	0.585
Thighs	32	0.198	(0.1470)	0.128	-	0.403	13	0.258	(0.0333)	0.258	-	0.360
Lower legs	32	0.207	(0.0379)	0.093	-	0.296	13	0.194	(0.0240)	0.165	-	0.229
Feet	32	0.112	(0.0177)	0.0611	-	0.156	13	0.0975	(0.00903)	0.0834	-	0.115
TOTAL		1.94[c]	(0.00374)[d]	1.66	-	2.28[e]		1.69[c]	(0.00374)[d]	1.45	-	2.09[e]

[a] number of observations.
[b] standard deviation.
[c] median (see Table 6-2).
[d] standard error.
[e] percentiles (5th - 95th).
Source: Adapted from U.S. EPA, 1985.

exposure at the legs due to frequent contact with spray nozzles during loading of pesticides.[38] In a separate study, high exposure at the thighs was reportedly due to workers rubbing their hands against their pants at the thigh area during pesticide operations.[39] The Pesticide Handlers Exposure Database (PHED) has been designed to facilitate estimations of pesticide exposures during mixing/loading, application, and flagging operations, based on application method, formulation type, procedure, equipment, clothing scenario, and other factors.[34]

FQ (Frequency of Events per Day)

Care must be taken in interpreting the number of exposure events occurring per day when using this model because it assumes that skin loading is always proportional to the number of events. Other equations (provided below) will provide more realistic ways to address the loading and subsequent absorption of an agent through the epidermis. In using this equation, the industrial hygienist or risk assessor must carefully consider whether each "event" will result in additional loading of the skin. The majority of experimental studies investigating dermal transfer and retention have been documented after a single exposure event. However, in this equation, it is presumed that during repeated events (or contacts), the amount of the agent retained on the skin will increase additively with each event in proportion to the number of contacts; this approach is likely to overestimate skin loading. The actual amount of agent contacting and subsequently retained on the skin may vary over several orders of magnitude depending on the way specific activities are performed by the worker and the physical nature of the agent matrix. If one assumes an unreasonably large number of events or contacts for the purposes of this model input, it will in turn result in an unreasonably high estimate of dermal loading over the period of interest. A more realistic scenario is that eventually the exposed area of skin would become "saturated" when the skin's capacity to retain the substance reaches its maximum level. After this point, additional exposure events would be unlikely to result in additional skin loading.[40,41] One way to address this issue when using this particular model is to consider assigning the number of exposure events (FQ) for periods of work separated by changes in work routine or skin decontamination.

Q (Quantity Deposited on Skin)

The amount of a substance deposited on the skin has been shown to vary by activity; for solids, parameters such as particle size may be the primary factor in the deposition rate.[42] A variety of methods can be used to measure the amount of contaminant deposited on the skin, and numerous summaries have been prepared.[7,42–48] Methods that can be used for measuring dermal exposure or the exposure source include absorbent pad or clothing sampling, glove and hand wash sampling, surface wipe sampling, fluorescent tracers, and other techniques. Additional information on dermal sampling techniques can be found in Appendix II of *A Strategy for Assessing and Managing Occupational Exposures*, 3rd edition.[49]

In the absence of dermal exposure monitoring data, potential default values for the amount retained on the skin per event (Q) are presented in Table 13.3. These values can be used for industrial scenarios to develop screening level estimates of potential dermal exposure. Note that these default values are based on a limited amount of data. The data can generally be used to represent exposure received by workers wearing no personal protective equipment. Some of the samples were collected on surrogate skin (e.g., cotton gloves), which may result in upwardly biased estimates of actual exposure. The reported physical skin loading is 1.3–10.3 mg/cm^2 for liquids.[50] A study of the amount of liquid insect repellent applied to the skin found that for most persons, the average application quantity was approximately 2 mg/cm^2. The upper limit was restricted to about 4 mg/cm^2 due to runoff from the skin.[51] Analysis of collected field data estimate median soil loadings to be 0.25 mg/cm^2 (50th percentile) and up to 1.7 mg/cm^2 (95th percentile) for most situations.[52] Studies also confirm that the amount of soil retained on the skin is influenced by the individual tasks.[42,45,46] The properties of the soil (e.g., particle size, moisture) may also influence soil adherence rates. The EPA Exposure Factors Handbook provides recommendations for soil adherence values for a variety of exposure scenarios.[26]

A loading of about 5 mg/cm^2, assuming particles ≤ 35 mm in diameter and a density of 2.6 g/cm^3, will completely cover the skin, forming a "monolayer" of coverage. Finer soils, like clay, will form a monolayer of loading with only about 2.5 mg/cm^2.[53] Any skin surface loading of particles that does not completely cover the skin can be presumed to have a mass loading equal to the above monolayer loadings times the fraction (or area) of skin covered by the particles. A loading of particles greater than this monolayer loading will probably not contribute significantly to additional percutaneous absorption.

ABS (Fraction of applied dose absorbed)

In this model, a constant fraction of the deposited contaminant is used to approximate the amount absorbed. In reality, however, the fraction of chemical absorbed into the skin is dependent on both concentration applied to the skin and duration of exposure. Thus, the absorbed fraction can vary substantially from one study to another and will be specific to the scenario under evaluation. The mass of chemical that ultimately penetrates the skin is typically determined by how much of the chemical is absorbed into the stratum corneum within the first 30 minutes. However, for some chemicals, such as those which are very lipophilic, penetration may continue for several days.[54–56] In practice, default fractional absorption assumptions are based on a very limited number of 24-hour *in vivo* animal exposure studies, primarily with pesticides.[57] Thus, the use of these absorption fractions may overestimate the

Table 13.3 — "Q" Values: Amount of Contaminant Retained on the Skin per Event that can be assumed for Screening-Level Assessments of Dermal Exposure to the Hands[50]

Type of Contact[1]	*Typical Examples*	*S*[2] *(cm^2)*	*Q*[3] *(mg/cm^2)*	*Resulting Dermal Contact (mg)*
Routine, direct handling of solids – 2 hands	• Filling/dumping containers of powders, flakes, granules • Weighing powder/scooping/mixing (e.g., dye weighing) • Handling wet or dried material in a filtration and drying process	840	(3.7)[5]	up to 3100[4]
Routine contact with surfaces – 2 hands – solids	• Handling bags of solid materials (closed or empty)	840	(1.3)[5]	up to 1100[4]
Routine immersions, 2 hands – liquids	• Handling wet surfaces • Spray painting	840	1.3 – 10.3	up to 8700
Routine contact, 2 hands – liquids	• Maintenance • Manual cleaning of equipment • Filling drum with liquid	840	0.7 – 2.1	up to 1800
Incidental contact, 2 hands – liquids	• Connecting transfer line	840	0.7 – 2.1	up to 1800
Incidental contact, 1 hand –liquid	• Sampling • Ladling liquid/bench scale liquid transfer	420	0.7 – 2.1	up to 900

Table 13.3 Notes:

[1] The terms "routine" and "incidental" reflect typical EPA/Office of Pollution Prevention and Toxics/Chemical Engineering Branch (CEB) judgments on likelihood of contact for the example activities.

[2] Values of the skin surface area of the hands taken from: EPA Exposure Factors Handbook, 1997 and are for the mean values of men.

[3] Selected ranges of "Q" values for liquid handling activities taken from: A Laboratory Method to Determine the Retention of Liquids on the Surface of Hands, Exposure Evaluation Division, Office of Pollution Prevention and Toxics, US EPA, EPA 747-R-92-003, September, 1992.

[4] Values for dermal contact for solids handling activities were taken from: Lansink, C.J.M., M.S.C. Breelen, J. Marquart, and J.J. van Hemmen: Skin Exposure to Calcium Carbonate in the Paint Industry. Preliminary Modeling of Skin Exposure Levels to Powders Based on Field Data (TNO Report V 96.064). Rijswijk, The Netherlands: TNO Nutrition and Food Research Institute, 1996.

[5] Estimated; not provided by CEB.

penetrated dose when the exposure duration is shorter than 24 hours. Keep in mind that some environmental agents will be absorbed through the skin during the exposure event, some will be absorbed after the event (if not removed before the next event), and some will be removed before absorption occurs. Thus, as discussed above, the exposure value used may lead to an overestimate of actual systemic absorption. Some regions or divisions of the U.S. EPA have recommended using ABS = 0.10 as the default absorption factor for organic chemicals when no empirical data are available, whereas others have preferred to assume complete absorption with ABS = 1.0.[50,58] As a result, default values, in the absence of empirical data, may differ by a factor of nearly 10-fold depending on the assumption.[35] In reality, there can be a wide range of actual absorption fractions experienced with different compounds and over various durations of skin contact. If empirical skin absorption values specific to the scenario under evaluation are available, they should be used where applicable. Empirical skin absorption values can be found in a number of original studies in the peer reviewed literature; a selected number can also be found in the U.S. EPA's report on dermal exposure assessment.[32]

WF (Weight fraction of the substance in the mixture, unitless)

When a chemical is present in a mixture, its concentration, or weight fraction, must be considered to estimate dermal exposure. A useful approximation is to assume that the mass of a specific chemical penetrating the skin after dermal exposure to a mixture is roughly proportional to the concentration of the chemical within the mixture. For example, if an individual is dermally exposed to a mixture containing a chemical at a concentration of 10 parts per million (ppm) by weight, then WF = 10 mg chemical / 1,000,000 mg mixture = 0.00001. If an individual is exposed to a neat liquid chemical, then WF = 1.

Model Application

When applying this model to estimate dermal exposure, all five terms in Equation 13-2 are multiplied together, arriving at the dermal absorbed dose rate in units of mg/day. Therefore, when applying this model, the exposure assessor implicitly assumes that dermal exposure is proportional to each of the following: the surface area of skin exposed, the quantity of mixture deposited, exposure frequency, the fraction of chemical absorbed, and the weight fraction of chemical in the mixture. He or she must consider this fact and understand potential biases and limitations of the model as described. Other models based on Fick's law of diffusion (as discussed below) may provide more realistic estimates of dermally absorbed dose rates. Example 1 below demonstrates the potential application of this model.

13.3.3 Dermal Exposure Estimation Using Average Daily Dose

The U.S. EPA in the Risk Assessment Guidance for Superfund (RAGS) Supplemental Guidance for Dermal Risk Assessment (as well as in the Exposure Factors Handbook and the Dermal Exposure Assessment Guidance documents) describes a dermal exposure estimation model in which the absorbed dose can be derived using either a permeability coefficient (for aqueous solutions) or a fraction of absorbed dose (for non-aqueous and non-steady state conditions such as exposure to soil).[6,26,35] In this model, the skin is considered to have two discrete layers with different penetration potentials: the stratum corneum and the viable epidermis. The model is based on Fick's law of diffusion, which describes the concentration of a chemical within the skin as functions of both the distance traveled into the skin and time. Ultimately, Fick's law is used here to predict the rate at which chemicals are absorbed by the stratum corneum under steady conditions. Methods for estimating percutaneous absorption using permeability coefficients have been developed for aqueous solutions or neat liquids and for vapors and gases. These methods require information about the chemical's permeability coefficient (distance per time) in water. This model can be used to evaluate both organic and non-organic chemicals.

Note that the model below is only applicable to non-carcinogens and calculates the average daily dose (ADD); the Lifetime Average Daily Dose (LADD) should be used to evaluate carcinogens. In the main conceptual equation, the dose rate is averaged over the period of exposure and a daily dose is expressed as mass per unit of body weight per day[6]:

$$ADD = \frac{DA_{event} \times EV \times ED \times EF \times SA}{BW \times AT} \quad (13\text{-}3)$$

where

- ADD = average daily dose, mg/kg-day
- DA_{event} = absorbed dose per event, mg/cm^2-event
- EV = event frequency, events/day
- ED = exposure duration, years
- EF = exposure frequency, days/year
- SA = skin surface area available for contact, cm^2
- BW = body weight, kg
- AT = averaging time, days,
  - for non-carcinogenic effects, AT = ED
  - for carcinogenic effects, AT = 70 years = 25,550 days

Absorbed Dose per Event for an Aqueous Vehicle (DA_{event})

For aqueous solutions, a permeability coefficient is used to calculate the mass of chemical absorbed per exposure event per unit area of skin (i.e., mg/cm^2-event). The permeability coefficient, K_p, is a function of (1) the distance of chemical diffusion (i.e., the thickness of the stratum corneum), (2) the membrane/vehicle partition coefficient (defined in this equation as the octanol/water solubility partition coefficient, K_{ow}), and (3) the effective diffusion coefficient, D_{sc}, cm^2/hr. K_p can be estimated simply from the chemical's molecular weight, MW, and K_{ow} using one of several empirical regression equations. Additionally, the RAGS Supplemental Guidance for Dermal Risk Assessment provides hundreds of K_p values for common chemical substances in Appendix B.[6] The K_p value can also be estimated using the SkinPerm tool, a simple computer-based model that will also estimate quantitative absorption values for dermal exposures to both liquids and vapors. This tool is available for free download from the internet (http://home.planet.nl/~wtberge/).[59] The website includes a description of the assumptions underlying the revised Robinson equation (Equation 13-8), which is used by SkinPerm to estimate K_p. The physicochemical data needed to estimate K_p are stored in an associated database for a number of common chemicals. If the appropriate chemical properties are available to the user, K_p can be estimated for chemical substances not contained in the SkinPerm database. Alternatively, K_p can be obtained using modeling tools on both the NIOSH and the Syracuse Research Center websites.[60,61]

Given the appropriate value for the permeability coefficient K_p, one can estimate the dermal absorbed dose rate per event using Equation 13-4:

$$DA_{event} = K_p \times C_w \times t_{event} \quad (13\text{-}4)$$

where

- DA_{event} = dermal absorbed dose rate per event, mg/cm^2-event
- K_p = appropriate permeability coefficient in water, cm/hr
- C_w = concentration of the substance in water, mg/cm^3
- t_{event} = duration of event, hr/event

Alternative Equation for Absorbed Dose per Event (DA_{event}) for Single Events and for Neat Chemicals

An alternative equation for DA_{event} can be used in the event that only the dose of interest is for a single specific event. This equation can also be used to estimate the dermally absorbed dose for substances in a vehicle other than water or as a neat chemical:

$$DA_{event} = K_{p,veh} \times C \times SA \times t_{event} \quad (13\text{-}5)$$

where

DA_{event} = dermal absorbed dose rate per event, mg/event
$K_{p,veh}$ = appropriate permeability coefficient in the vehicle of interest, cm/hr
C = concentration of the substance in vehicle of interest, mg/cm^3
(note that for neat liquids, $C = \rho$, the density of the liquid)
SA = surface area of skin, cm^2
t_{event} = duration of event, hr/event

The permeability coefficient $K_{p,veh}$ may be difficult to determine for most chemicals in specific nonaqueous vehicles. In these cases, K_p can be approximated by:

$$K_{p,veh} = K_p \times \frac{C_w^{sat}}{C_{veh}^{sat}} \quad (13\text{-}6)$$

where

K_p = appropriate permeability coefficient in water, cm/hr
C_w^{sat} = the saturation concentration of the vehicle of interest in water
C_{veh}^{sat} = the saturation concentration of the chemical of interest in the vehicle of interest
K_p = appropriate permeability coefficient in water, cm/hr

Note that if the chemical is neat, then C_{veh}^{sat} is equivalent to the chemical's density. The above equation is based on two main approximations: (1) that the stratum corneum is not significantly altered by the chemical in a vehicle other than water, and (2) that the chemical activity of the chemical at its saturation concentration in water is roughly the same as its activity when saturated in the second vehicle. Note the somewhat counter-intuitive prediction that the dermal flux of a chemical saturated at a relatively high concentration in a particular vehicle, or present as a neat liquid, will be equivalent to the predicted flux of the same chemical at its relatively low saturation concentration in water. This prediction is in agreement with theoretical predictions[62] as well as the empirical observations of Bunge and Romonchuk, who demonstrated that phenanthrene in water at its saturation concentration exhibited nearly the same dermal flux as phenanthrene saturated in silicone oil, even though the saturation concentration in the oil was several orders of magnitude higher.[63] Other empirical studies have shown that the dermal flux for skin absorption from a saturated aqueous solution of benzene and for neat benzene were similar, a phenomenon that is true for several other solvents, but may not be true for all.[3,64]

Derivation of K_p for use in DA_{event}: Diffusion through the Skin from Aqueous Solutions or Neat Liquids

In its simplest form, if concentration remains steady, then the permeability coefficient, K_p, is essentially the speed at which a chemical diffuses through the skin with units of cm/hr. The flux of a chemical is then the permeability coefficient multiplied by the concentration of chemical on the surface of the skin, a result of Fick's Law of Diffusion:

$$\text{Flux } (mg/cm^2/hr) = K_p \times \text{Concentration } (mg/cm^3)$$

A number of equations have been used to define K_p for aqueous solutions. For example, part E of the EPA's risk assessment guidance for superfund document provides a relatively simple correlation[6]:

$$\log K_p = -2.80 + 0.66 \times \log K_{ow} - 0.0056 \times MW \quad (13\text{-}7)$$

where

K_{ow} = octanol-water partition coefficient of the chemical
MW = molecular weight of the chemical, g/mol

The current revised Robinson model provides another, more comprehensive, means to estimate the skin permeation and lag time for highly hydrophilic and highly lipophilic chemicals. This method has the smallest residual variance relative to similar equations.[65,66] A quantitative structure-activity relationship (QSAR) for the skin permeability coefficient from aqueous solution (K_p) has been developed on the basis of the following starting points.

- Resistance against mass transfer is assumed to be only in the stratum corneum, the outermost layer of the dermis.
- Permeation is assumed to occur via two pathways in the stratum corneum, the transcellular route through the protein corneocytes (dead, flattened skin cells) (K_{pro}) and the intercellular route through the extracellular lipid matrix which holds the corneocytes in place (K_{lip}).
- Transfer from the stratum corneum to the viable epidermis is assumed to be controlled by the water solubility of the permeating substance. The water solubility may be enhanced by the presence of water soluble organic solvents like ethanol in the permeating solution.
- The viable epidermis is not assumed to be a limiting factor for permeation. For water soluble

compounds the viable epidermis does not form a barrier compared to the stratum corneum. The permeability coefficient of the stratum corneum increases with increasing lipophilicity, but the permeability coefficient of the viable epidermis will also increase. The reason for this is that the blood solubility of lipophilic compounds is higher than the water solubility, due to the presence of about 1% blood lipids. The removal rate of a substance from the viable epidermis by skin blood increases with increasing lipophilicity, but the amount in the viable epidermis available for permeation decreases due to decreasing water solubility.

The revised Robinson model is based on the rate of chemical diffusion through lipids and proteins in the stratum corneum, along with the subsequent rate of diffusion through the aqueous viable epidermis. As shown in Equation 13-8, if the rate of diffusion through the aqueous viable epidermis is rapid (i.e., K_{aq} is large), then diffusion is limited by the lipids and proteins of the stratum corneum.

$$K_p = \frac{1}{\dfrac{1}{K_{lip} + K_{pro}} + \dfrac{1}{K_{aq}}} \qquad (13\text{-}8)$$

where

K_p = overall aqueous permeability coefficient, cm/hr
K_{lip} = permeability coefficient of the lipid matrix fraction of the stratum corneum, cm/hr
K_{pro} = permeability coefficient of the protein fraction of the stratum corneum, cm/hr
K_{aq} = permeability coefficient of the viable layer of the epidermis, posing an aqueous barrier, cm/hr

The estimation of K_p for this model considers two parallel pathways (for either lipophilic or hydrophilic compounds) via the lipid and protein fractions of the stratum corneum, in addition to the viable epidermis as an aqueous barrier. As mentioned above, K_p multiplied by the chemical concentration (mg/cm^3) results in the flux of chemical through the skin (mg/cm^2-hour). Note that the chemical concentration can never exceed the chemical's solubility in water. A database of measured permeability coefficients and physicochemical properties of numerous substances was used to evaluate the revised Robinson model and provide estimates of K_p for a wide range of chemicals in an aqueous vehicle.[67] Note that the following equations are used to estimate K_p in an aqueous vehicle only. To estimate K_p for other vehicles, appropriate adjustments must be made, as described above.

The permeability coefficient of the lipid fraction of the stratum corneum, K_{lip}, in cm/hr can be estimated using Equation 13-9:

$$\log K_{lip} = -2.59 + 0.7318 \times \log K_{ow} - 0.006832 \times MW \qquad (13\text{-}9)$$

where

K_{ow} = octanol-water partition coefficient of the chemical, and
MW = molecular weight, g/mol.

The permeability coefficient of the protein fraction of the stratum corneum (K_{pro}) in cm/hr can be estimated using Equation 13-10:

$$K_{pro} = \frac{0.043}{MW^{1.36}} \qquad (13\text{-}10)$$

To estimate the permeability coefficient of the viable layer of the epidermis, Bunge and Cleek (1995) treated this layer as a pseudo-liquid in which chemicals would theoretically partition equally between the viable epidermis and water.[68] Their estimates assume that diffusivity scales with MW$^{-1/2}$ and that a chemical with MW = 50 g/mol would exhibit a diffusivity of 10^{-6} cm^2/s (ten-fold slower than a comparable molecule would likely diffuse in water). Equation 13-11 results from this formulation, using a viable epidermis thickness of 100 μm:

$$K_{aq} = \frac{2.5}{\sqrt{MW}} \qquad (13\text{-}11)$$

Note that the permeability coefficients outlined above are influenced by the molecular volume of the chemical. Smaller molecules tend to diffuse faster, and larger molecules tend to diffuse more slowly. To a very good approximation for most chemicals, MW is a reasonable indication of the molecular volume, as chemicals with larger molecular weights tend to be more voluminous. However, the molecular volumes of chemicals containing very dense atoms such as halogenated compounds tend to be smaller than anticipated using correlations with other chemicals.

Derivation of K_p for use in DA_{event}: Diffusion through the Skin from Vapors and Gases

At least one study has provided recommendations for evaluating diffusion through the skin from vapors and gases.[69] To estimate the skin permeability coefficient from gases, the permeability coefficient from aqueous solutions can be modified through multiplication with the water/air partition coefficient. The water/air partition coefficient (K_{wa}) is defined as follows in Equation 13-12:

$$K_{wa} = \frac{R \times T \times W_{sb}}{VP \times MW} \qquad (13\text{-}12)$$

where

R = gas constant in 8.314 Joule/mol/degree Kelvin
T = temperature in degree Kelvin

W_{sb} = Chemical solubility in water (g/m^3)
VP = Chemical vapor pressure at temperature T in degrees Kelvin
MW = molecular weight

Multiplication of K_p and K_{wa} produces the permeability coefficient of the skin, $K_{p,sk\text{-}air}$ (cm/hour) for gaseous compounds:

$$K_{p,sk-air} = K_{wa} \times K_p \quad (13\text{-}13)$$

For gaseous compounds permeating the skin quickly, the diffusion from the air to the skin surface may become the rate-limiting factor. To take this effect into account it is assumed that a layer of stagnant air is present between the skin surface and workroom air. The thickness, δ, of this air layer is estimated to be 3 cm for light work clothing, based on a 1993 study.[70]

The air diffusion constant (D_{air}, in cm^2/hour) and the permeability coefficient of the stagnant-air boundary layer on the skin ($K_{p,air}$, cm/hour) are formulated as follows:

$$D_{air} = 360\sqrt{\frac{76}{MW}} \quad (13\text{-}14)$$

and:

$$K_{p,air} = \frac{D_{air}}{\delta} \quad (13\text{-}15)$$

where

δ = 3 cm

Finally, the overall $K_{p,sk\text{-}air,air}$ (in cm/hour) is controlled by the diffusion through the air boundary layer on the skin ($K_{p,air}$) and diffusion through the skin ($K_{p,sk\text{-}air}$):

$$K_{p,sk-air,\,air} = \frac{1}{\dfrac{1}{K_{p,sk-air}} + \dfrac{1}{K_{p,air}}} \quad (13\text{-}16)$$

As with the dermal flux for aqueous solutions, the dermal flux of gaseous chemicals is estimated by multiplication of the $K_{p,sk\text{-}air,air}$ with the atmospheric concentration of the chemical, Cair (mg/cm^3):

$$\text{Flux (mg/cm}^2\text{/hr)} = K_{p,sk\text{-}air,\,air} \times C_{air}\ \text{(mg/cm}^3\text{)}$$

The atmospheric concentration is limited by its vapor pressure: $C_{air,max} = MW \times VP / R \times T$, where T is in degrees Kelvin.

The advantage of this simple approach is that it provides an estimate of the skin permeation of vapors when the permeability coefficient from aqueous solutions is known. This model will estimate a high permeation rate for gaseous aniline, nitrobenzene, and glycol ethers and a very low permeation rate for hydrocarbons such as benzene and toluene. Thus, this model's estimates are consistent with real-world observations. Ignoring the air boundary layer, $K_{p,air}$, in the equation for the permeability coefficient of gases will result in larger estimates of the skin permeation rate.

Although some mathematical models can be used to estimate the amount of a contaminant in an aqueous media permeating the skin, there is still much to be learned about the process. The revised Robinson Model, which is useful for estimating the permeability coefficient, is recommended over the other models that have been evaluated.

Estimation of Lag Time

The permeability coefficients discussed above describe the permeation rate at steady state. In reality, it may take some time before steady state has been achieved. The following equations can be used to better characterize this phenomenon:

$$K_p = \frac{P_{scw} \times D_{sc}}{h_{sc}} \quad (13\text{-}17)$$

$$t_{lag} = \frac{h_{sc}^{\,2}}{6D_{sc}} = \frac{h_{sc}P_{scw}}{6K_p} \quad (13\text{-}18)$$

$$D_{sc} = \frac{h_{sc}^{\,2}}{6t_{lag}} = \frac{h_{sc}K_p}{P_{scw}} \quad (13\text{-}19)$$

where

t_{lag} = lag time (hour)
P_{scw} = partition coefficient stratum corneum/water
D_{sc} = diffusivity stratum corneum (cm^2/hour)
h_{sc} = thickness stratum corneum (cm)

The lag time and the diffusivity may be estimated from the steady state permeability coefficient and the partition coefficient between stratum corneum and water, which is specific for each substance. A database of measured partition coefficients between wet stratum corneum and water has been compiled.[71] The wet stratum corneum/water partition coefficient P_{scw} was approximated by:

$$P_{scw} = 0.72 \times K_{ow}^{0.43} \quad (13\text{-}20)$$

The lowest P_{scw} of the database was about 0.2, which is the experimental wet stratum corneum/water partition coefficient of water. Because substances with smaller values of K_{ow} were not part of the database, it is not advisable to use the above equation to estimate P_{scw} for these substances, since it is assumed that the same partition coefficient applies. In these cases, a value of P_{scw} = 0.2

should selected. For the estimation of the lag time, the thickness of the hydrated stratum corneum should be used, with h_{sc} = 0.002 cm.

Fraction of dermal absorption of a volatile liquid deposited on the uncovered skin

When a volatile liquid chemical is deposited on the skin, a fraction of the volume will volatilize and a fraction may be absorbed into the stratum corneum. For risk assessments, the absorbed fraction may need to be estimated. This requires an understanding of the initial rate of absorption (dermal flux) of neat liquid into the stratum corneum in mg/cm²/hour and an understanding of the initial rate of evaporation from the skin in mg/cm²/hour. The fraction of dermal absorption when a volatile liquid is deposited on uncovered skin, such as hands and forearms is given by:

$$\text{Fraction absorbed} = \frac{J_{dermal,neat}}{J_{dermal,neat} + J_{evap,neat}} \qquad (13\text{-}21)$$

The initial dermal flux into the stratum corneum might be approximated by:

$$J_{dermal,neat} = K_{p,veh} \times \rho = \left[K_p \frac{C_w^{sat}}{\rho}\right] \rho = K_p \times C_w^{sat} \qquad (13\text{-}22)$$

where

$J_{dermal,neat}$ = Dermal flux of neat chemical (mg/cm²/hr)
ρ = density of the neat liquid (mg/cm³)
K_p = aqueous permeability coefficient (cm/hour)
C_w^{sat} = the saturation concentration of the vehicle of interest in water
$k_{p,veh}$ = appropriate permeability coefficient in the vehicle of interest, cm/hr

The evaporation rate of a liquid film[72] might be approximated by:

$$J_{evap,neat} = \frac{MW \times VP}{9.57 \times T} \qquad (13\text{-}23)$$

where

$J_{evap,neat}$ = Evaporative flux of neat chemical (mg/cm²/hr)
MW = Molecular weight
VP = Vapor pressure in Pascal at the temperature T in degrees Kelvin
T = Temperature in degrees Kelvin

It is not necessary to use the equations presented in this section if the evaporative flux is small relative to the dermal flux. However, it provides a useful means of estimating a more realistic dermal absorption when working with volatile chemicals in the open air (e.g., cleaning and painting) when a fraction is likely to volatilize.

13.3.4 Dermal Exposure Estimation Using Acceptable Skin Exposure Duration

Dermal exposure criteria do not need to be thought of only in terms of surface concentrations or mass loading. An alternative approach is to think of exposure in terms of duration.[73] The premise here is that longer contact time results in greater skin absorption and potential for toxic effect. As an acceptable toxicologically-based criterion, the previously introduced concepts can be used to derive an OEL equivalent for chemicals that are systemically toxic. This model estimates the amount of time required to exceed the OEL equivalent dose when exposure is via the skin route.

Skin absorption time (hr) =

$$\frac{\text{Total absorption at OEL (mg)}}{\text{Skin area (cm}^2) \times \text{Flux (mg/cm}^2 - \text{hr)}} \qquad (13\text{-}24)$$

If the skin absorption time is larger than the lag time, this equation is valid. In the event that the skin absorption time is found to be smaller than the lag time, the accuracy of this equation is diminished. One should realize that in the case of an infinite dermal dose, only 60% of its steady state flux is achieved after a period that will be approximately equal to the lag time.

Total Absorption at OEL (mg)

This estimate can be calculated using the OEL conversion described in Equation 13-1.

Skin area (cm²)

The combined surface area of the skin can be estimated for the hands, hands and forearms, or for any other extent of skin surface that is expected to be potentially exposed. These areas can be taken from the table of skin surface areas presented above.

Flux (mg/cm²-hr) and K_p (cm/hr)

The flux of chemical through the skin can be determined empirically or taken from the literature if available, or else estimated. As discussed above, flux is defined as: K_p × Concentration. K_p is measured in cm/hr and expresses the speed of penetration of chemicals through skin. The Potts and Guy and revised Robinson equations are presently the most commonly used regression equations for this purpose, with the revised Robinson equation being used most commonly.[32,74]

Once K_p is obtained, it and the concentration (which must be in units of mg/mL) of the chemical can be substituted for the flux in Equation 13-25 as follows. A scenario demonstrating this approach is also provided in Example 3 at the end of this chapter.

Skin Absorption Time (hr) =

$$\frac{\text{Total absorption at OEL (mg)}}{K_p \text{ (cm/h)} \times \text{Concentration (mg/ml)} \times \text{skin area (cm}^2\text{)}} \quad (13\text{-}25)$$

It is important to note that K_p can be significantly affected by the type of solution carrying the chemical contaminant of concern, for example whether the agent is in an aqueous or organic solvent. The K_p used in Equation 13-25 must be the appropriate value for the occupational situation that one is trying to model (i.e., aqueous, nonaqueous). However, the Potts and Guy and revised Robinson equations are simply multiple regression equations fitted to empirical human *in vitro* skin permeation data obtained from experiments of chemicals in aqueous solutions only.(66,74) Fortunately for simplicity's sake, for single chemicals not in a mixture or solution, the permeation rate is often about the same as for a saturated aqueous solution, as long as the components of the solution itself do not alter the skin.(75) However, for complex mixtures, there might be appreciable differences in K_p and there are no equations yet that have been successful in estimating these situations. The most reliable method to determine K_p for nonaqueous solutions is to test the compound mixture using *in vitro* or *in vivo* test methods.

13.3.5 Estimating Hand-to-Mouth Contact Related to Dermal Exposure

Hand-to-mouth contact is a path of dermal exposure that applies to substances which are ingested as a result of handling a product or chemical. Exposure via this route occurs by "direct" and "indirect" transfers of chemical. Direct transfer occurs when the chemical in the product is transferred to the hand, which is then followed by transfer from the hand to the mouth, usually by touching the mouth or lips with the hand, by nail biting, or finger or thumb sucking. Indirect hand-to-mouth transfer occurs when contaminated products are handled followed by handling or touching of food, cigarettes, drinking straws, silverware, clothing, door knobs, or other objects. The contribution of exposures via this route to total exposure has been considered recently in at least one review of occupational exposures.(76)

A study by Camann, et al. (2000) assessed the fraction of three pesticides removed from the hands of three adult volunteers using gauze wipes moistened with human saliva, artificial saliva, and a surfactant (1.3% dioctyl sulfosuccinate or DSS). Results showed that the average removal efficiency for all three pesticides was 48.3% for human saliva, 42.1% for artificial saliva, and 55% for DSS. Based on these data, the removal efficiency for hand-to-mouth transfer was estimated to be approximately 50%.(77)

The U.S. Consumer Product Safety Commission (CPSC) has developed an estimate of the percent of dislodgeable residue on the hands that is ingested daily by children, based on soil, in order to estimate exposures to copper chromate arsenic (CCA) on playground equipment. The CPSC found large variability and uncertainty in the available data, and therefore combined reasonable upper and lower bound values reported for soil ingestion and skin adherence to estimate a percent residue on the hands that may be ingested. The estimates for children ranged from 3% to 700% per day, with an average of 43% for both direct and indirect hand-to-mouth activities together.(78)

The U.S. CPSC has used the hand-to-mouth transfer efficiency of 50% measured by Camann, et al.(77) to estimate lead exposures to children's polyvinyl chloride products. The U.S. Environmental Protection Agency's Office of Pesticide Programs has also used this value for estimating hand-to-mouth exposures to pesticides.(79) More recently, the California Environmental Protection Agency's (CalEPA) Office of Environmental Hazard Assessment (OEHHA) has proposed using a default value of 50% for direct dermal transfer and 25% for indirect dermal transfer when assessing exposure to lead in fishing tackle products under Proposition 65, in their guidelines for hand-to-mouth transfer of lead exposure from fishing tackle products.(80) OEHHA cautions that the guidelines are intended to be specific to fishing tackle, but the concepts and equations provided are valuable and may be applicable to other scenarios following appropriate analysis.

OEHHA provides equations to estimate intake associated with both direct and indirect dermal exposure associated with hand-to-mouth activity. The guidance document defines a direct hand-to-mouth transfer factor as the mass transferred to the mouth divided by the mass on the surface area of that part of the hand which is in contact with the mouth, per contact. An indirect hand-to-mouth transfer factor is defined as the fraction of mass on the portion of the hand that comes in contact with an intermediate object and subsequently comes in contact with the mouth. These transfer factors are reported as a fraction or percent of the total available mass. Intake associated with hand-to-mouth activity is defined as follows:

$$\text{Intake}_{HM} = \text{Intake}_{HM\ direct} + \text{Intake}_{HM\ indirect} \quad (13\text{-}26)$$

Intake associated with direct hand-to-mouth exposure is defined as follows:

$$\text{Intake}_{HMdirect} = L_{hand\text{-}D} \times SA_D \times f_{direct} \times \lambda_D \times t \quad (13\text{-}27)$$

where

$L_{hand\text{-}D}$ = the loading on the part of the hand touching the mouth, in units of mass per surface area of skin, $\mu g/cm^2$

SA_D = the surface area of the part of the hand in direct contact with the mouth, cm^2

f_{direct} = the direct hand-to-mouth transfer factor, presented as a fraction or percentage

λ_D = the frequency of direct hand-to-mouth contact, e.g., #/hr
t = the number of hours of activity

The indirect transfer factor, $f_{indirect}$, is defined as the transfer of mass to an intermediate object and then subsequent introduction of the mass from the intermediate object into the mouth by the hand, with the potential for some loss of mass from the hand before contact with the mouth. It can be estimated as follows:

$$f_{indirect} = f_{direct} \times (1 - f_{loss}) \quad (13\text{-}28)$$

where

f_{loss} = the fraction of mass loading lost during intermediate contacts

Intake associated with indirect hand-to-mouth exposure is therefore defined as follows:

$$Intake_{HMdirect} = L_{hand\text{-}I} \times SA_I \times [f_{direct} \times (1 - f_{loss})] \times \lambda_I \times t \quad (13\text{-}29)$$

where

$L_{hand\text{-}I}$ = the loading on the part of the hand touching the intermediate object, in units of weight per surface area, µg/cm²
SA_I = the surface area of the hand in contact with the material reaching the mouth
λ_I = the frequency of indirect hand-to-mouth object contact, e.g., #/hr
t = the number of hours of activity

Hand-to-mouth contact is an area of dermal exposure assessment where additional research is needed to accurately estimate exposures resulting from hand-to-mouth transfer, although the above equations help to conceptually frame the exposure potential. Example 4 below demonstrates a potential application of this approach.

13.3.6 Derivation of risk-based wipe surface screening levels

The Risk Assessment Guidance for Superfund (RAGS) equations allow for the assessment of total risk from exposure via inhalation, incidental ingestion, and skin contact.[(6)] These equations can be modified to allow for calculation of screening level surface concentration limits that are health-based. It should be noted, however, that these equations are intended to compute risk estimates based on a 24-hour exposure time rather than the more typical 8-hour exposure time for most occupational scenarios. Input variables include exposure factors that are based on default assumption values or preferably are empirically known. Models for both carcinogens and non-carcinogens will be presented. The general risk equation is:

$$\text{Total Risk} = Risk_{dermal} + Risk_{ingestion} + Risk_{inhalation} \quad (13\text{-}30)$$

Risk for Carcinogens

For carcinogenic chemicals, the estimated increase to an individual's lifetime background risk for cancer can be calculated from the carcinogenic slope factor (CSF_p) for exposure pathway *p* (mg/kg-day)$^{-1}$ and the lifetime average daily intake ($LADI_p$) for exposure pathway *p* (mg/kg-day):

$$Risk_p = CSF_p \times LADI_p \quad (13\text{-}31)$$

LADI for each pathway of exposure is calculated using the following equation:

$$LADI_p = \frac{C_{surface} \times CR_p \times EF \times ED}{BW \times AT_{lifetime} \times 365 \text{ days/yr}} \quad (13\text{-}32)$$

where

$C_{surface}$ = surface concentration (mg/m²)
CR_p = exposure pathway specific contact rate (m²/day)
EF = exposure frequency (days/yr)
ED = exposure duration (yr)
BW = body weight (kg)
$AT_{lifetime}$ = carcinogenic averaging time (yr)

By factoring the components of LADI into Equation 13-32 for each pathway and selecting an acceptable target risk (TR), the equation (13-33) now appears below.

Acceptable target risks vary depending on the population and regulatory agency assigning the risk level, but generally range from 1×10^{-4} to 1×10^{-6}. Note that each contact rate for each pathway is in units of m²/day and the cancer slope factors are each in units of (mg/kg-day)$^{-1}$. Algebraically rearranging Equation 13-33 so that $C_{surface}$ is separated out, and its units adjusted from mg/m² to µg/100cm², the equation becomes 13-34 below.

Next, the contact rates must be calculated. Exposure variables are included, along with prerequisite assumptions

$$TR = \frac{C_{surface} \times [(CR_{dermal} \times CSF_{dermal}) + (CR_{ingestion} \times CSF_{ingestion}) + (CR_{inhalation} \times CSF_{inhalation})] \times (EF \times ED)}{(BW \times AT_{lifetime} \times 365 \text{ days/yr})} \quad (13\text{-}33)$$

$$C_{surface} = \frac{TR \times BW \times AT_{lifetime} \times 365 \times 10}{(CR_{dermal} \times CSF_{dermal} + CR_{ingestion} \times CSF_{ingestion} + CR_{inhalation} \times CSF_{inhalation}) \times EF \times ED} \quad (13\text{-}34)$$

about their magnitude to calculate separate contact rates for a dermal risk equation (CR_{derm}), the ingestion risk equation ($CR_{ingestion}$), and for inhalation (CR_{inhale}). For more details about this modeling approach, refer to the reference publication.[81] To calculate CR_{dermal}, it consists of the following factors:

$$CR_{dermal} = SA_d \times F_d \times EV \times FT_{ss} \times DAF \qquad (13\text{-}35)$$

where

CR_{dermal} = dermal contact rate, m^2/day
SA_d = dermal surface area available for absorption, m^2
F_d = fraction of available dermal area contacted on a given day
EV = contact frequency with surface, day^{-1}
FT_{ss} = fraction of dust transferred from surface to skin
DAF = dermal absorption efficiency

To calculate CR_{ingest}, the following equation is used:

$$CR_{ingest} = SA_g \times F_g \times EV \times FT_{ss} \times FT_{ftm} \times HTME \qquad (13\text{-}36)$$

where

CR_{ingest} = ingestion contact rate, m^2/day
SA_g = dermal surface area available for ingestion, m^2
F_g = fraction of available dermal area that contacts mouth
EV = contact frequency with surface, day^{-1}
FT_{ss} = fraction of dust transferred from surface to skin
FT_{ftm} = fraction of dust transferred from skin to mouth
HTME = number of hand-to-mouth events

Calculating CR_{inhale} considers the resuspension of surface contaminants into the air as a source of inhalation exposure. Empirical data for resuspension of dust from surfaces due to typical physical activities and air movement presently is very limited, and is considered site-specific. Available data used for extrapolation to other scenarios therefore should be considered rough estimates.[82]

$$CR_{inhale} = IR_i \times K \qquad (13\text{-}37)$$

where

CR_{inhale} = inhalation contact rate, m^2/day
IR_i = inhalation rate, m^3/day
K = re-suspension factor, m^{-1}

Cancer slope factors and dermal absorption efficiency fractions can be found for many compounds online at the U.S. EPA.[58]

Risk for Non-Carcinogens

For non-carcinogenic effects of chemicals, a slightly different approach is used for calculating risk. A hazard quotient of risk (HQ) is simply the ratio of the estimated average daily intake for each pathway (ADI_p) to the acceptable intake. Thus, when the hazard quotient is less than 1, the exposure does not exceed the acceptable dose. Acceptable dose estimates are provided by the U.S. EPA for many chemicals and are referred to as reference doses for each pathway (RfD_p). They can be found at the same site that was provided above in reference to cancer slope factors for carcinogens. This relationship is expressed as:

$$HQ = ADI_p / RfD_p \qquad (13\text{-}38)$$

As with the $LADI_p$, the ADI_p is an estimate of an average daily intake. However, the averaging time used for the ADI_p is the average for chronic exposure duration, not the lifetime average. Equation 13-39 is used to estimate the ADI_p:

$$ADI_p = \frac{C_s \times CR_p \times EF \times ED}{(BW \times AT \times 365 \text{ days/year})} \qquad (13\text{-}39)$$

where

ADI_p = average daily intake for pathway p, mg/kg-day
C_s = surface concentration, mg/m^2
CR_p = route specific contact rate, m^2/day
EF = exposure frequency, day/year
ED = exposure duration, year
BW = body weight, kg
AT = averaging time, year = exposure duration, year

Setting the target hazard quotient (THQ) to 1 (the maximum acceptable ratio of ADI_p/RfD_p) and including the sum of all pathways, Equation 13-38 can be expanded by including the factors shown in Equation 13-39 that equal ADI_p. Algebraically rearranging this expanded equation for all pathways so that C_s is separated out provides Equation 13-40:

$$C_s = \left[\frac{THQ}{\frac{CR_{dermal}}{RfD_{dermal}} + \frac{CR_{ingest}}{RfD_{ingest}} + \frac{CR_{inhale}}{RfD_{inhale}}}\right] \times \left[\frac{BW \times AT \times 365 \text{ days/yr}}{EF \times ED}\right] \times CF \qquad (13\text{-}40)$$

where

C_s = surface concentration (mg/100cm^2)
CF = 10 - conversion factor: 10 (mg/100 cm^2) / (mg/m^2)

Example 5 below uses the above approach for polychlorinated bisphenyls to illustrate how to calculate surface concentrations for carcinogens.

13.4 Exposure Modeling — What to Do with Dermal Exposure and Risk Estimates

One of the primary reasons for estimating dermal exposure is to identify exposures that are unacceptable and need to be controlled. This can be quite challenging considering the uncertainties associated with attempting to quantify dermal exposure or absorbed dose. However, methodologies are available to begin the effort of evaluating the potential for an unacceptable skin exposure. Where possible, use of an OEL equivalent, Biological Exposure Index (BEI®), or Biological Environmental Exposure Level (BEEL) to evaluate dose estimates should be used. Although the results of the model estimations might be uncertain, they are still useful in making determinations about prioritizing resources for managing skin exposures.

13.5 Evaluation of Methodologies and Approaches to Dermal Exposure Modeling

Various approaches can be used to assess skin exposure and absorbed dose, and though several of these approaches have been validated for specific scenarios, the general applicability of these methods to industrial settings has not been fully evaluated. The permeability coefficient approach is useful to estimate skin permeation in aqueous media, but it might have limited utility in many industrial settings where solvent mixtures and other nonaqueous media are common. This is because the Kp value is not easily calculated for non-aqueous solutions. If this value can be experimentally determined, it may be possible to apply the models presented in this chapter to determine flux and mass absorbed. Because the equations imply that the exposed skin is fully immersed in an infinite volume of liquid (i.e., the exposed concentration and surface area are assumed to remain steady), in some cases the application of this approach to industrial activities can misrepresent the exposure scenario. A far more realistic scenario is to consider a finite volume of chemical deposited on the skin that is subsequently removed by one or more mechanisms, such as washing or evaporation. In some cases, the chemical may be reloaded onto the skin in periodic intervals.

It can be difficult to describe the many nuances of an individual's daily work activities through models. As with any model, the use of the appropriate input or parameter values is critical and can significantly affect the final model outcomes. The default absorption approach described in Equation 13-2 may be relevant to a wider range of scenarios, but it is based on numerous simplifying assumptions and is used most appropriately for screening level assessments.

13.5.1 Standardization of Data Collection and Reporting

Standardization dermal exposure data collection and reporting of data would help to make modeling inputs and results more reliable. In many studies, the data have been normalized by time or quantity, and some studies present individual data points, ranges, or statistical descriptors without data points. If exposure has been normalized by time or quantity, the time or quantity should be presented with the data. Some investigators report only total exposure for a given body section or combination of sections, while others report data as unit loading rate (such as from pad samples). Again, data reporting requirements must be linked to data interpretation and the need for increased transparency.

Many factors contribute to the magnitude of the dermal exposure, but factors such as duration of exposure, quantity handled, effect of washing, chemical loss through evaporation, effect of skin hydration (e.g., migration of contaminants, contaminants dripping off the skin) and effect of skin damage have not received sufficient evaluation. A more developed understanding of these factors will aid in improving our capabilities to appropriately assess dermal exposure.

13.5.2 Data Interpretation

Numerous assumptions and uncertainties are inherent in the approaches currently used to estimate workplace-related dermal exposure. The impact of many of these assumptions has not been fully evaluated. Some of these assumptions and uncertainties include:

- Extrapolating the deposition on absorbent sampling pads to assume uniform deposition within a specific area of the body
- Normalizing exposure measurements by quantity of chemicals handled or by exposure time
- Assuming that the retention rate on absorbent pads is equivalent to that of the skin
- Inadequately characterizing variability due to the selection of collection media and handling conditions used
- Selecting widely varying exposure duration times for dermal exposure monitoring studies
- Using varying standards for quality assurance between studies
- Extrapolating from a specific contaminant (e.g., active ingredient) to total mass
- Differing workplace conditions, work practices, etc., which may significantly impact the dermal exposure level measured

13.5.3 Workplace-Related Dermal Exposure Activity Patterns

Information and data on activity patterns in industrial environments are limited. Although data have been collected for some activities in the pesticides industry and in a few non-industrial environments, additional data are needed for a variety of activities and age groups. Tools such as DREAM and RiskofDerm may help to standardize dermal exposure and activity pattern characterization. Dermal exposure for a given body area can vary over a range of several orders of magnitude. Individual practices and activities can significantly impact the exposure received, and little information is available with which to characterize the worker activities and to estimate the dermal exposure for any given task in a workplace environment. Information on personal protective equipment (PPE) and other control measures used to mitigate dermal exposure also should be collected when gathering information on activity patterns.

13.5.4 Assessment of the Effectiveness of Control Measures

Although data on permeation of single contaminants through glove materials is becoming more commonplace, the effectiveness of gloves and other control measures during actual use is still poorly characterized. Information on barrier effectiveness has been collected in many pesticides studies, but interpretation and use of this information may be unreliable because of glove contamination due to hygiene practices and use practices, as well as other factors. The actual protection afforded by chemical protective clothing may be much lower than that demonstrated in laboratory evaluations of glove permeation, because factors such as elevated glove temperature due to the skin temperature, the pressure exerted on the glove material during use of the glove (e.g., stretching of glove), as well as cross-contamination may impact the protection provided in the field. Many types of engineering and process controls will presumably reduce or mitigate dermal exposures in addition to inhalation exposures, but the efficiency of these measures in the reduction of dermal exposure potential has not been routinely investigated. In some cases, administrative controls and modifications of work practices may also aid in reducing exposures due to surface contamination, secondary contamination, etc.

13.6 Dermal Modeling Research Needs

Additional monitoring data to better estimate dermal exposure in workplace settings are needed, and standardization of data collection and assessment methods for measuring indoor surface contamination or estimating human exposure from contact with these surfaces is recommended. While hand exposure often constitutes the majority of the total body exposure, exposure to other parts of the body does occur, and non-hand exposure may be an important contribution to the total dose. Finally, additional information on worker activities and practices and guidance in interpretation of exposure monitoring data would greatly enhance our ability to evaluate dermal exposures in a consistent and meaningful fashion.

13.7 Examples of Dermal Exposure Assessments

13.7.1 Example 1 — Screening Assessment of Dermally Absorbed Dose

An industrial hygienist was contacted to investigate an incidence of skin disease among maintenance workers responsible for repairing water service lines for a facility. The initial basic characterization and information gathering revealed that workers reported some "tingling" sensation in their hands, but the primary symptom was dryness and peeling of the skin of the hands. Acrylamide grouting compounds were used by the workers, and the potential for dermal exposure had likely been reduced over time by various process changes. However, glove use was still required for portions of the grouting task and skin contact was possible due to misuse or inadequacy of the gloves. The ACGIH® TLV® for acrylamide (0.03 mg/m^3 as a time-weighted average or TWA) was selected as the OEL to be used for judging whether exposures were acceptable. Exposure during the grouting operation was believed to occur primarily through skin contact.

Acrylamide is an animal carcinogen as defined by ACGIH® and has been identified as a suspected human carcinogen.[83] The TLV® carries a skin notation which cautions that the potential for significant contribution to the overall exposure by the dermal route exists. There are no quantitative dermal exposure limits for acrylamide, and ACGIH® has not established a BEI®. Although there was no prior sampling data to measure skin exposure, the industrial hygienist determined that the health hazard was potentially significant based on the health complaints and his personal observations. The industrial hygienist proceeded to use modeling to try to make a more definitive exposure judgment. Converting the airborne inhalation OEL to total body dose using Equation 13-1 provided an equivalent total body dose of 0.3 mg/day:

Total Body Dose OEL: (0.03 mg/m^3) x (10 m^3/day) =
0.3 mg/day total body dose

Because no dermal monitoring data were available, a screening level estimate of potential dermal dose was developed using Equation 13-2 and default values from Tables 13.2 and 13.3. Specific activities included mixing the two-part acrylamide product (acrylamide as powder with activator and other materials), evaluating the consistency of the acrylamide gel (manually pouring a small amount of liquid mixture from one cup to another), assembling the grouting equipment, and using the equipment to fill the leak in the water line.

Using Equation 13-2:

$$DA = S \times Q \times FQ \times ABS \times WF$$

where

- DA = dermal absorbed dose rate, mg/day
- S = surface area of skin available for contact, cm^2
- Q = quantity of mixture deposited on the skin per event, mg/cm^2 — event
- FQ = contact frequency, number of events per day, 1/day
- ABS = fraction of applied dose absorbed through the skin per event, unitless
- WF = weight fraction of the substance in the mixture, unitless

The potential dermal absorbed dose estimates summarized in Table 13.4 were developed. The skin absorption of acrylamide has been determined to be approximately 25%, based on laboratory studies conducted by several researchers.[84] When compared with the computed dermal OEL of 0.3 mg/day, the screening level estimates (column DA in Table 13.4) identified a potentially unacceptable exposure, even though it was acknowledged that the screening level estimates were believed to overestimate the true dose. The industrial hygienist took immediate action to reduce dermal exposures by instituting improvements in the glove practices, housekeeping, and prescribed work practices controls. Skin exposure monitoring also was performed to confirm that the work practices and protective equipment were effective. Also, the bags of powdered acrylamide were replaced with a liquid form of acrylamide, further reducing the potential contribution of exposure due to inhalation.

Follow-Up Information: Following the implementation of health hazard controls, the measured dermal exposure using hand washes and wipes ranged between 0.017–0.090 mg/day for the various job activities. The surface contamination measurements ranged from 0.0062–0.234 mg/100 cm^2 on top of the acrylamide mixing tank. The major sources of exposure were now contact with contaminated work surfaces, equipment, and tools. Comparison of the measured skin exposures with the 0.3 mg/day (after adjusting for fraction dermally absorbed) indicated that exposures could be effectively controlled (i.e., 0.004–0.02 mg/day for the absorbed dermal exposures).

The industrial hygienist appreciated that the skin surface sampling measurements probably underestimated the actual exposure to some extent, but the greater than 10-fold difference in absorbed dose when compared to the OEL provided a reasonable margin of safety. Nevertheless, the surface contamination measurements indicated that exposures could be reduced further. The industrial hygienist sought additional ways of reducing dermal exposures, such as evaluating possible substitutes for acrylamide grout.

13.7.2 Example 2 — Assessment of Dermal Exposure Potential and Hazard Assessment for Benzene

Benzene is a common organic solvent that can affect the nervous and hematopoietic systems and can cause leukemia. A skin designation is given by the ACGIH® TLV® committee, indicating dermal absorption is a potentially significant route of exposure and may lead to systemic effects.[83] Benzene is present in small quantities (~1%) in vehicle fuel because of its antiknock properties in unleaded fuel. This concentration was confirmed experimentally by Adami et al., who found benzene concentrations in gasoline of 0.39%, 0.74%, and 1.06%.[85] Nevertheless, concerns have been raised that the benzene in vehicle fuel may result in a potentially significant exposure to automotive repair workers who are exposed to fuel during the disassembly of carburetors, cleaning of fuel injectors, changing of fuel filters, replacing and repairing of fuel tanks and fuel lines, as well as cleaning of parts and the skin using gasoline. A risk assessment was performed to estimate the daily exposure to workers through unprotected skin among automotive repair professionals specializing in such work at a large service repair shop.

Table 13.4 — Screening Level Estimates of Acrylamide Dermal-Absorbed Dose During Grouting Activities

Activity	*S, cm²*	*Q, mg/cm²*	*WF*	*FQ, 1/day*	*ABS Fraction*	*DA, mg/day*
Pour acrylamide into tank	840 (two hands)	3.7 (dump bags of powder)	1	1	0.25	777
Manually test for consistency	420 (one hand)	0.7–2.1 (sampling)	0.05	1	0.25	4–11
Assemble equipment 7.4–22	840 (one hand)	0.7–2.1 (connecting transfer line)		0.05	1	0.25
Performgrouting	840 (two hands)	1.3–10.3 (assume similar to handling wet surfaces)	0.05	2	0.25	27–216

Air samples were also taken to measure potential inhalation exposure during some actual repair work. The geometric mean 8-hr TWA concentration for benzene during personal monitoring was 0.06 mg/m^3.[86] These concentrations are less than 10% of the ACGIH® TLV® of 0.5 ppm (1.6 mg/m^3), indicating a low concern for benzene inhalation exposure from gasoline during repair work.

However, a study using both biological monitoring and air sampling of benzene exposure postulated that a large percentage of the total exposure to garage repair professionals was through the skin, which raised concern for dermal overexposure among the workers in the present situation.[87] Benzene is relatively volatile and much evaporates from the skin upon contact, but there is potential for absorption of a small fraction of the applied dose. One in vivo study found that that 0.14% of an applied dose of pure benzene is absorbed through the skin of rhesus monkeys, and 0.05% through human skin on the forearm following application of a thin layer of benzene.[88] Another *in vivo* human study showed that an applied dose of pure benzene to the forearm resulted in an average total absorption of 0.07 ± 0.04% and an applied dose to the palm resulted in an average total absorption of 0.13 ± 0.04% of the applied dose.[89] However, it has also been noted that absorption can increase through damaged skin by about 5-fold, which could be the case among these workers.

Using the above information and the following assumptions, a hazard assessment of dermal exposure was conducted for these workers. First, an estimated total body dose OEL equivalent was calculated using Equation 13-1:

Exposure dose (mg/day) = OEL (mg/m^3) x inhaled air volume (m^3/day)

It was determined that the vehicle maintenance tasks resulted in a light to moderate activity level during the work day, and therefore a value of 11 m^3/day was selected (Table 13.1). The TLV® of 0.5 ppm (1.6 mg/m^3) was used as the OEL equivalent:

Exposure dose (mg/day) = 1.6 mg/m^3 x 11 m^3/day = 17.6 mg/day

It should be noted that the use of this method to calculate the dermal OEL may not leave as sufficient a margin of error as might be found for inhalation exposures, since although an assumption is made that 100% of an inhaled dose is subsequently absorbed, in reality it may be only a small fraction of this dose which is actually absorbed. For example, in the case of benzene, it has been reported that 46% of inhaled benzene is absorbed.[90] This "safety factor" should be considered when using equivalent dose methods for evaluating dermal exposures. It was assumed from observations, interviews and work records that each worker had the potential for 20 skin contacts with benzene per day (FQ = 20/day). Because of the low viscosity of benzene in gasoline, which could allow it to drip and easily spread over contacted skin, it was assumed that approximately 150 cm^2 per hand, or 300 cm^2 of skin surface area could be contaminated during each episode (S = 300 cm^2). To estimate skin loading (Q), a value of 2.1 mg/cm^2 was used, based on the values in Table 13.3 for routine or incidental contacts with liquids and according to the EPA's 1992 guidance document on determining the retention of liquids on the surface of the hands.[91] For the absorption estimate (ABS), a conservative value based on the empirical studies of 1% absorption was used. This absorption value is also well over 5 times the absorption rate of 0.13% for benzene reported by the two studies above, and might be an appropriate conservative value to select if the industrial hygienist is interested in accounting for the potential for increased absorption due to damaged skin. It was assumed that the concentration of benzene on the skin was a function of its concentration in gasoline (~1%), as measured empirically by Adami, et al. (2006).[85] Using Equation 13-2 as a screening calculation, the amount of benzene potentially absorbed through the skin was calculated:

DA = S x Q FQ x ABS x WF

DA = 300 cm^2 x 2.1 mg/cm^2 x 20/day x 0.01 x 0.01 = 1.26 mg/day

It should be noted that this screening calculation is conservative because it does not take into account the mass of benzene which could evaporate from the skin prior to absorption. This screening calculation indicates preliminarily that exposures to benzene for this vehicle repair professional scenario may not reach the daily OEL for benzene based on dermal exposure alone, although it likely would be a more significant route of exposure than inhalation, which, based on the exposure measurements cited above, would result in a dose of approximately 0.06 mg/m^3 × 11 m^3/day = 0.66 mg /day. Therefore, the more robust equation for DA_{event} and a K_p value for benzene were also used to confirm this initial screening calculation in a more quantitative manner using Equation 13-5.

$$DA_{event} = K_{p,veh} \times C \times SA \times t_{event}$$

or

$$DA_{event} = Flux \times SA \times t_{event}$$

Adami et al. reported a measured K_p value for benzene in gasoline on human skin *in vitro* of K_p = 0.00044 cm/hr. They reported a dermal flux for benzene in gasoline of 0.002 mg/cm^2-hr.[85] These values were used to calculate the estimated dermal absorption dose in the equations that follow. Similar K_p values for benzene have been derived experimentally in other *in vitro* studies of pure benzene on human skin.[88,92,93] Another study, Blank, et al. (1985), also experimentally measured the K_p of benzene in gasoline on human skin in vitro and reported a value of K_p = 0.0014 cm/hr, which was significantly higher than the values reported in the other four studies.[94] It is unclear why Blank et al. measured a K_p value that was so much higher than other studies, although it is possible that the time in which the *in vitro* skin was in contact with the solvents used in the study may have caused damage to the skin. As a result, this K_p value was not used in this example. However, this reported K_p value might be appropriate for use in situations where, for example, the industrial hygienist is concerned about generating a conservative estimate of skin uptake where it is believed that worker skin damage has occurred.

Using the reported flux value in the above equation for DA_{event}, and again using an assumed skin surface area of SA = 300 cm^2 and an assumed direct contact time with gasoline on the hands of 20 contacts/day × 6 min/contact × 1 hr / 60 min = 2 hr/day, the following dermal absorbed dose can be estimated:

$$DA_{event} = 0.002 \text{ mg/cm}^2\text{-hr} \times 300 \text{ cm}^2 \times 2 \text{ hr/day} = 1.2 \text{ mg/day}$$

Using the empirical K_p value provided and a concentration of 1%, the flux is equal to:

$$\text{Flux} = K_p \times \text{Concentration} = 0.00044 \text{ cm/hr} \times 10 \text{ mg/cm}^3 = 0.0044 \text{ mg/cm}^2\text{-hr}$$

This flux is twice that reported in the Adami study because the highest weight concentration of benzene in gasoline reported in the study was used (10 mg/cm^3 or 1%), resulting in a dermal absorbed dose of:

$$DA_{event} = 0.0044 \text{ mg/cm}^2\text{-hr} \times 300 \text{ cm}^2 \times 120 \text{ minutes} \times 1 \text{ hr/60 minutes} = 2.6 \text{ mg/day.}$$

Both values (using the parameters selected) were well below the calculated dermal OEL equivalent of 17.6 mg/day. It should also be noted that even if an assumption of 8 hours/day of continuous contact with benzene in gasoline were used in this equation, the results of the calculation would still indicate that the total absorbed dose of benzene via the dermal route would be below the calculated dermal OEL equivalent.

$$DA_{event} = 0.0044 \text{ mg/cm}^2\text{-hr} \times 300 \text{ cm}^2 \times 8 \text{ hours} = 10.6 \text{ mg/day}$$

For comparison, K_p was also estimated using the revised Robinson model equation (Equation 13-8). The molecular weight and K_{ow} for benzene were obtained from the SRC web site (MW = 78.11 g/mol; log K_{ow} = 2.13) for use in the revised Robinson model for calculating K_p.[60] The revised Robinson model, as shown in Equation 13-8, was used for this calculation:

$$K_p = \frac{1}{\frac{1}{K_{lip} + K_{pro}} + \frac{1}{K_{aq}}}$$

where

K_p = overall permeability coefficient, cm/hr
K_{lip} = permeability coefficient of the lipid matrix fraction of the stratum corneum, cm/hr
K_{pro} = permeability coefficient of the protein fraction of the stratum corneum, cm/hr
K_{aq} = permeability coefficient of the viable layer of the epidermis, posing an aqueous barrier, cm/hr

The permeability coefficient of the lipid fraction of the stratum corneum, Klip, in cm/hr can be estimated using Equation 13-9:

$$\log K_{lip} = -2.59 + 0.7318 \times \log K_{ow} - 0.006832 \times MW$$

$$\log K_{lip} = -2.59 + 0.7318 \times 2.13 - 0.006832 \times 78.11 = -1.565$$

$$K_{lip} = 10^{-1.565} = 0.027$$

The permeability coefficient of the protein fraction of the stratum corneum (K_{pro}) in cm/hr can be estimated using Equation 13-10:

$$K_{pro} = \frac{0.043}{MW^{1.36}}$$

$$K_{pro} = 0.043/(78.11)^{1.36} = 0.00011$$

The permeability coefficient of the viable epidermal layer (K_{aq}) in cm/hr can be estimated using Equation 13-11:

$$K_{aq} = \frac{2.5}{\sqrt{MW}}$$

$$K_{aq} = 2.5/(78.11)^{1/2} = 0.283$$

Substituting values for K_{lip}, K_{pro}, and K_{aq} into the revised Robinson equation yields:

$$K_p = 0.025 \text{ cm/hr}$$

Alternatively, K_p can be estimated directly using the correlation in RAGS part E (Equation 13-7):

$$\log K_p = -2.80 + 0.66 \times \log K_{ow} - 0.0056 \times MW$$

$$K_p = 10\text{^}(-2.80 + 0.66 \times \log K_{ow} - 0.0056 \times MW) = 0.015 \text{ cm/hr}$$

Because the above estimations for K_p are based on an aqueous vehicle, K_p must be adjusted. The solubility of benzene in water is C_w^{sat}= 1.8 mg/mL. Benzene is completely miscible with gasoline, and therefore C_w^{sat}= 880 mg/mL. As described in Section 13.3.3., K_p is adjusted using Equation 13-6:

$$K_{p,veh} = 1.8 \text{ mg/mL} / 880 \text{ mg/mL} \times 0.025 \text{ cm/hr} = 5.1\times10^{-5} \text{ cm/hr.}$$

Although this predicted value of $K_{p,veh}$ is smaller than empirically observed (about 4.4×10^{-4} cm/hr), RAGS Part E reports a 95% UCL on K_p of 0.37 cm/hr, which, when adjusted to account for gasoline as the vehicle equates to:

$$K_{p,veh} = 1.8 \text{ mg/mL} / 880 \text{ mg/mL} \times 0.37 \text{ cm/hr} = 7.6 \times 10^{-4} \text{ cm/hr}$$

which is greater than both the empirical measurement and the predicted value of K_p. The estimated flux can then be substituted into the equation for DA_{event}, assuming 20 events per day:

$$DA_{event} = 0.00076 \text{ cm/hr} \times 10 \text{ mg/cm}^3 \times 300 \text{ cm}^2 \times 120 \text{ min/day} \times 1 \text{ hr/60 min} = 4.56 \text{ mg/day}$$

or, assuming a full 8-hour work day of continuous exposure:

$$DA_{event} = 0.00076 \text{ cm/hr} \times 10 \text{ mg/cm}^3 \times 300 \text{ cm}^2 \times 8 \text{ hr/day} = 18.24 \text{ mg/day}$$

This estimate is likely higher than the estimates developed using the empirical data because of the use of the 95th percentile UCL adjustment for the modeled K_p value. Overall, the estimations presented in Example 2 show that model results can be quite variable and have the potential to be significantly higher or lower than empirical measurements, depending on parameter selection. Further, these estimates do not take into account skin condition or skin damage, which may affect results substantially. Although dermal exposure and risk modeling can be valuable for characterizing the potential for dermal absorption and systemic toxicity from specific chemicals, the results in this example underscore the uncertainty inherent in modeling dermal exposures. This is particularly true for chemicals in non-aqueous solutions, and highlights the importance of careful selection of the model inputs. It is possible that for some industrial dermal exposure scenarios, the appropriate data will not be available to determine with sufficient certainty the quantitative potential for dermal exposure through modeling alone. Empirical data should always be used for as many model inputs as possible.

13.7.3 Example 3 — Estimation of Acceptable Skin Exposure Duration

For comparison, Equations 13-24 and 13-25 using the acceptable skin absorption time (SAT) can also be used to determine whether benzene in gasoline exposures might be acceptable under the scenario for vehicle mechanics described in Example 2. Using either form of this equation, one can estimate how long workers might be able to work unprotected with benzene-containing gasoline if gloves were not handy or the work could not easily be performed while wearing gloves. Equation 13-25 was used to estimate an acceptable skin exposure time for this work.

$$\text{Skin Absorption Time (hr)} = \frac{\text{Total absorption at OEL (mg)}}{K_p \text{ (cm/h)} \times \text{Concentration (mg/ml)} \times \text{skin area (cm}^2\text{)}}$$

Using the calculated dermal OEL equivalent of 17.6 mg/day and the empirical values for flux and Kp for benzene in gasoline reported in Adami et al.[85], and the reported concentration of benzene in the gasoline (1% or 10 mg/cm³), the acceptable skin absorption time would be in the range of 13.3–29.3 hours. The difference in SAT values seen when using the empirical K_p value versus the empirical flux to calculate SAT is due in large part to the use of a more conservative default value of 1% for benzene by weight in gasoline, when in actuality the measured average concentration in the study was 0.73%. In any case, these calculations appear to confirm the results obtained using the equation for DA_{event}, indicating that the dermal OEL equivalent for benzene exposure in the vehicle repair scenario described above is unlikely to be reached during an 8-hour work day:

$$\text{Skin Absorption Time} = \frac{1.6 \text{ mg/m}^3 \times 11 \text{ m}^3}{0.00044 \text{ cm/hr} \times 10 \text{ mg/cm}^3 \times 300 \text{ cm}^3} = 13.3 \text{ hrs}$$

Or, using the reported flux value,

$$\text{Skin Absorption Time} = \frac{1.6 \text{ mg/m}^3 \times 11 \text{ m}^3}{0.002 \text{ mg/cm}^2 - \text{hr} \times 300 \text{ cm}^2} = 29.3 \text{ hrs}$$

This calculation seems to confirm the possibility in this scenario that overexposure to benzene through the dermal route would be relatively unlikely if dermal flux is similar to that reported in the study by Adami et al. However, as noted above, if the skin is damaged by exposure to other solvents or by repeated exposures to gasoline over time, the resulting dermal exposures may be significantly higher than those estimated here. Care should always be taken when working with solvents, especially when the work occurs on a routine basis or when significant direct contact with the skin occurs.

13.7.4 Example 4 — Assessment of Dermal Exposure to Lead by Dermal Absorption and Hand-to-Mouth Exposure

Exposure estimates for all routes of exposure for a jewelry product were developed as part of an evaluation of compliance with the California Proposition 65 Maximum Allowable Dose Level (MADL) of 0.5 µg/day, and the U.S. CPSC policy of 15 µg/day. The jewelry product was a bracelet containing 12 silver beads with a measurable extractable lead content. Both dermal absorption and ingestion via hand-to-mouth contact following dermal exposure were evaluated for the product. Table 13.5 presents key default assumptions and variables used in the estimates.

First, a dermal loading of lead on the skin (based on research involving skin contact with lead-containing objects) was determined. These values were then used to estimate the potential ingestion exposure from hand-to-mouth contact and dermal absorption based on normal wear and use of the bracelet.

Estimated Dermal Loading of Lead from the Silver Beads

A dermal exposure assessment for lead conducted by the Institute of Occupational Medicine (IOM) in Edinburgh, United Kingdom was used to estimate these dermal loading values.(95) In this study, researchers collected hand wipe samples after six volunteers handled pure lead materials with various geometries. A total of 18 data points were collected, based on a single contact with lead, ranging from less than the limit of detection (<LOD) to 1.84 μg lead / cm^2 skin, with a median value of 0.225 μg/cm^2 and a 95th percentile of 1.04 μg/cm^2.

A total of three laboratory experiments resulted in soluble/extractable lead masses for the silver bracelet beads of 5100 μg, 7800 μg and 9500 μg. Each silver bracelet bead weighed approximately 0.69 g, therefore containing between 0.74% and 1.4% extractable lead by weight.

The results of the IOM study were extrapolated to estimate dermal lead exposure from the silver bracelet beads by multiplying the IOM study results by the extractable lead content measured in the beads. Estimates of the total potential dermal lead loading onto the skin, Lskin, for the silver beads therefore ranged from:

$L_{skin} = 0.225\ \mu g/cm^2 \times 0.74\% = 0.0017\ \mu g/cm^2$

to

$L_{skin} = 1.04\ \mu g/cm^2 \times 1.4\% = 0.015\ \mu g/cm^2$

Estimated Hand-to-Mouth Exposure to Lead for the Silver Beads

This exposure route involved the direct transfer of lead from the skin to the mouth from touching a lead-containing material and then performing hand-to-mouth activities such as touching the mouth, nail biting, or consumption of food. Hand-to-mouth transfer can be estimated using Equation 13-27(80):

$$M_{hand\text{-}mouth} = L_{skin} \times SA \times f_{direct} \times \lambda_D \times t,$$

where

$M_{hand\text{-}mouth}$ is the daily mass of lead transferred to the mouth (μg/day),
L_{skin} is the loading of lead on the hands (μg/cm^2),
SA is the portion of the surface area of the hand in contact with the silver beads on the bracelet (cm^2),
f_{direct} is the hand-to-mouth transfer factor,
λ_D is the rate of hand-to-mouth contact (#/hr), and
t is the amount of time the hand remains loaded with lead (hr/day).

L_{skin} is estimated above for the bracelet and ranges from 0.0017 μg/cm^2 to 0.015 μg/cm^2. One half of the surface area of the 12 beads on the bracelet was used to estimate the area of dermal contact with the hand, palm and fingers during normal wear and periodic repositioning of the bracelet. Thus, $A = 12 \times ½ \times 4\pi \times (½ \times 0.5\ cm)^2 = 4.7\ cm^2$. A value of 50% was selected for f_{direct}, based on the study by Camann, et al.(77) who studied the dermal removal of pesticides. A value for hand-to-mouth contact of $\lambda_D = 9$ /hr for adults was selected, based on Table 13.5. A value of t = 16 hours/day was selected to represent the number of waking hours in a day for adults in which the hand-to-mouth activity might occur. Therefore, hand-to-mouth estimates using this method range between:

$M_{hand\text{-}mouth} = 0.0017\ \mu g/cm^2 \times 4.7\ cm^2 \times 50\% \times 9/hr \times 16\ hr/day = 0.58\ \mu g/day$

$M_{hand\text{-}mouth} = 0.015\ \mu g/cm^2 \times 4.7\ cm^2 \times 50\% \times 9/hr \times 16\ hr/day = 5.1\ \mu g/day$

Table 13.5 — Assumptions Used in the Exposure Estimates for a Bracelet with Silver Beads

Parameter	*Value(s)*	*Unit*	*Note / Reference*
Diameter of one silver bead	5	mm	Measurement
Silver bead weight	0.69	g	Measurement
Silver bead extractable lead	5100–9500	μg	Laboratory results
Dermal loading of pure lead	0.225–1.04	μg/cm^2	IOM,2006(95)
Frequency of hand contact with bracelet for adults (medium and large bracelets), λ_A	5–10	1/day	Professional judgment
Dermal absorption fraction for Lead, ABS_d	0.001	—	Extrapolated from cadmium (Washington state, 2004)(96)
Permeability Coefficient for Lead, K_p	1×10^{-4}	cm/hour	U.S. EPA, 2004(6)
Hand-to-mouth transfer factor for lead, f_{direct}	50%	—	Camann et al., 2000(77)
Rate of hand-to-mouth contact for adults, λ_D	9	1/hr	Cal EPA OEHHA, 2007(80)
Waking hours for adults	16	hr	Professional judgment

However, as will be shown, the above methodology provides estimates that are unrealistically high, as they exceed the mass of lead originally loaded onto the skin. Bounding estimates of exposure via this pathway are therefore made by estimating the mass of lead originally loaded onto the skin and conservatively assuming that all of the lead that was transferred to the hand touching the bracelet over the course of the day is ultimately transferred to the mouth. Therefore, $M_{hand\text{-}mouth}$ is limited by λ_A:

$$M_{hand\text{-}mouth} = L_{skin} \times A \times \lambda_A$$

where

$M_{hand\text{-}mouth}$ is the mass of lead transferred to the mouth (μg),
L_{skin} is the loading of lead on the hands (μg/cm²).
A is the portion of the surface area of the hand in contact with the silver beads on the bracelet (cm²), and
λ_A is the rate of bracelet-to-hand contact (#/day) for adults.

Using these values, the estimates of $M_{hand\text{-}mouth}$ for adults wearing the medium or large size bracelets range between:

$$M_{hand\text{-}mouth} = 0.0017\ \mu g/cm^2 \times 4.7\ cm^2 \times 5/day = 0.040\ \mu g/day$$

$$M_{hand\text{-}mouth} = 0.015\ \mu g/cm^2 \times 4.7\ cm^2 \times 10/day = 0.71\ \mu g/day$$

The above estimates better reflect a day's hand-to-mouth transfer mass, as they are based on the more realistic assumption that lead is not constantly contaminating the transfer hand from the bracelet throughout the day. The previous estimates were larger, but relied on the assumption of continuous lead contamination over the 16-hour period in which hand-to-mouth activities might occur. The bounding estimates calculated here for the silver beads on the bracelet exceed the 0.5 μg/day MADL for lead under Proposition 65 but are below the U.S. CPSC policy recommendation for maximum chronic daily ingestion of lead by children of 15 μg/day.

Estimated Dermal Absorption of Lead from the Silver Beads

Once loaded onto the skin via dermal contact, the mass of lead absorbed dermally at the site of the bracelet was estimated using the U.S. EPA's Risk Assessment Guidance for Superfund (RAGS), Volume I: Human Health Evaluation Manual, Part E, Supplemental Guidance for Dermal Risk Assessment methodology for inorganic compounds in an aqueous solution.(6) This mass, DA_{event}, is given by:

$$DA_{event} = K_p \times t \times SA \times C_w,$$

where

K_p is the permeability coefficient in cm/hr,
t is time in hours,
SA is the surface area of skin in contact with the beads, and
C_w is the chemical concentration in water or a thin film of sweat on the dermal surface in μg/cm³.

The U.S. EPA has reported the permeability coefficient for lead as $K_p = 1\times10^{-4}$ cm/hr.(6) As a bounding estimate, t = 24 hours was selected to characterize the amount of time the skin was likely to be in contact with the beads of the bracelet and remain loaded with lead in a typical day.

One fourth the surface area of a sphere with the same diameter as a silver bracelet bead (diameter = 0.5 cm) was used to estimate the area of dermal contact. For the 12 silver metal beads on the bracelet, it was estimated that approximately ¼ of the surface area of each bead would be continuously in contact with the skin when the bracelet is worn. Thus, the surface area of skin in contact with the beads was equal to:

$$A = 12 \times \tfrac{1}{4} \times 4\pi \times (\tfrac{1}{2} \times 0.5\ cm)^2 = 2.4\ cm^2$$

C_w was estimated based on the dermal loading estimates (§13.7.4) using a film thickness of 0.006 cm to characterize the thickness of sweat on the dermal surface.(97)

Therefore, estimates for C_w range from:

$$C_w = 0.0017\ \mu g/cm^2 / 0.006\ cm = 0.28\ \mu g/cm^3$$

to:

$$C_w = 0.015\ \mu g/cm^2 / 0.006\ cm = 25\ \mu g/cm^3.$$

Dermal exposure estimates for the bracelet therefore range from:

$$DA_{event} = 1 \times 10^{-4}\ cm/hr \times 24\ hr \times 2.4\ cm^2 \times 0.28\ \mu g/cm^3 = 0.0016\ \mu g$$

to:

$$DA_{event} = 1 \times 10^{-4}\ cm/hr \times 24\ hr \times 2.4\ cm^2 \times 2.5\ \mu g/cm^3 = 0.014\ \mu g.$$

The above is the estimated range of dermal absorption from wearing a bracelet with twelve 5-mm beads over one 24-hour period. In this instance, the lower and upper bound estimates calculated were below both the 0.5 μg/day MADL for lead under Proposition 65, and the U.S. CPSC policy recommendation for maximum chronic daily ingestion of lead by children of 15 μg/day.

13.7.5 Example 5 — Derivation of acceptable surface concentrations for carcinogens

Initial surface wipe samples taken at the Pentagon following the crash of a hijacked airliner on September 11, 2001 identified polychlorinated biphenyls and assorted dioxins and furans. Site-specific, risk-based surface screening levels were needed in order to assure protection

$$C_{surface} = \frac{TR \times BW \times AT_{lifetime} \times 365 \times 10}{(CR_{dermal} \times CSF_{dermal} + CR_{ingestion} \times CSF_{ingestion} + CR_{inhal} \times CSF_{inhal}) \times EF \times ED} \quad (13\text{-}34)$$

of emergency response and remediation workers, as well as office workers once the building was cleared for normal occupancy. An approach that would include all possible routes of exposures specific to the Pentagon, such as provided in Equation 13-30, was needed.[98]

Because PCBs, dioxins and furans are considered carcinogenic, the first step was to obtain cancer slope factors (CSF) and decide on input variable values needed to calculate lifetime average daily intake (LADI) for each route of exposure using the following equations 13-31 and 13-32:

$$Risk_p = CSF_p \times LADI_p$$

$$LADI_p = \frac{C_{surface} \times CR_p \times EF \times ED}{BW \times AT_{lifetime} \times 365 \text{ days/yr}}$$

Some of the values used for site-specific exposure assumptions came from the USEPA Exposure Factors Handbook and some assumption were based on professional judgment.[26] Calculation of an acceptable surface concentration come from Equation 13-34 (see above).

A target risk (TR) rate of 1×10^{-4} was selected. Although it is at the least protective end of the range recommended by the USEPA for National Priority List site remediation, it is often more protective than standards recommended by the Occupational Safety and Health Administration for workers. To compute each of the three contact rate factors a few assumptions about the variables have to be made. The first contact rate for dermal exposure is from Equation 13-35:

$$CR_{dermal} = SA_d \times F_d \times EV \times FT_{ss} \times DAF$$

The following assumptions relative to dermal exposure were made:

SA_d = 0.198 m^2 (total surface area of hands and forearms of an adult male)[26]
F_d = 0.25 (fraction of dermal surface area available for contact that comes in contact each day)
EV = 3 (contact frequency per day)
FT_{ss} = 0.1 (fraction of dust transferred to surface to skin and remains on the skin)
DAF = 0.14 (dermal absorption efficiency, in this case based on empirical data)

Thus, CR_{dermal} = 0.0021 m^2/day.

Next, the contact rate for ingestion exposure is calculated using Equation 13-36:

$$CR_{ingest} = SA_g \times F_g \times EV \times FT_{ss} \times FT_{ftm} \times DAF$$

The additional variables used in this equation include:

F_g = 0.1 (fraction of available skin area that contacts mouth)
FT_{ftm} = 0.3 (fraction of dust transferred from skin to mouth)
HTME = 3 (hand-to-mouth events per day)
SA_g in this case equals 0.084 m^2 which is the surface are of two hands.[26]

Thus, CR_{ingest} = 0.0023 m^2/day.

Finally, an inhalation contact rate is computed from Equation 13-37:

$$CR_{inhale} = IR_i \times K$$

The variables and values used here include:

IR_i = 15 m^3/day (inhalation volume for a moderately active worker)[26]
K = 5×10^{-8} m^{-1} (this dust re-suspension factor is at the lower range representing lighter or less dust producing activities such as walking)[82]

Thus, $C_{Rinhale} = 7.5 \times 10^{-7}$ m^2/day.

Incorporating the CR_p values into the overall equation for calculating site-specific surface limits:

$$C_{surface} = \frac{TR \times BW \times AT_{lifetime} \times 365 \times 10}{(CR_{dermal} \times CSF_{dermal} + CR_{ingestion} \times CSF_{ingestion} + CR_{inhal} \times CSF_{inhal}) \times EF \times ED}$$

where

TR = 1×10^{-4}
BW = 71.8 kg (body weight)[26]
$AT_{lifetime}$ = 70 years
EF = 7 days for remediation worker and 250 days per year for an office worker
ED = 1 year for remediation worker and 25 years for an office worker

After substituting all the values for the variables in the above equation, surface concentration resulting in a risk of cancer health effects at 1×10^{-4} was computed to be 0.012 µg/100cm^2 for remediation worker exposure patterns. All 12 of the initial wipe samples taken were less than this concentration. For the returning Pentagon office worker, the surface concentration resulting in a 1 in 10,000 cancer risk was computed to be 1.4×10^{-5} µg/100cm^2. A few of the wipe samples taken contained concentrations greater than this limit. But because the sampling locations were selected to represent surfaces appearing to have the most contamination, that the exposure assumptions used for the risk-based evaluation were considered protective, and that the concentrations detected were either below or only slightly higher than available screening levels, it was decided that additional sampling was not necessary and that adverse health effects among exposed Pentagon personnel were unlikely.

Upon careful consideration of the variables in each equation above, it should be clear that the value chosen for each one is typically not a hard and empirically determined measure. Some of the estimates are based on a precedent established by prior experience and judgment, while others are based solely on the professional judgment of the current user. The range of possibilities for each variable are potentially broad, and choosing the incorrect value might have significant impact on the determination of risk potential and the actions that are taken to prevent exposure. Thus, these computations must be done with care and with full knowledge that they are only estimates, and should always be supplemented by additional tools and approaches for monitoring the exposure and health of potentially affected populations. Additional advanced techniques, such as Monte Carlo analysis (see Chapter 19), can produce a probability distribution range based on thousands of possibilities for choosing variable parameters.

13.8 References

1. **Elias, P.M. and K.R. Feingold (eds.):** *Skin Barrier.* London: Informa HealthCare, 2005.
2. **Chilcott R. and S. Price (eds.):** *Principles and Practice of Skin Toxicology.* San Francisco, CA: Wiley, Inc., 2008.
3. **Leung, H. and D.J. Paustenbach:** Techniques for Estimating the Percutaneous Absorption of Chemicals Due to Occupational and Environmental Exposure. *Appl. Occup. Environ. Hyg. 9(3)*:187–197 (1994).
4. **Grandjean, P.:** "Skin Penetration: Hazardous Chemicals at Work." A report prepared for the Commission of the European Communities, Directorate-General Employment, Industrial Relations and Social Affairs. London, England: Taylor & Francis, 1990.
5. **Burstyn, I. and K. Teschke:** Studying the determinants of exposure: a review of methods. *Am. Ind. Hyg. Assoc. J. 60(1)*:57–72 (1999).
6. **U.S. Environmental Protection Agency (EPA):** Risk Assessment Guidance for Superfund (RAGS). Volume I: Human Health Evaluation Manual (Part E, Supplemental Guidance for Dermal Risk Assessment). EPA/540/R/99/005. Office of Superfund Remediation and Technology Innovation. Washington D.C.: U.S. Environmental Protection Agency, July 2004.
7. **Fenske, R.A.:** Dermal exposure assessment techniques. *Ann. Occup. Hyg. 37*:687–706 (1993).
8. **Marquart, H., S. Maidment, J.L. McClaflin, and M.C. Fehrenbacher:** Harmonization of future needs for dermal exposure assessment and modeling: a workshop report. *Appl. Occup. Environ. Hyg. 16*:218–227 (2001).
9. **American Chemistry Council:** Methods for Assessing Matrix Effects on Oral and Dermal Absorption and Dermal Contact and Transfer. White Paper HHEA-3. Report prepared by ICF Consulting. Arlington, VA, 2001.
10. **Schneider, T., R. Vermeulen, D.H. Brouwer, J.W. Cherrie, H. Kromhout, and C.L. Fogh:** Conceptual model for assessment of dermal exposure. *Occup. Environ. Med. 56*:765–73 (1999).
11. **Marquart, J., et al.:** Determinants of Dermal Exposure Relevant for Exposure Modeling in Regulatory Risk Assessment. *Ann. Occup. Hyg. 47(8)*:599–607 (2003).
12. **Goede, H.A., et al.:** Classification of dermal exposure modifiers and assignment of values for a risk assessment toolkit. *Ann. Occup. Hyg. 47(8)*: 609–618 (2003).
13. **Warren, N., H.A. Goede, T. Scha, R. Oppl, H.J. Schipper, and J.J. Van Hemmen:** Deriving Default Dermal Exposure Values for Use in a Risk Assessment Toolkit for Small and Medium-Sized Enterprises. *Ann. Occup. Hyg. 47*:619–627 (2003).
14. **Brouwer, D.H., et al.:** A dermal model for spray painters. Part I: subjective exposure modelling of spray paint deposition. *Ann. Occup. Hyg. 45(1)*:15–23 (2001).
15. **Semple, S., et al.:** A dermal model for spray painters. Part II: estimating the deposition and uptake of solvents. *Ann. Occup. Hyg. 45(1)*:25–33 (2001).
16. **Van-Wendel-de-Joode, B., D.H. Brouwer, R. Vermeulen, J.J. Van Hemmen, D. Heederik, and H. Kromhout:** DREAM: a method for semi-quantitative dermal exposure assessment. *Ann. Occup. Hyg. 47*:71–87 (2003).
17. **European Chemicals Agency:** Technical Guidance Documents on Registration, Evaluation, Authorisation, and Restriction of Chemicals (REACH), 2008. http://guidance.echa.europa.eu/ guidance_en.htm.
18. **Rajan-Sithamparanadarajah, R., et al.:** Patterns of dermal exposure to hazardous substances in European Union workplaces. *Ann. Occup. Hyg. 48(3)*:285–97 (2004).
19. **Oppl, R., F. Kalberlah, P.G. Evans, and J.J. Van Hemmen:** A Toolkit for Dermal Risk Assessment and Management: an Overview. *Ann. Occup. Hyg. 47*:629–640 (2003).
20. **Schuhmacher-Wolz, U., F. Kalberlah, R. Oppl, and J. J. Van Hemmen:** A Toolkit for Dermal Risk Assessment: Toxicological Approach for Hazard Characterization. *Annals of Occupational Hygiene 47*: 641-652, 2003.

21. **Eurofins:** Eurofins Product Testing — RISKOFDERM project, European Community, RTD Project : QLK4-CT-1999-01107. http://product-testing.eurofins.com/services/research—development/projects-on-skin-exposure-and-rotection/riskofderm—-skin-exposure-and-risk-assessment.aspx.
22. **Fenske, R.A., and J.J. van Hemmen:** Occupational Skin Exposure to Chemical Substances: Setting Limits. *Ann. Occup. Hyg. 38(4)*:333–336 (1994).
23. **Bos, P. M., D.H. Brouwer, H. Stevenson, P.J. Boogaard, W.L. de Kort, and J.J. van Hemmen:** Proposal for the assessment of quantitative dermal exposure limits in occupational environments: Part 1. Development of a concept to derive a quantitative dermal occupational exposure limit. *Occup. Environ. Med. 55*:795–804 (1998).
24. **Brouwer, D.H., L. Hoogendoorn, P. M. Bos, P. J. Boogaard, and J. J. van Hemmen:** Proposal for the assessment to quantitative dermal exposure limits in occupational environments: Part 2. Feasibility study for application in an exposure scenario for MDA by two different dermal exposure sampling methods. *Occup. Environ. Med. 55*:805–11 (1998).
25. **Sartorelli, P., et al.:** Percutaneous penetration studies for risk assessment. *Environ. Toxicol. Pharmacol., 8*:133–152 (2000).
26. **U.S. Environmental Protection Agency (EPA):** Exposure Factors Handbook. EPA/600/P-95/002Fa. Office of Research and Development, Washington, D.C.: National Center for Environmental Assessment, August 1997.
27. **Adams, W.C.:** Measurement of breathing rate and volume in routinely performed daily activities, Final Report. California Air Resources Board (CARB) Contract No. A033-205. June 1993. 185 pgs.
28. **Layton, D.W.:** Metabolically consistent breathing rates for use in dose assessments. *Health Physics 64(1)*:23–36 (1993).
29. **Linn, W.S. and D.A. Shamoo:** Hackney, J.D. Documentation of activity patterns in "high-risk" groups exposed to ozone in the Los Angeles area. In: *Proceedings of the Second EPA/AWMA Conference on Tropospheric Ozone*, Atlanta, Nov. 1991. pp. 701–712. Pittsburgh, PA: Air and Waste Management Association, 1992.
30. **Fiserova-Bergerova, V., J.T. Pierce, and P.O. Droz:** Dermal Absorption Potential of Industrial Chemicals: Criteria for Skin Notation. *Am. J. Ind. Medicine 17*:617–635 (1990).
31. **DeCock, J., D. Heedrik, H. Kromhout, and J.S.M. Boleij:** Strategy for Assigning a "Skin Notation": A Comment. *Ann. Occup. Hyg. 40*:611–614. (1996).
32. **U.S. Environmental Protection Agency (EPA):** Dermal Exposure Assessment: Principles and Applications; Interim Report (EPA/600/8-91/011B). Washington, D.C.: U.S. Environmental Development, 1992.
33. **Leung, H. and D.J. Paustenbach:** Techniques for Estimating the Percutaneous Absorption of Chemicals Due to Occupational and Environmental Exposure. *Appl. Occup. Environ. Hyg. 9(3)*:187–197 (1994).
34. **Health and Welfare Canada, U.S. Environmental Protection Agency, and National Agricultural Chemicals Association:** Pesticide Handlers Exposure Database (PHED), February, 1992. [PHED Version 1.1 update, 1995].
35. **McDougal, J.N. and M. Boeniger:** Methods for Assessing Risks of Dermal Exposures in the Workplace. *Crit. Rev. Tox. 32*:291–327 (2002).
36. **VanRooij, J.G.M., J.G.C. de Roos, M.M. Bodelier-Bade, and F.J. Jongeneelen:** Absorption of Polycyclic Aromatic Hydrocarbons through the Human Skin: Differences Between Anatomical Sites and Individuals. *J. Toxicol. Environ. Health 38*:355–368 (1993).
37. **Knaak, J.B., M.A. Al-Bayati, O.G. Raabe, J.L. Wiedmann, J.W. Pensyl, J.H. Ross, A.P. Leber, and P. Jones:** Mixer-Loader-Applicator Exposure and Percutaneous Absorption Studies Involving EPTC Herbicide. Biological Monitoring for Pesticide Exposure, Measurement Estimation, and Risk Reduction. Washington, D.C.: American Chemical Society, 1992.
38. **Knarr, R.D., G.L. Cooper, E.A. Brian, M.G. Kleinschmidt, and D.G. Graham:** Worker Exposure During Aerial Application of a Liquid and a Granular Formulation of Ordram Selective Herbicide to Rice. *Arch. Environ. Contam. Toxicol. 14*:523–527 (1985).
39. **Lavy, T.L., J.S. Shepard, and D.C. Bouchard:** Field Worker Exposure and Helicopter Spray Pattern of 3,4,5-%, Bull. *Environ. Contam. Toxicol. 24*:90–96 (1980).
40. **Roff, M. and J. Wheeler:** Dislodgeable residues from hard surfaces: Interim report, IR/ECO/00/11. Health and Safety Laboratory, Biomedical Sciences Group. United Kingdom, 2000.
41. **Brouwer, D.H., et al.:** Transfer of contaminants from surface to hands: experimental assessment of linearity of the exposure process, adherence to the skin, and area exposed during fixed pressure and repeated contact with surfaces contaminated with a powder. *Appl. Occup. Environ. Hyg. 14(4)*:231–39 (1999).
42. **Kissel, J., K. Richter, and R. Fenske:** Field Measurements of Dermal Soil Loading Attributes to Various Activities: Implications for Exposure Assessment. *Risk Anal. 16*:116–125 (1996).
43. **McArthur, B.:** Dermal Measurements and Wipe Sampling Methods: A Review. *Appl. Occup. Environ. Hyg. 7(9)*:599–606 (1992).
44. **Ness, S.A.:** *Surface and Dermal Monitoring for Toxic Exposures*. New York: Van Nostrand Reinhold, 1994.
45. **Kissel, J. K. Richter, R. Duff, and R. Fenske:** Factors Affecting Soil Adherence to Skin in Hand-Press Trials. Bull. Environ. *Contamin. Toxicol. 56*:722–728 (1996).

46. **Driver, J.H., J.J. Konz, and G.K. Whitmyre:** Soil adherence to human skin. Bull. Environ. *Contam. Toxicol. 43*:814–820 (1989).
47. **Que Hee, S.S., B. Peace, C.S. Clark, J.R. Boyle, R.L. Bornschein, and P.B. Hammond:** Evolution of efficient methods to sample lead sources, such as house dust and hand dust, in the homes of children. *Environ. Res. 38*:77–95 (1985).
48. **Brouwer D., M. Boeniger, and J. van Hemmen:** Hand Wash and Manual Skin Wipes. *Ann. Occup. Hyg. 44*:501–510 (2000).
49. **Ignacio, J.S. and W.H. Bullock (eds.):** *A Strategy for Assessing and Managing Occupational Exposures*, 3rd edition. Fairfax, VA: American Industrial Hygiene Association, 2006.
50. **U.S. Environmental Protection Agency (EPA):** Options for revising CEB's Method for Screening-Level Estimates of Dermal Exposure., Final Report, June 2000, Chemical Engineering Branch, Economics, Exposure and Technology Division, Washington DC.
51. **Rutledge, L.C.:** Some Corrections to the Record on Insect Repellents and Attractants. *J. Am. Mosquito Cont. Assoc. 4*:441–425 (1988).
52. **Finley, B.L., P.K. Scott, and D.A. Mayhall:** Development of a standard soil-to-skin adherence probability density function for use in Monte Carlo analyses of dermal exposure. *Risk Analysis 14*: 555–569 (1994).
53. **Kromhout, H., W. Fransman, R. Vermeulen, M. Roff, and J.J. van Hemmen:** Variability of Task-Based Dermal Exposure Measurements from a Variety of Workplaces. *Ann. Occup. Hyg. 48*:187–196 (2004).
54. **Rougier, A., D. Dupuis, C. Lotte, and R. Roguet:** The measurement of the stratum corneum reservoir. A predictive method for in vivo percutaneous absorption studies: influence of application time. *J. Invest. Dermatol. 84(1)*:66–68 (1985).
55. **Rougier, A., C. Lotte, and D. Dupuis:** An original predictive method for in vivo percutaneous absorption studies. *J. Soc. Cosmetic Chemists 38*:397–417 (1987).
56. **Rougier, A.:** Predictive Measurement of In Vivo Percutaneous Absorption. In *Prediction of Percutaneous Penetration Methods, Measurements, Modeling*. Scott, R.C., R.H. Guy, and J. Hadgraft (eds.). London: IBC Technical Services, 1990.
57. **Maibach, H.I. and R. Feldman:** Systemic Absorption of Pesticides Through the Skin of Man, In: Occupational Exposure to Pesticides, Report to the Federal Working Group on Pest Management. From the Task Group on Occupational Exposure to Pesticides. Washington, DC: White House Council on Environmental Quality, Federal Working Group on Pest Management, 1974.
58. **U.S. Environmental Protection Agency (EPA):** Region IX Preliminary Remediation Guidelines, Toxicity Tables (http://www.epa.gov/Region9/waste/sfund/prg/index.htm). [Accessed on April 1, 2009].
59. **ten Berge, W.:** QSARS of Skin Permeation for Chemicals Webpage. http://home.planet.nl/~wtberge/qsarperm.html. [Accessed on April 1, 2009].
60. **Syracuse Research Center:** Physical Properties Database (PHYSPROP). http://www.syrres.com/what-we-do/product.aspx?id=133. [Accessed on April 1, 2009].
61. **National Institute for Occupational Safety and Health:** Skin Exposures & Effects Webpage, Skin Permeation Calculator. Centers for Disease Control and Prevention, http://www.cdc.gov/niosh/topics/skin/skinPermCalc.html. [Accessed on April 1, 2009].
62. **Theeuwes, F., R. Gale, and R. Baker:** Transference: A comprehensive parameter governing permeation of solutes through membranes. *J. Membrane Sci. 1*:3–16 (1976).
63. **Bunge, A.L. and W.J. Romonchuk:** Measuring Skin Permeability Coefficients of Highly Lipophilic Compounds, Final Report, Award No. R-82951801. Submitted to the U.S. Environmental Protection Agency (EPA), July 31, 2006.
64. **Jakasa, I., N. Mohammadi, J. Kruse, and S. Kezic:** Percutaneous Absorption of Neat and Aqueous Solutions of 2- Butoxyethanol in Volunteers. *Internatl. Arch. Occup. Env. Health 77*:79–84 (2004).
65. **ten Berge, W.:** A Simple Dermal Absorption Model: Derivation and Application. *Chemosphere*, 2009. [In Press].
66. **Wilschut, A., W.F., ten Berge, P.J. Robinson, and T.E. McKone:** Estimating Skin Permeation. The Validation of Five Mathematical Skin Permeation Models. *Chemosphere 30*:1275–1296 (1995).
67. **Vecchia B.E. and A.L. Bunge:** (Chapter 3) Skin absorption databases and predictive equations. In *Transdermal Drug Delivery.* Guy, R.H. and J. Hadgraft (eds.). New York: Marcel Dekker, 2002.
68. **Bunge, A.L. and R.L. Cleek:** A New Method for Estimating Dermal Absorption from Chemical-Exposure .2. Effect of Molecular-Weight and Octanol Water Partitioning. *Pharma. Res. 12*:88–95 (1995).
69. **Wilschut, A. and W.F., ten Berge:** Two Mathematical Skin Permeation Models for Vapours. Perspectives in Percutaneous Penetration (PPP) Conference, 4B: La Grande Motte, France (1997).
70. **Lotens, W.A. and L.J.A. Wammes:** Vapour Transfer in Two Layer Clothing due to Diffusion and Ventilation. *Ergo. 36(1)*:1223–1240 (1993).
71. **Vecchia B.E. and A.L. Bunge:** Partitioning of Chemicals into Skin: Results and Predictions (Chapter 4). In *Transdermal Drug Delivery.* Guy, R.H. and J. Hadgraft (eds.). New York: Marcel Dekker, 2002.

72. **European Chemicals Agency:** Guidance on information requirements and chemical safety assessment Chapter R.14: Occupational Exposure Estimation, Appendix R 14.1, May 2008. (http://echa.europa.eu/reach_en.asp).
73. **Walker, J.D., C. Whittaker, and J.N. McDougal:** Role of the TSCA Interagency testing committee in meeting the U.S. government data needs: designating chemicals for percutaneous absorption rate testing. In *Dermatotoxicology*, 5th edition. Marzulli, F.N. and H.I. Maibach (eds.). Bristol, PA: Taylor & Francis, 1996. pp. 371–381.
74. **Potts, R.O. and R.H. Guy:** Predicting skin permeability. *Pharm. Res. 9*:663–669 (1992).
75. **Bunge, A.L., G.L. Flynn, and R.H. Guy:** Predictive Model for Dermal Exposure Assessment. In *Water Contamination and Health: Integration of Exposure Assessment, Toxicology, and Risk Assessment.* Wang, R. (ed.). New York: Marcel Dekker, 1994.
76. **Cherrie, J.W., S. Semple, Y. Christopher, A. Saleem, G.W. Hughson, and A. Phillips:** How important is inadvertent ingestion of hazardous substances at work? *Ann. Occup. Hyg. 50(7)*: 693–704 (2006).
77. **Camann, D.E., T.K. Majumdar, and P.W. Geno:** Evaluation of saliva and artificial salivary fluids for removal of pesticide residues from human skin. Southwest Research Institute, San Antonio, TX and ManTech Environmental Technology, Inc., Research Triangle Park, NC for U.S. EPA. National Exposure Research Laboratory, Research Triangle Park, NC. EPA/600/R00/041, 2000.
78. **U.S. Consumer Product Safety Commission:** Briefing Package. Petition to ban chromate copper arsenate (CCA)-treated wood in playground equipment (Petition HP01-3). Washington, DC, February, 2003.
79. **U.S. Environmental Protection Agency (EPA):** Policy paper from the Science Advisory Council for Exposure, Policy Number 12. Recommended revisions to the Standard Operating Procedures (SOPs) for residential exposure assessments, Washington, DC, 2001.
80. **California Environmental Protection Agency, Office of Environmental Health Hazard Assessment (OEHHA):** Safe Drinking Water and Toxic Enforcement Act off 1986 (Proposition 65). Draft: Guideline for Hand-to-Mouth Transfer of Lead Exposure from Fishing Tackle Products. Interpretive Guideline No. 2007-001. May 2007.
81. **May, L.M., B. Gaborek, T. Pitrat, and L. Peters:** Derivation of risk based wipe surface screening levels for industrial scenarios. *Sci. Total Environ. 288*:65–80 (2002).
82. **Sansone, E.B.:** Redispersion of Indoor Surface Contamination and its Implications. *Treatise on Clean Surface Technology*, Vol. 1. New York: Plenum Press, 1987. pp 261–290.
83. **American Conference of Governmental Industrial Hygienists (ACGIH®):** Threshold Limit Values (TLVs®) for Chemical Substances and Physical Agents and Biological Exposure Indices (BEIs®). Cincinnati, Ohio: American Conference of Governmental Industrial Hygienists, 2007.
84. **Ramsey, J., J. Young, and S. Gorzinski:** "Acrylamide: Toxicodynamics in Rats." 1984. [Unpublished Dow Chemical Company Report.] In Assessment of Health Risks from Exposure to Acrylamide. Washington, D.C.: U.S. Environmental Protection Agency/Office of Toxic Substances, 1990.
85. **Adami, G., F. Larese, M. Venier, P. Barbieri, F. Lo Coco, and E. Reisenhofer:** Penetration of benzene, toluene, and xylenes contained in gasolines through human abdominal skin *in vitro*. *Tox. in Vitro (20)*:1321–1330 (2006).
86. **Javelaud, B., et al.:** Benzene Exposure in Car Mechanics and Road Tanker Drivers. *Int. Arch. Occup. Environ. Health 71*:277–283 (1998).
87. **Laitinen, J., J. Kangas, K. Pekari, and J. Liesivuori:** Short-term Exposure to Benzene and Gasoline at Garages. *Chemosphere 28(1)*:197–205 (1994).
88. **Franz, T.J.:** Percutaneous Absorption of Benzene. (Chapter 5). In *Adv. Modern Environ. Toxicol. Applied Toxicology of Petroleum Hydrocarbons*, vol. 6. Princeton, N.J.: Princeton Scientific Publishers, Inc., 1984. pp. 61–70.
89. **Modjtahedi, B.S. and H.I. Maibach:** In vivo percutaneous absorption of benzene in man: Forearm and palm. *Food Chem. Tox. (46)*:1171–1174 (2008).
90. **Rusch, G.M., B.K. Leong, and S. Laskin:** Benzene metabolism. *J. Toxicol. Environ. Health 2(suppl)*:23–26 (1977).
91. **Cinalli, C., C. Carter, A. Clark, and D. Dixon:** A Laboratory Method to Determine the Retention of Liquids on the Surface of Hands. U.S. Environmental Protection Agency. EPA 747-R-92-003, September 1992.
92. **Hanke, J., T. Dutkiewicz, and J. Piotrowski:** The Absorption of Benzene through Human Skin. *Int. J. Occup. Environ. Health. 6*:104–111 (2000).
93. **Loden, M.:** The in Vitro Permeability of Human Skin to Benzene, Ethylene Glycol, Formaldehyde, and n-Hexane. *Acta Pharmacol. et Toxicol. 58*:382–389 (1986).
94. **Blank, I.H. and D.J. McAuliffe:** Penetration of Benzene through Human Skin. *J. Invest. Dermatol. 85*:522–526 (1985).
95. **Sleeuwenhoek, A. and M. van Tongeren:** "Assessment of dermal exposure to inorganic lead caused by direct skin contact with lead sheet and moulded PVC profiles." Institute of Occupational Medicine, Research Report TM/06/04. Edinburgh, United Kingdom: December 2006.

96. **Washington State Department of Ecology:** Response to Questions from the Science Advisory Board May 2004. pp. 5–10. "Lead Exposure from Dermal Contact with Lead-Contaminated Soils." 2004.
97. **U.S. Environmental Protection Agency (EPA):** A Laboratory Method to Determine Retention of Liquids on the Surfaces of Hands. Washington D.C., Office of Pollution Prevention and Toxics., EPA/747/R-92/003. 1992.
98. **Gaborek, G.J., J.M. Mullikin, T. Pitrat, L. Cummings, and L.M. May:** Pentagon surface wipe sampling health risk assessment Presented at 2002 Toxicology and Risk Assessment Conference, Cincinnati, Ohio. U.S. Army Center for Health Promotion and Preventive Medicine, Aberdeen Proving Ground, Maryland (2002).

14

REACh — A New and Important Reason to Learn Modeling

By
Ron Pearson, MS, CIH
Michael Jayjock, PhD, CIH
Susan F. Arnold, CIH

14.0 Overview

In the 1990s, at meetings of the European Union's (EU) Council of Environment Ministers, concerns were raised that European chemical policies didn't provide enough protection to chemical product users. An evaluation of existing rules revealed that new substances were heavily regulated, but they made up only a tiny fraction of the total chemicals in commerce. On the other hand, existing substances, which made up the vast majority of the total volume of chemicals in use, were essentially unregulated. Out of this, REACh was conceived.(1,2)

This lack of risk assessment attention for existing chemicals has also been recognized on this side of the Atlantic. It has been explicitly noted by the U.S. Environmental Protection Agency (EPA) Board of Scientific Councilors that any comprehensive assessment of human consumer exposure to chemicals has not occurred and that the vast majority of exposures have not been systematically and proactively addressed.(3) Indeed, it is reasonably well established at this point that the risks of most types of consumer chemical exposures in modern society have not been assessed.

The types of exposure mentioned here are the exposures to chemicals that result predominately from residential sources. Because of the leadership of the European Union and the development of REACh, we are on the cusp of a change in which regulatory mandates playing out in the rest of the world are literally driving the overall scientific development of human health exposure assessment for the multitude of common and relatively unstudied substances to which humans are exposed.

14.1 What then is REACh all About at its Base Level and What Does it Mean to the Industrial Hygienist?

REACh stands for **R**egistration, **E**valuation, **A**uthorisation and restriction of **C**hemical Substances. It is a complex piece of rulemaking that lawmakers have been hammering out in Europe for some time. REACh became the law of the land in the EU on June 1, 2007. Its primary purpose is to evaluate and control the human health risk from chemicals. It places the burden of proof for chemical safety on manufacturers, removing it from regulatory authorities. Because of the complexity and broad scope of REACh, it is driving the need to screen and assess many chemicals and substances in a relatively short (but rather immediate) timeframe.

So it is thus clear that REACh was conceived in Europe to regulate the majority of chemicals that exist in commerce. What then is the upshot of REACh relative to industrial hygiene? Professionals anticipate REACh will actually spur innovations and opportunities in estimating exposures to consumer products, which would positively affect the field of IH. Toxicologists are also facing the same issues as the second half of any risk assessment team.

On an administrative level the law presents a lot of details surrounding thresholds, implementation deadlines, information requirements, etc. It is an enormously multifaceted and complicated piece of legislature; however, one needs to understand that it essentially replaces about 40 different pieces of existing laws and regulations within the EU. Thus, this brief chapter is certainly not an exhaustive review, but instead intended to provide a relatively high-level (10,000 meter high) overview.

Let's first break down some of the detail within the REACh acronym:

Registration:

Manufacturers of covered substances in the EU (and importers) will be required to submit a registration "dossier" to the newly created European Chemicals Bureau (ECB). The dossier will contain information on the properties, uses, classification of a substance, and guidance on safe use. This will be phased in over an 11-year period, starting with those substances shipped in quantities of more than 100 tons per year, eventually decreasing to 1 ton-per-year products. You also should be aware that quite a number of exemptions from this regulation will come into play, such as pharmaceuticals, food additives, polymers, etc.

Evaluation:

The European Chemicals Bureau (ECB) will evaluate the dossiers and individual substances. Depending on the potential risks posed by a substance, the dossier packages are expected to vary by product. They may include additional data from toxicological testing, ecotoxicological testing, and studies of human exposure potential qualifying safe-use scenarios. Many of the requirements remain to be fully detailed, but it is a given that they will lead to much "back and forth" before decisions are made about restrictions on substance use, if any.

Authorization:

Manufacturers that produce substances of very high concern will need to identify them, as they may eventually require authorization. These substances of very high concern (SVHCs), at a minimum, likely will include three classes of substances to include: 1. certain categories of carcinogens, mutagens and reproductive toxins 2. persistent, bioaccumulative and toxic chemicals and 3. substances that are very persistent and very bioaccumulative.

All of this, of course, is applied to existing **Ch**emicals, hence the name **REACh**.

14.2 What REACh Should Mean to the Industrial Hygienist

To help better understand the practical implications of REACh, it may help to discuss REACh implementation projects. In 2006, the EU began issuing draft technical guidance documents that started to address some of the issues surrounding these types of questions by means of REACh Implementation Projects (RIPs). The Arona Network, named for a city in Northern Italy near the European Commission's Joint Research Centre, addresses issues surrounding exposure scenarios. It is in this forum where the technical details of the evaluation of human exposure to these chemicals are supposed to be worked out.

Basically, there are likely to be three types of users for which REACh Exposure Scenarios (ESs) will need to be defined: industrial, professional and personal/consumer. All of the ESs have the end goal of defining what constitutes safe use. This sounds fairly straightforward, but even a simple example can develop into relatively high levels of complexity very quickly.

For example, suppose company X in the U.S. sells a solvent-based adhesive to several EU manufacturers, and the solvent contains 1 percent toluene, which has been identified as a substance of very high concern (SVHC). EU manufacturer 1 sells to an industrial customer that primarily sprays the material to bond automobile door panels to steel. Manufacturer 2 sells to a professional customer that uses the adhesive infrequently to bond damaged insulation in HVAC systems. Manufacturer 3 sells to consumers for use as a general purpose bonding agent. Think about the most basic variables that would differ among the three different user types. At a minimum, these variables would include:

- Application method: spraying vs. brushing vs. dabbing
- Use temperature: anywhere from 50°F to 90°F
- Use volume: hundreds of gallons vs. 1 liter vs. a few milliliters
- Frequency of use: daily vs. weekly vs. biannually
- Ventilation: specific LEV vs. commercial building exchange rates vs. household or even outdoor use
- Availability of respiratory protection: Airline respirator vs. organic vapor air-purifying respirator vs. no knowledge of what a respirator is

Consider the aspects of both dermal and inhalation exposure potential; for residential exposures, we must also consider the potential for secondary ingestion by children.

Also consider the task of quantifying all of these with a reasonable degree of confidence. Think about what happens when the amount of toluene in our example is doubled in the formulation or replaced with a less toxic substitute. Eventually, safety data sheets (the EU equivalent of MSDSs) will be appended with information specific to each exposure scenario. Thus, a picture is clearly emerging that will require a lot of technological input by persons skilled in the science of exposure assessment.

Pre-registration for existing substances began June 1, 2007. As more implementation deadlines arrive, U.S. corporations will be compelled to react. If they do not, they run the risk of not gaining authorization for one of their product components and the real loss of overseas business. The provisions of REACH will require a different mindset and a much stronger proactive approach in order to comply. Several key issues that U.S. companies need to know are:

- REACh will mandate increased cooperation among manufacturers in data sharing, even though it purports to protect commercially sensitive information.
- The EU, in general, has a stronger aversion toward toxicological data obtained from animal tests. Although this is a political/scientific issue, it could also have a number of practical implications for those wanting to sell into these markets.
- Requests for additional information must be approved by all EU member states.

These and other changes will mean significant transformations in the manner of doing business for many companies. Perhaps the most controversial aspect of REACh, which involved extended discussion and study in Europe over the last decade, addressed the overall cost-benefit. The political conclusion of that work resulted in the decision that the "light" of REACh was worth the "candle" of the cost. Furthermore, and perhaps more importantly, the responsibility and cost for proving "safety" have been shifted from EU government bodies to manufacturers and the private sector.

As mentioned above, numerous studies have been published trying to quantify the potential financial impact on manufacturers, with some conflicting results. The "green" organizations argue that it will foster a lot more innovation in chemical substitution at a nominal cost while the companies claim it will be economically harmful. Only time will tell who is more correct, but the fact remains that REACh is the law of the land, and in a global economy it will affect everyone.

14.3 First Comes Assessment and Modeling

If REACh requires a comprehensive and scientifically valid and rational evaluation of human exposure to substances, then modeling is an indispensable element of that assessment. Since it has been established that this level of comprehensive assessment of human exposure to chemicals has not occurred and that the vast majority of exposures have not been systematically and proactively addressed, a considerable amount of scientific model development needs to occur. For example, personal chemical exposures to consumers from predominately residential sources have been ignored by most chemical modelers. Given this commitment within REACh to assessment, exposure model utilization and development should come to the forefront.(4)

Exposure models are critical to any commitment for comprehensive exposure assessment because we will never be able to monitor or measure every exposure everywhere. The need for models is particularly acute with REACh because it increases proportionally with the growing universe of chemicals under consideration. Also, as a technical expert, an industrial hygienist has a critical need for objective and rational scientific tools for analysis. These facts necessitate both continuing education in and the ongoing scientific development of the discipline.

To comply with this landmark legislation, non-EU companies wanting to do business in Europe must first assess their current and planned chemical substance volumes that could be impacted by REACh. Indeed, concurrent efforts need to be under way that will refine tools and standardized approaches for:

- Prioritizing chemical/exposure scenario (ES) combinations
- Screening/quantifying all routes of human exposure
- providing questionnaires to gather information from end users
- interfacing health effect benchmarks from toxicologist with estimated exposure from exposure assessors

Specific research needs relative to exposure model development and validation have been outlined and presented(5); unfortunately, to date the authors are unaware of any concerted and coordinated effort to follow up on these recommendations.

Also, tools are being developed and forwarded for the evaluation and prioritization of large numbers of chemicals within the context of occupational and consumer product use.(6)

14.4 Conclusion

Stripping away all the complexity, the "bottom line" for those responsible for conducting risk assessments under REACh is to provide enough quality information about any substance's exposure pathway(s) and toxic effects in the real world to pass the "red face" test of accountability. The size of the challenge is clearly daunting and the need for tool development is critical. However, if properly addressed, the end result will be a more rational and proactive approach to human health risk assessment from chemicals.

14.5 References

1. **Europa:** What is REACh? http://ec.europa.eu/environment/chemicals/reach/reach_intro.htm. [Accessed January 9, 2008].
2. **Ex-European Chemicals bureau (European Commission Joint Research Centre):** Explanation of REACh. http://ecb.jrc.it/REACH/. [Accessed January 9, 2008].
3. **U.S. Environmental Protection Agency (EPA):** Human Health Research Program Review: A Report of the U.S. EPA Science Advisory Board, Final Report of the Subcommittee on Human Health, EPA Office of Research and Development. May 18, 2005, revised July 18, 2006. http://www.epa.gov/osp/bosc/pdf/hh0507rpt.pdf. [Accessed January 9, 2008].
4. **Jayjock, M.A, C.F. Chaisson, S. Arnold, and E.J. Dederick:** Modeling framework for human exposure assessment. *J. Exp. Sci. Env. Epidemiol. 17*:S81–S89 (2007).

5. **Kephalopoulos, S,. A. Arvanitis, and M.A. Jayjock (eds.):** Global CEM Net Report of the Workshop no. 2 on "Source Characterization, Transport and Fate", Intra (Italy), June 20–21, 2005. Available online: http://www.jrc.ec.europa.eu/pce/documentation/eur_reports/Global%20CEM%20Net%20Workshop%202%20SOURCES.pdf [Accessed April, 23, 2009].

6. **Jayjock, M.A., C.F. Chaisson, C.A. Franklin, S. Arnold, and P.S. Price:** Using publicly available information to create exposure and risk-based ranking of chemicals used in the workplace and consumer products. *J. Exp. Sci. Env. Epidemiol.* (2009). [In Press].

Computational Fluid Dynamics Modeling

15

By
T. Renée Anthony, PhD, CIH, CSP

15.0 Overview

This chapter changes gears from previous modeling chapters where contaminant concentrations are estimated using chemical properties, such as vapor pressure and evaporation rate, and simple estimates of room conditions, such as room dimensions, air exchange rates, and diffusion rates. Computational fluid dynamics (CFD) modeling uses first principles to solve both equations of fluid flow and contaminant transport for a computer representation of the physical space. When using CFD, estimates of contaminant concentration anywhere within the modeled space can be generated.

As we know, exposure is a function of contaminant generation and transport. CFD modeling focuses on fluid transport, the second half of this exposure equation. The objective of CFD is to represent the behavior of a continuous fluid, in this case air, as discrete elements and to numerically approximate solutions to fluid dynamics equations (partial differential Navier-Stokes equations). Not only can air motion be simulated, but contaminant transport and thermodynamic gradients can be evaluated to examine environmental exposure potential.

CFD techniques have historically been used to examine flow and drag to improve the design of automobiles, airplanes, and engines. Because fluid dynamics is ideal for studying the transport phenomenon, CFD tools have been adopted by environmental health and industrial hygiene engineers to investigate environmental flow and to investigate occupational ventilation in order to evaluate contaminant transport. As computing resources improve and costs decline, the speed and complexity of simulations has allowed researchers to examine more realistic representations of workspaces and associated human interactions. The end of this chapter guides readers to a snapshot of publications that have used CFD to investigate contaminant transport and exposures.

CFD modeling is useful in many aspects of exposure investigation. Control decisions to minimize exposures can be optimized using accurate exposure estimations given a physical description of the geometry and flow conditions. CFD is also useful in projects targeted at improving control solutions: changing parameters, either physical geometry or flow, can allow for what-if analysis of operating conditions and allow for ranking of alternative solutions. In addition, the flow visualizations generated from CFD simulations are ideal to share with employees to communicate the effects of proper and improper operation of control technologies. Finally, CFD can be useful to aid in exposure assessments. As with other modeling methods, the generation rate of a contaminant is still required as an input variable, but the transport of contaminants through a workplace or through a control system is well represented with proper selection of CFD equations and constants.

The goal of this chapter is to provide an overview of the CFD process. This text is not intended to provide exhaustive detail of CFD methodology, but will provide basics to allow readers to understand the benefits and costs associated with CFD modeling, comprehend CFD articles that may apply to a process at one's own plant, and evaluate whether CFD simulations may be beneficial in engineering the control of exposures. Details for additional reading for more detailed information are provided at the end of this chapter. This chapter primarily focuses on Reynolds-Averaged Navier-Stokes (RANS) models, as commercially available software is available for these methods. Other computational models, including Large Eddy Simulations (LES), Detached Eddy Simulations (DES) and Discrete Vortex Methods (DVM) are available but are beyond the scope of this chapter.

15.1 CFD Basics

A CFD model requires users to develop a computer representation of the physical space of interest: this space is referred to as the *computational domain*. The air within this space is divided into smaller, discrete volumes by generating a "*mesh*" of intersecting lines: nodes are defined as the position where these lines intersect. The known velocities into or out of this space are assigned, such as the velocity at the ventilation return duct on a wall of the room. For all solid surfaces within the room, zero velocity is assigned. Finally, an initial guess must be assigned to all other nodes within the space.

Note that the Discrete Vortex Method (DVM) is an exception to the concept, above. Instead of treating the flow field as discrete elements within a volume, the DVM is a time-stepping algorithm in which vorticity is descretized and transported by the velocity field that it induces. DVM is *Lagrangian* in approach, versus the *Eulerian* approach of the RANS model discussed in this chapter.

Next, a system of equations must be selected and applied to each node within the space. These equations include the continuity equation (conservation of mass) and the Navier-Stokes momentum equations. If thermal effects are important to the problem, an energy equation is also attached to each node. Although these equations are derived from first principles, the methods used to solve them are approximations, as assumptions are used in solving the partial differential equations numerically.

These equations are solved at each node, with up to 8 unknowns per node within the computational domain. The 8 unknowns are referred to as "degrees of freedom," consisting of horizontal velocity (U), lateral velocity (V), vertical velocity (W), turbulence kinetic energy (κ), dissipation of turbulence kinetic energy (ε), pressure (P), temperature (T), and concentration (C). Simplifications to the model could include the following:

- An isothermal study (no variation in temperature) would eliminate the energy equation and, hence, temperature (T) in the solution matrix.
- A simplified two-dimensional representation of the physical space would result in only two of the three velocity components.
- Simulations can uncouple the concentration and velocity field for separate analyses, eliminating the concentration (C) from the fluid simulations, requiring post-processing to evaluate concentration transport from a source.
- Laminar instead of turbulent equations can be used in few cases, and laminar simulations result in the elimination turbulence kinetic energy (κ) and dissipation of turbulence kinetic energy (ε).

Steady-state analyses report time averaged results. In some cases, time dependency is of interest and solving accurate velocity field at each time step greatly increases the computational resources needed to simulate the fluid dynamics.

Typically, realistic simulations will include all three dimensions of velocity, turbulent flow patterns, and pressure, at a minimum, for 6 degrees of freedom (U, V, W, κ,ε, and P). The solution relies on matrix algebra defined by:

$$\mathbf{A}x = b \qquad (15\text{-}1)$$

where

$\mathbf{A}$ = the coefficient matrix
x = the matrix of unknowns
b = the boundary condition matrix

The size of the **A** matrix is:

(# nodes x # degrees of freedom) by
(# nodes x # degrees of freedom).

For a physical space represented by 1,000,000 nodes and 6 degrees of freedom (3-dimensions of velocity, turbulence, isothermal and uncoupled concentration), the size of the **A** matrix is $36\text{x}10^{12}$. Indeed, this solution matrix is huge, likely exceeding the computation resources of today.

To make these solutions tractable, finite element and, later, finite volume methods have been developed in which only information from the neighboring nodes are used to solve the equations. Mathematically, this places a lot of zeros in the **A** matrix, above, simplifying the computational storage requirements and making even very refined meshes easily solvable by today's computers. In addition, there are tools available to further reduce the size of the matrix by decoupling the simultaneous solution of all degrees of freedom. The popular *segregated solver* decouples pressure from the above solution matrix, solving pressure by enforcing the mass balance equation after other components are solved. This also reduces the size of the matrix and, hence, improves the ability to solve the equations at moderate computational expense.

Because these methods require an iterative process to solve the equations, users of CFD software must determine when the solution is converged to an acceptable tolerance. The computer generates estimates for each degree of freedom from each solution iteration and compares them to the previous system of solutions. The differences between these two solutions are typically reported in terms of error norms, discussed in detail later. Typically, users specify a tolerance level for this value (such as 0.001, the default in most packages). When this error norm, over all nodes in the simulation, reaches this predetermined value for all degrees of freedom, calculations cease and the simulation ends, yielding estimates of all degrees of freedom at all nodes within the physical space modeled. Evaluation of iterative convergence and mesh independence are required to ensure that the system of equations have been solved correctly.

Finally, as with any model development, the model requires validation with experimental data or, rarely available, analytic solutions. This requires comparing your estimates with experimental measurements and determining whether the two agree, within the uncertainty inherent in both methods. This answers the question of whether the correct equations were solved, providing insight into whether the initial simplifications or the selected representation of the flow conditions were adequate.

15.2 CFD Modeling Sequence

The sequence of steps for CFD modeling are:

1. **Define the fluid dynamic problem:** Identify a conceptual simplification of the real world problem.
2. **Generate the computational domain:** Construct physical geometry and fluid mesh; consider three meshes needed to assess mesh independence.
3. **Formulate a well-posed problem:** Select appropriate equations, boundary conditions and initial conditions.
4. **Solve the equations:** Select the most appropriate and accurate method to solve your flow field, to successively decreasing tolerance levels and multiple mesh densities.
5. **Assess convergence:** Verify that equations were solved correctly; examine iterative tolerance and mesh independence, at a minimum.
6. **Validate the model with data:** If convergence is assured, determine whether the correct equations were solved: do the estimates agree with reality?

Although the process appears to be linear, you may find that you have to revisit earlier steps along the way. For example, if you are modeling wind tunnel experiments, in Step 2 you have placed the exit of the computational wind tunnel in the same position as the real wind tunnel, and in Step 3 you specified the boundary condition of "no acceleration" through the back of the wind tunnel; however, in Step 4 you realized convergence problems because your problem wasn't well posed: you need to revise Step 2 and extend the computational domain.

The last two steps in CFD modeling require reevaluation of your previous work. If convergence is not achieved, you need to examine whether your mesh is simply too coarse or examine whether you selected an inappropriate model, e.g., used a high Reynolds number when a low Reynolds number model is more appropriate. Changing models may now require a different mesh density to adequately solve the equations. If convergence is achieved (Step 5) but does not reflect experimental values (Step 6), users need to consider whether the conceptual simplification was too big of an abstraction from reality (Step 1) or whether some numerical assumptions were inadequate (Step 3).

To follow are more detailed descriptions of each step in this process.

15.2.1 Step 1: Define the Fluid Dynamic Problem

Defining the problem requires more than just coming up with a statement. Users must consider the problem they are interested in solving and identify how the complexity associated with the real-world problem can be simulated in the computer. Simplification of the problem requires the user to identify the key parameters to represent the computational domain: can we simulate the problem with two dimensions or will the model require three dimensions (a more difficult and time-consuming solution); can we solve steady-state equations or is the time-dependent phenomenon significant; can we make simple assumptions regarding flow into and out of our system or do we need to develop subroutines to more accurately reflect a flow profile? Often, the physical geometry is considered first, but understanding the set-up of the equations you will use to solve the fluid flow field often requires adjustments to the physical representation. Example 15.1 illustrates the discussions involved in defining the computational abstraction of a real-world scenario.

Considerations for defining the computational abstraction include the following:

Geometry:
How much detail is actually needed? Can we highly refine our geometry in areas of interest and simplify elsewhere?

2D or 3D:
Is a two-dimensional representation sufficient or is the dynamic behavior of the air or contaminant a three-dimensional phenomenon? 2D problems are easier to solve, but is this an oversimplification of a real world situation?

Time-dependent:
Is steady-state simulation sufficient or are temporal changes important?

Effect of Contaminant on Air:
Can you apply the assumption that the contaminants do not induce changes to the air?

Flow Regime:
Is the Reynold's number high (turbulent) or low (laminar)? This will affect not only the equations you use, but the thickness of your boundary layer surrounding solid surfaces.

Validation Data:
Can you obtain experimental data to validate your model? If so, what are the experimental conditions you need to match in your model?

Particle Simulation Simplifications:
Are particle bounce and gravitational effects important (large particles in low velocity air)? Is laminar particle trajectory sufficient, or do you need to assess variability and simulate turbulent particle motion?

Example 15.1

You want to look at flow past a human torso. Specifically, you want to know if concentration builds up in the near wake when a person stands with her back to the oncoming wind. What simplifications can you consider?

Discussion:

(1) A human torso is very complex, with head, torso, arms, and legs. You decide to model the human with a simple geometry. A circular cylinder is presented in literature, but you think that the shape of the human is more elliptical, with persons being wider shoulder to shoulder than from front to back.

Benefits: Simple to generate the shape.
Possible Problems: Air moving between the torso and arms may contribute to turbulence in the wake not accounted for in the simple model.

(2) Using the elliptical cylinder, you now need to decide if what you want to study can be modeled in two dimensions or if you need to model in 3D. Figure 15.1a illustrates the two-dimensional model, where air can only slow down or turn laterally as it approaches the cylinder; this 2D simulation would represent flow past an infinitely long cylinder. Figure 15.1b illustrates the three-dimensional model, where air can slow down, move laterally or move upward / downward. Airflow past a person may have downwash as it moves over the head, and this may change the wake concentration near the mouth, where we really are interested.

Figure 15.1a: 2D Figure 15.1b: 3D

Benefits: Easier to solve a 2D problem, but 3D better reflects what we want to evaluate.
Possible Problems: Top of human head is rounded: this may affect flow and result in differences from real world.

(3) A human with her back to the wind will be breathing in the near wake of the head. Will you include breathing and, if so, how will you simulate it? What mouth opening shape is appropriate? Will breathing be represented by continuous inhalation or a cyclical breathing pattern? You decide to position an elliptical orifice to represent the mouth on the cylinder and have it continuously inhale at a moderate breathing rate. This way, you can use steady-state equations (no time dependency).

Benefits: Mildly addresses breathing impact on the system, but not completely. You want to look at steady-state simulations, so no time-dependent breathing or air flow in the model.
Possible Problems: Representation of vortex shedding behind the cylinder and cyclical breathing indicate true time-dependent conditions that are not well represented with the model. Differences between modeled estimates and experimental studies may be due to non-breathing time-dependencies, even if the experiment has continuous inhalation.

These are only a few of the decisions to be made in Step 1, but documentation of the implications of your decision is recommended.

15.2.2 Step 2: Generate the Computational Domain

In order to simulate air and contaminant transport throughout a room or to a breathing zone with a RANS model, the next step requires generating a computer representation of the space through which this air travels. For this discussion, a simulated "room" is used to describe our geometry. Obviously, this computational domain need not be limited to the workroom discussed here.

The simulated room can be a two- or three-dimensional space, such as that expressed with four walls, a floor and ceiling. Objects can be placed within this computational room, including walls, equipment, or persons. All of these types of objects are solid surfaces or "walls," through which no air is allowed to pass. Sources of ventilation, either exhaust or supply, can be indicated within the room: these physical structures are not solid but allow air to pass into or out of the room through these openings. Typically, vertices are used to indicate the relative position of each "edge" of surfaces within our room. Vertices are connected by lines, and lines are connected to define surfaces. In three-dimensions, such as our room, surfaces are connected to form volumes.

Defining the surfaces within our model gives us an idea of where, for the most part, air is NOT allowed to pass through. Next, we have to subdivide the volume of the room where air is in motion. This is accomplished by using a routine to first generate a two-dimensional (2-D) grid on the surfaces of each object used to define the room. Finally, these surface grids are then projected into the three dimensional space that defines the air. Where these grid lines intersect, a "*node*" exists: it is at these nodes where the equations of fluid flow are solved. Example 15.2 provides readers with an illustration and discussion of a computational domain generated to solve a specific problem.

Once the volume that defines our fluid is generated and meshed, the positions of the nodes needed to solve the equation are established. We then identify our system of equations and apply boundary or initial conditions to each of these nodes. The equations and conditions we select must formulate a well-posed problem.

Example 15.2

We want to simulate particulate sampling on a person's torso. A 2D approximation was used to generate the geometry and mesh using an elliptical cylinder with a sample cassette positioned on the front. We want to examine room velocities that represent workplace conditions (0.4 m s^{-1}), but the velocity is much larger near and within the sampler. Hence, a graded mesh is needed to model this system, illustrated by Figure 15.2a.

Discussion: Figure 15.2a is a close up of the mesh illustrating the torso. A quadrilateral meshing scheme was used (4-sided elements), but due to the complexity of shapes, a paving algorithm was used. A mapped mesh results in a uniform grid throughout the area; a paved mesh contains less regular quadrilateral elements, as seen at the torso edge in the close-up Figure 15.2b. Especially when paving a mesh onto a surface, modelers should evaluate the quality of the mesh to eliminate highly skewed mesh elements.

The physical walls of the domain are beyond the image captured, as this is a fairly dense grid.

The blacker the region, the denser the grid: hence, the darkest region is in the area of the sampler, where we are most interested in the flow. Figure 15.2 is a closer look at the mesh, now focused on the sampler. The inside area of the sampler is fairly refined, necessary to address the high velocity through the sampler itself. To the left of the sampler, the mesh transitions from coarser to finer as you move from freestream to the sampler. The moderately refined mesh in the rectangular area just upstream of the sampler was needed to control mesh refinement.

15.2.3 Step 3: Formulate a Well-Posed Problem

A well-posed problem requires that a solution exists and that it is unique. There is yet to be proof that Navier-Stokes equations in three dimensions and high Reynolds numbers have a unique solution. Other requirements are (1) the solutions depend continuously on boundary conditions and (2) small changes in the initial conditions result in small changes to the resulting solution.

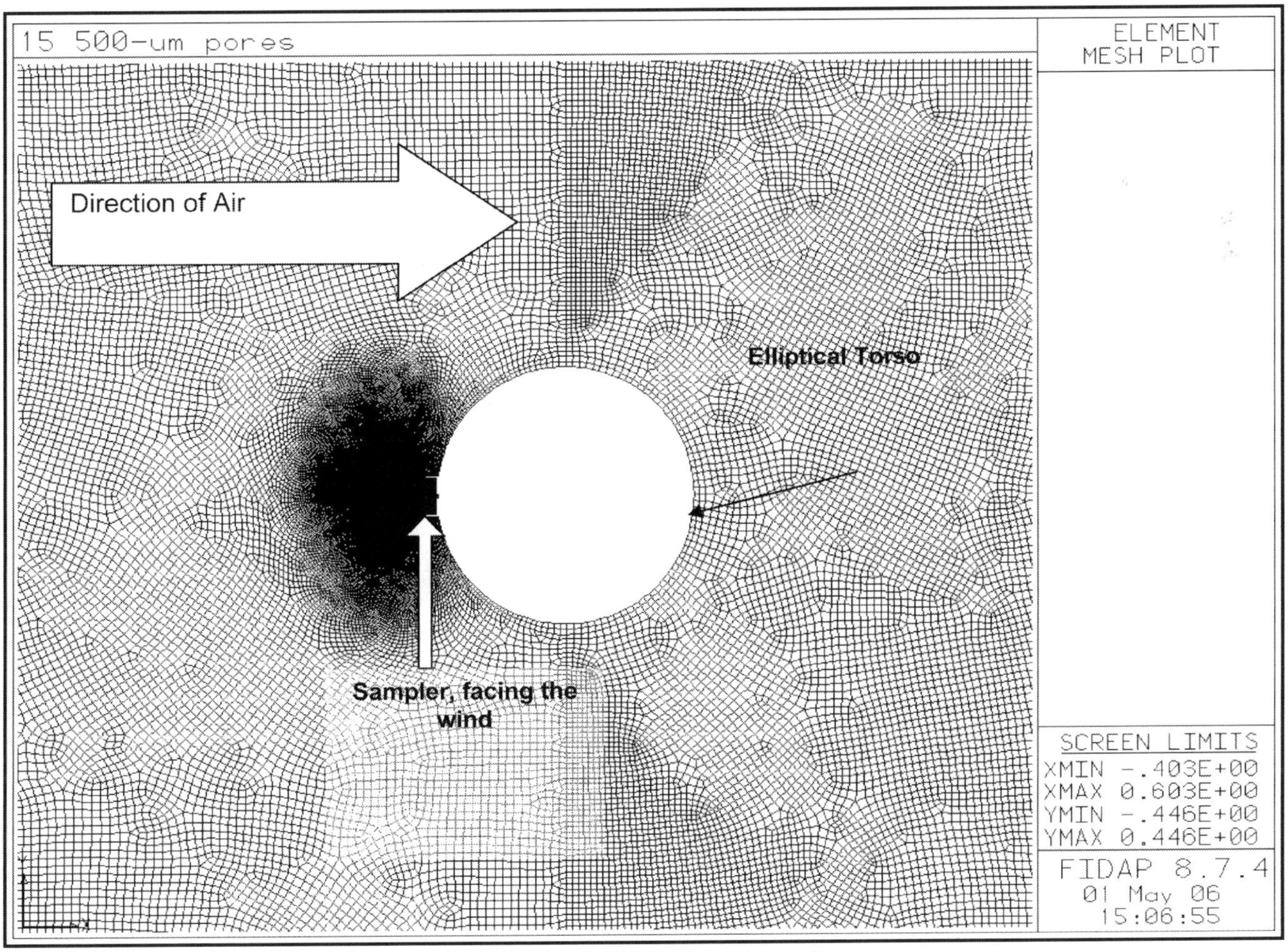

Figure 15.2a — Example of mesh surrounding the elliptical cylinder and sampler.

The formulation of a well-posed CFD problem requires the selection of the fluid dynamic model, boundary conditions, and initial conditions, all of which are necessary to run the subsequent simulation. Without simplifications, the full version of continuity and momentum equations are given as:

$$\text{Continuity: } \partial \frac{\partial}{\partial t}(\rho) + \nabla \bullet (\rho u) = 0 \quad (15\text{-}2)$$

$$\text{Momentum: } \frac{\partial}{\partial t}(\rho u) + \rho u \bullet \nabla u + u(\nabla \bullet \rho u) = \nabla \bullet [\Pi_{ij}] + S \quad (15\text{-}3)$$

These equations include density (ρ), velocity in vector notation (u), and time (t). The term ∇u represents the velocity gradient; the $\nabla \bullet$ operator indicates the "divergence" of the term that follows. The $\prod_{ij}$ term represents the stress tensor, and S indicates the surface forces, such as gravity (ρg). Note that the equation provided in (15-3) is a set of *three* momentum equations, one for each component of velocity. Further explanation of vector calculus needed to describe these equations in full, which is beyond the scope of this chapter. An excellent explanation is given by Schey.[1]

The complexity of these equations is reduced when flow conditions are simplified. For example, the condition of incompressible flow reduces the continuity equation to:

$$\nabla \bullet u = 0 \quad (15\text{-}4)$$

When flow is incompressible but we allow for turbulence, the continuity equation in (15-2) is used along with the simplified momentum equations from (15-3):

$$u \bullet \nabla u = \nabla \bullet \left[\frac{1}{\rho}(\mu + \mu_t)(\nabla u + \nabla u^T)\right] - \frac{1}{\rho}\nabla P \quad (15\text{-}5)$$

If we have incompressible flow and constant density, viscosity, and thermal conductivity, we can add the energy Equation (15-7) to our system of equations:

$$\rho c_v \frac{\partial T}{\partial t} + u \bullet \nabla T = \kappa \nabla^2 T + \Theta \quad (15\text{-}6)$$

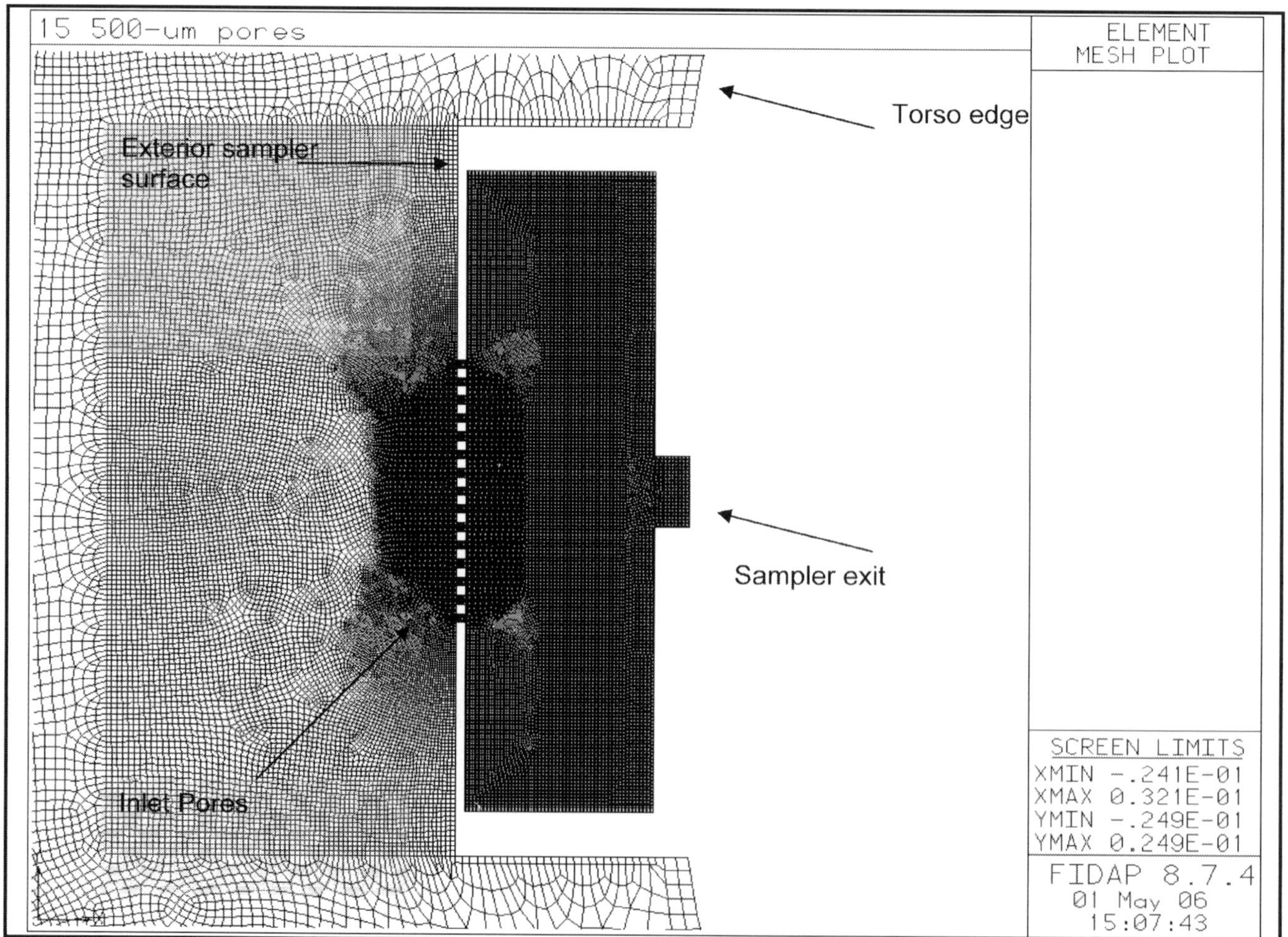

Figure 15.2b — Close-up of the sampler mesh.

Commercially available CFD software allows users to select, apply, and solve appropriate versions of these partial differential equations. Although the options and terminology may vary with software package, considerations of flow type are necessary to properly set-up and run the simulation of fluid flow. The level of detail provided in the following discussion may seem daunting, but it gives readers insight into the complexity of decisions needed to set-up the computational problem. Users should understand that although CFD software is available off-the-shelf, using the software is not as simple as selecting a few parameters and running it. Selection of one parameter to define the type of problem you wish to solve will require additional choices in assumptions and methods later on. Standard recommendations are provided for *k*-epsilon turbulence modeling, but users must know that particular problems associated with their fluid dynamic problem may be better solved with other equations, assumptions, and solvers.

Problem Type

Determining your "problem type" results in selecting what simplifications can be made to the continuity and momentum equations. These decisions go hand-in-hand with your decisions from Step 1 in the CFD modeling process. Simplifications should be specific to the flow conditions you are modeling. A list of considerations is given, to help you understand how much detail you need to know about your flow.

Selection guidance is provided for generic exposure modeling scenarios.

Steady-state vs Transient: For steady-state analyses, the time-derivative terms in each equation are discarded. For transient systems, time integration schemes must be selected.

Incompressible vs Compressible: Is density constant? Incompressible assumptions can be applied when Mach numbers (velocity/speed of sound) are less than 0.3 (velocities are less than 100 m s^{-1}). In most exposure simulations, we are dealing with incompressible flow, with the possible exception of scenarios with high velocity jets (e.g., spray painting).

Laminar vs Turbulent: Specify whether laminar equations (turbulence kinetic energy and dissipation of turbulence kinetic energy are not solved) or turbulent equations are to be applied. More discussion on selecting the appropriate turbulence model is provided later.

Newtonian vs Non-Newtonian: Air and water are Newtonian fluids. Blood flow through veins is an example of a non-Newtonian fluid, and a non-Newtonian viscosity model must be selected.

Momentum vs No Momentum: The default is typically for the momentum equation to be solved. If you are solving a heat transfer or a multi-species diffusion problem, you may not need the momentum equations; you would, however, be required to select at least one other transport equation, such as energy.

Isothermal: If the system has no temperature-induced buoyancy effects, the energy equation is not solved. A comprehensive examination of contaminant transport near human body in low flow environments would involve investigating the effect of temperature gradients and would not be isothermal.

Free vs Fixed Surface: If the system has a free surface, such as investigations of open channel flow of water, a melting surface, or a liquid-vapor interface within the domain, there is a free-surface boundary. Algorithms to solve these equations must incorporate either a local perturbation method onto the surface, a method of spines to generate line guides for the free surface, or other methods that introduce an additional degree of freedom into each node on the free surface. Most de-coupled industrial hygiene problems can be accomplished using fixed surface.

Single or Two-Phase Flow: This is different from studying flows of two species of fluids within a room. Rather, two-phase flow modeling is required when there is a non-negligible transfer of momentum, heat, and/or mass transfer between two phases in the model. To solve this, Eulerian (each phase is treated as a continuum and is modeled at each point of the computational domain) or Lagrangian (incorporation of additional particle tracking equations into the simulation) techniques have been developed. In many industrial hygiene cases, however, the fluid can be solved separately as a single phase, with particles or contaminants solved later using two-phase Lagrangian techniques. This requires the assumption that the dispersed phase does not affect the conveying fluid. This is a reasonable assumption when the dispersed phase is very dilute, the temperatures of the two are the same, and when the interfacial forces between the two are negligible.

Turbulence

When turbulent flow is modeled, users are attempting to represent a random, three-dimensional motion of "fluid particles" superimposed on the mean motion.[2] Figure 15.3 illustrates a representation of turbulence in the velocity term: there is a mean value for the velocity ($\bar{u}$) and a random component at time *t*, indicated by u', which indicates the fluctuation from the mean velocity. In Figure 15.3, the value of u' for the last measurement of velocity is illustrated. For each time velocity was measured, there is a unique value of u'. Measurements of this fluctuation in velocity around the mean are often reported in terms of *turbulence intensity*, which is a measure of the root mean square of velocity normalized by the mean velocity:

$$TI = \sqrt{(u')^2} / \bar{u} \qquad (15\text{-}7)$$

To represent turbulence, we need to add Reynolds stress terms to the momentum equations (15-3). An example of the x-momentum equation from equation (15-3) is provided as:

$$\rho \frac{d\bar{u}}{dt} = -\frac{\partial p}{\partial x} + \frac{\partial}{\partial x}\left[\mu \frac{\partial \bar{u}}{\partial x} - \overline{\rho u'^2}\right] +$$

$$\frac{\partial}{\partial y}\left[\mu \frac{\partial \bar{u}}{\partial y} - \overline{\rho u'v'}\right] + \frac{\partial}{\partial z}\left[\mu \frac{\partial \bar{u}}{\partial z} - \overline{\rho u'w'}\right] \qquad (15\text{-}8)$$

The new terms (e.g., $\overline{\rho u'^2}$) are referred to as Reynolds stresses that result in increased transfer of momentum due to correlations in velocity fluctuations. Boussinesq method is by far the most commonly used method to represent turbulent stresses, but Speziale and Launder methods estimate normal turbulent stresses better, with Launder providing better estimates of turbulence in some cases of swirling flow.

Once the eddy-viscosity model is selected, users must select the appropriate turbulence model. Table 15.1 highlights common two-equation turbulence models available. By far, the standard k-epsilon model is the most popular, but modelers should consult with literature to understand known limitations of this model before selecting it as a default (Richmond-Bryant, et al., 2006).

By specifying the type of fluid in your model (e.g., incompressible, Newtonian), its flow conditions (single phase, fixed surfaces, isothermal, momentum, and turbulent), and the methods you want to use to represent turbulence (Boussinesq eddy-viscosity method and standard *k*-epsilon two-equation turbulence model), users now have to apply conditions to the boundary and initial guesses to all other nodes within the computational space to complete the formulation of a well-posed problem.

Boundary and Initial Conditions

This portion of Step 3 requires users to assign conditions to all surfaces in the computational domain.

Typical boundary conditions include:

- No slip at walls: all velocity components are zero at all nodes
- Plane of symmetry: zero normal velocity; tangential velocity allowed
- Conditions at all inlets: usually normal velocity is the known value; tangential velocities are set to zero (or some other known value)

At boundaries where no velocity is specified, the first derivative of that value ("acceleration") is set to zero. Simulations attempt to solve for the values of velocity at these unspecified planes.

Table 15.1 — Two-equation turbulence models

Model	*Variants*	*Comments*
k-epsilon	Standard	Uses standard constants. It tends to predict flatter flow profiles for swirling flows and shorter re-circulating zones in separated flows. For flows with strong anisotropy, use Speziale or Launder relationships.
k-epsilon	Extended	Updated constants, claims of better predictions for complex flows with recirculation, streamline curvature, swirl.
k-epsilon	RNG (Renormalized group)	Includes the ratio of characteristic time scales of turbulence and the mean flow fields in the dissipation equation.
k-epsilon	RNG revised	Same as RNG, but with different coefficients; better than RNG-original with recirculation; less diffusive than original RNG.
k-epsilon	Anisotropic	For flows of strong curvature, swirl and body force. (Cannot use Boussinesq relation, only Speziale or Launder.)
Wilcox low-Re *k*-omega	—	Uses turbulent frequency (ϖ) rather than dissipation of turbulence kinetic energy (ε). Used to estimate the variation of turbulence through the viscous sublayers to the wall. Used for *globally* low Re *internal* flow. *Highly* sensitive to free-stream values of ϖ.

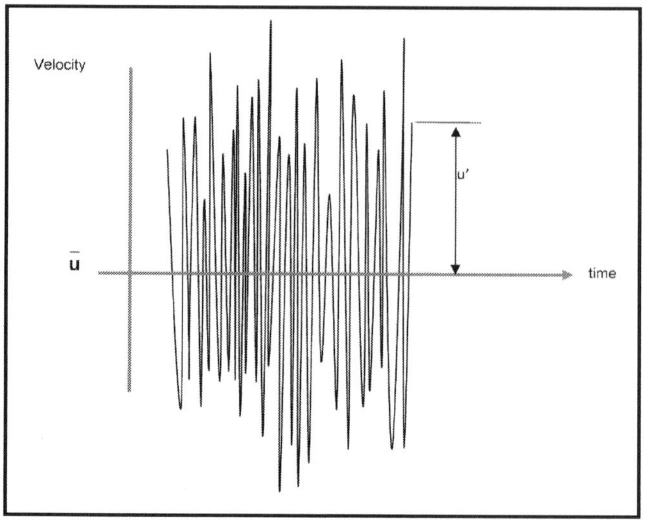

Figure 15.3 — Illustration of Turbulence

Care should be exercised when assigning velocity values to *all* sources or sinks within a domain. For instance, consider the simulation of a room with two air supplies and one exhaust, with no other air entering or leaving the system. If you assign specific values to all three surfaces, you run the risk of violating the mass conservation equation, partly because of the flow rate and mesh dependency described above. It is best to specify one surface as a "no acceleration" surface and allow the velocity through that surface to be computed. In addition, enforcing a uniform velocity profile across an exhausting duct may not be the best representation, such as with an air exhaust in the corner of a room: enforcing no-acceleration through that surface will allow for an appropriate velocity distribution to be developed into that exhaust duct.

The above discussion focused only on "velocity", but in fact you need to consider *all* degrees of freedom when assigning boundary conditions. In is important to realize that, when solving the flow field, the boundary conditions you assign *do not change* during the iterative process used to solve the system of equations. If you want to change any boundary condition, you must change it in the code and re-solve your system of equations.

The final step before simulations can begin is to provide an initial guess for all nodes within your system. Often, some mean value of air flow can be assumed for your domain. Example 15.3 illustrates a well-posed problem for the geometry identified in Example 15.2.

Example 15.3

We wish to formulate the well-posed problem to evaluate particle transport and determine particle concentrations for the model developed in Example 15.2.

Discussion: Equations that will be iteratively solved to generate the air simulation were the steady-state, incompressible, turbulent, Navier-Stokes equations:

$$\nabla \cdot u = 0$$

$$u \cdot \nabla u = \nabla \cdot \left[\frac{1}{\rho}(\mu + \mu_t)(\nabla u + \nabla u^T) \right] - \frac{1}{\rho} \nabla P$$

The Boussinesq constitutive relationship will be used to model the Reynolds stresses, with the eddy viscosity defined by:

$$\mu_t = \rho c_\mu k^2 / \varepsilon$$

The standard two-equation k-epsilon model equations will be used:

$$\rho(u \cdot \nabla k) = \nabla \cdot (\mu + (\mu_t/\sigma_k)\nabla k) + \mu_t \Phi - \rho\varepsilon$$

$$\rho(u \cdot \nabla \varepsilon) = \nabla \cdot (\mu + (\mu_t/\sigma_\varepsilon)\nabla \varepsilon) + c_1 \frac{\varepsilon}{k} \mu_t \Phi + c_2 \frac{\varepsilon}{k} \Phi - c_2 \rho \frac{\varepsilon^2}{k}$$

where Φ is the viscous dissipation function and the standard constants are:

c_μ=0.09, σ_k=1.00, σ_ε=1.30, c_1=1.44, c_2=1.92

For fluid simulations, all surfaces will be assigned the no-slip condition (zero normal and tangential velocity). The initial conditions will match the wind tunnel freestream velocity, namely 0.4 m s^{-1} in the streamwise direction; zero lateral velocity; turbulence kinetic energy and dissipation of turbulence kinetic energy will be based on 8% turbulence intensity present in the wind tunnel and Ru=10, typical of wind tunnel studies. The inlet to the domain will also be set to these experimental conditions.

The no-acceleration rule will be applied to the exit of the domain. The velocity through the exit of the sampler will be specified to 1.044 m s^{-1} to correspond to a 2.5 L min^{-1} sampling rate.

15.2.4 Step 4: Solve the Equations

After the physical space is discretized in Step 2 and the equations and conditions are assigned in Step 3, modelers must now select the appropriate numerical methods to *solve* their system of equations. Remember that at each node, we need to solve for many degrees of freedom (u_x, u_y, u_z, k, ε, pressure, temperature, concentration, etc.). For small 2-dimensional problems (no u_z), fully coupled algorithms with direct Gaussian elimination are reasonable. However, as discussed earlier, CPU storage becomes an issue with larger matrices associated with 3-dimensional problems. Many algorithms are available, including successive substitution, Newton-Raphson, modified-Newton, quasi-Newton, and a segregated solver. For specific types of flows, selection of appropriate models is limited. For example, free-surface problems can use any of the Newton-based method or segregated solver, but compressible flow problems can only use the segregated solver.

Most CFD exposure studies rely on segregated algorithms, which decompose the large global system matrix into smaller sub-matrices associated with one conservation equation and successively solve the matrices to enforce conservation. Although the storage requirements are *reduced*, the number of iterations to converge to a solution is increased over the fully coupled algorithms. However, there is a tradeoff in increased number of iterations and decreased time to complete iterations due to the smaller matrix.

Because the segregated solver solves each degree of freedom sequentially and separately, different methods to solve *pressure* (or the increment of change in *pressure*) have been developed. A pressure *projection* method solves directly for pressure at each iteration with a mass adjustment of the velocity field at the end of each iteration; a pressure *update* method also solves directly for pressure but only ensures that the continuity equation is solved at convergence. A pressure *correction* method solves for an increment of pressure change per iteration rather than pressure itself. Most commonly, *pressure projection* is the default method for *segregated solvers*.

The above solver tools represent a small fraction of available methods. Consult the software you are using to be clear about what options work with the set of equations you have selected to apply to your model. Because these methods require an iterative process to solve the equations, users of CFD software must determine when the solution is converged to an acceptable tolerance. The software generates estimates for each degree of freedom at each solution iteration and compares them to the previous system of solutions. The differences between these two solutions are typically reported in terms of error norms, such as:

$$L_2 = \frac{[\Sigma\,(f_i - f_{i-1})^2]^{0.5}}{[\Sigma\,(f_i)^2]^{0.5}} \qquad (15\text{-}11)$$

where

L_2 = the error norm
f = the degree of freedom of interest
i = indicator of the ith iteration

Once the coding is complete, you are ready to specify the global solution error tolerance (e.g., 0.001) and the number of iterations to run. The code will run until the first of these criteria are met. The first time you run a simulation, it is a good idea to run *one* iteration and check the results to evaluate your set-up, specifically your initial and boundary conditions. Check for flow through walls, flow through edges that you didn't want flow through, and velocities that you specified in boundary conditions. It is good practice to check the flowrate through surfaces where you assigned non-zero boundary conditions: evaluate whether the nodal velocities should be changed to account for mesh density.

When all degrees of freedom are converged below the specified tolerance limit or the maximum number of iterations is reached, the solution matrix is posted. Graphical user interface options available with most CFD processors are useful to examine the flow information graphically.

Numerical solutions can be extracted anywhere within the computational domain to examine information pertinent to your study. Most software includes graphical user interface options to extract information, but data can be extracted and post-processed directly from the solution database. Consult with software reference manuals to determine the database format before extracting data. The solution matrix contains information for all degrees of freedom at each node within the computational domain: interpolation between nodal points allows for extracting this information at any location within the domain.

Once a solution is obtained, an evaluation of the quality of this solution is required.

15.2.5 Step 5: Assess Convergence

Previously, we discussed establishing a tolerance for the global solution error (GSE) to indicate cessation of iterative computation on the domain. Typical defaults are 0.001 (or 0.1%): when this error norm, over *all* nodes in the simulation, reaches this predetermined value for *all* degrees of freedom, calculations cease and the simulation ends, yielding estimates of all degrees of freedom at all nodes within the physical space modeled. However, this default tolerance may be insufficient: users then re-solve the matrix to successively lower tolerances until solutions in specific areas of interest are not changing with lowered tolerances. *Non-linear convergence* can be assessed by applying error norm calculations to solutions of successively lower tolerance levels. This is often referred to as *iterative convergence* studies. Using an error norm equation similar to equation (15-11), the iterative convergence is examined using an L_2 norm that is summed over positions of interest within your domain:

$$L_2 = \frac{\left[\sum_{\text{position}} (f_{\text{lower}} - f_{\text{higher}})^2\right]^{0.5}}{\left[\sum_{\text{position}} (f_{\text{lower}})^2\right]^{0.5}} \qquad (15\text{-}12)$$

where

L_2 = the error norm
f = the degree of freedom of interest
$_{\text{lower}}$ = indicates the set of data from solution with lower tolerance (e.g., 0.0001)
$_{\text{higher}}$ = indicates the set of data from solution with higher tolerance (e.g., 0.001)

The resulting L_2 indicates of the error magnitude over that set of positions you included in the analysis. An L_2=0.2 indicates 20% difference between the solutions, indicating lower tolerances should be investigated. An L_2=0.02, however, indicates a 2% difference in values between the solutions provided at the lower tolerance versus the higher tolerance limit at the positions you are examining. Reducing the global solution tolerance will likely yield only minimal improvement to the solutions you currently have. Depending on your need for accuracy, these solutions at the lower tolerance limit may be "acceptable," and future mesh densities should be solved to at least this GSE tolerance.

Once the non-linear convergence is assessed, CFD simulations must also ensure that the solution provided is independent of the *mesh density*. Users should ask whether the spacing of nodes affects the solution: if the distance between the nodes is, for example, halved, would the results be different? This question requires at least *three* successively refined meshes to be generated and solved to the same iterative tolerance levels, and compared with a three-mesh error norm, such as

$$R_2 = \frac{[\Sigma\,(f_{fine} - f_{mid})^2]^{0.5}}{[\Sigma\,(f_{mid} - f_{coarse})^2]^{0.5}} \qquad (15\text{-}13)$$

$$R_1 = \frac{[\Sigma\,(f_{fine} - f_{mid})^2]^{1}}{[\Sigma\,(f_{mid} - f_{coarse})^2]^{1}} \qquad (15\text{-}14)$$

Where f still refers to a given degree of freedom but the subscript now refers to the mesh density, from coarsest (the least number of nodes) to the finest (the most number of nodes).

The value of the R_2 error norm, summed over a series of locations of interest in the calculation domain, provides an indication of monotonic convergence over a region ($R_2 < 1$). The value of the R_1 error norm can be used for individual points to indicate whether the location is:

monotonically converging ($0<R_1<1$),
oscillatory ($R_1<0$), or
divergent ($R_1>1$).

See Stern *et al.* for detailed discussion of assessing convergence.[4]

15.2.6 Step 6: Validate the Model with Data

Estimated values for any modeled degree of freedom are available from anywhere within the computational domain. When queried, codes will interpolate results and provide data for positions not on an actual node. Uncertainty in the solution estimates for all degrees of freedom are obtained from *point-wise* evaluation of mesh convergence. Roache provides an excellent reference for the method summarized here.[5]

First, modelers should compare the order of accuracy for the computational method with the order of accuracy indicated from mesh convergence tests. For solutions from three meshes, converged to the same GSE tolerance (e.g., 10^{-5}), use the following equation to compute your order of accuracy:

$$p = \frac{1n[(f_{coarse} - f_{mid}) / (f_{mid} - f_{fine})]}{1n(r)} \qquad (15\text{-}15)$$

where

p = calculated order of accuracy
f = value of the degree of freedom at one location
r = mesh refinement ratio, defined earlier in Equation (15-2).

Subscript refers to the mesh density.

Compare the order of accuracy of your method (e.g., FIDAP's streamline upwinding method is nearly second order accurate, $1 \le p \le 2$).

Next, assess the point-wise three-mesh R_1 values, using Equation (15-13). Only where locations are demonstrate monotonic convergence ($0<R_1<1$), use the following to determine the grid convergence index (GCI):

$$GCI = C\left|\frac{f_{mid} - f_{fine}}{r^p - 1}\right| \qquad (15\text{-}16)$$

where

C = "safety factor."

Roach recommends C = 1.25 in locations where *both* monotonic convergence and correct order of accuracy have been demonstrated. For locations that are monotonically convergent but with the wrong order of accuracy, use a value of 3 as the safety factor. This GCI indicates the uncertainty associated with the estimated degree of freedom:

Uncertainty interval = estimate +/- GCI

Validation of the model with experimental data now merely requires the comparison of the numerical estimate's uncertainty interval with the uncertainty interval associated with the experimental measurement at paired locations.

The goal of both numerical and experimental tests should always focus on minimizing uncertainty. Recent validation studies have run into the problem of generating reasonable representations of the fluid dynamic field computationally but cannot be validated with experimental results due to very small confidence intervals on both measurement and simulated data. As such, traditional least-squares methods have been employed to "validate" the computational estimates with experimental data.[6] Data are fitted to the traditional least-squares form:

$$f_{experimental} = \alpha + \beta f_{numerical} \qquad (15\text{-}17)$$

The shape of the fitted line and the coefficient of determination (R^2) provide a more intuitive evaluation of the model's performance. When compared to the line of perfect agreement ($f_{experimental} = f_{numerical}$), the least-squares fit can indicate whether and where the estimates underestimate or overestimate the experimental data. These evaluations assist the evaluator in determining whether (and where) systematic biases may be occurring in one of the two systems.

15.2.7 Step 7: Post-Processing

If your model solves fluid and contaminant flow separately, you will have to perform post-processing to analyze contaminant concentration data. Work to investigate contaminant transport should only begin after the fluid field is solved, its convergence is assessed, and the results have been validated. The results from your post-processing activities should also include rigorous convergence testing and validation, similar to Steps 5 and 6 of the modeling process.

Contaminant transport in RANS models is typically performed using Lagrangian techniques, where a discrete "particle" is tracked through the computational domain. Gaseous or vapor phase contaminants are typically treated as massless "particles", as these move convectively with the air. Particulates can be assigned density and diameter, and gravitational effects can be evaluated.

Equation 15-18 details the particle momentum equation, commonly solved with a backwards implicit Euler method. The subscript *i* refers to the particle position, the P designates the particle, and τ is the particle relaxation time, which is a function of particle diameter, fluid viscosity, particle Reynolds number (which includes particle velocity term) and coefficient of drag (which is a function of particle Reynolds number).

$$\frac{du_i^P}{dt} = \frac{(u_i - u_i^P)}{\tau} + \frac{(\rho_p - \rho)g_i}{\rho_p} + f_i^P \qquad (15\text{-}18)$$

Simulations require the specification of a time step (how long the particle is allowed to move until another calculation is performed) for a set number of iterations. Sensitivity to the time step and number of iterations allowed per time step should be evaluated to ensure solutions are independent of the settings you use. This requires repeated simulations of the same particle release and a determination of the effect of different time steps. When the concentration, final location of particles, or whatever you are solving the equations to find does not change with successive reductions in the time step and increased number of iterations allowed per time step, independence of the solution to these variables is assured.

Particle simulations can rely on either laminar or turbulent particle trajectories. Laminar simulations of particles result in the exclusion of turbulence on particle behavior: repeated laminar simulations will yield the same results. Turbulent simulations incorporate stochastic model, originally described by Gosman and Ioannides.[7] The instantaneous values of the carrier fluid is combined with a random fluctuation and applied to the particle trajectory equations. Particles are moved through the carrier gas through hypothetical turbulent eddies. The duration over which a particle moves through one of these eddies is limited to the eddy lifetime or the time needed to move a particle through that eddy. After each interaction, a new fluctuation is assumed to act on the particle, resulting in a series different turbulence effects on the particle as it moves through the domain. The motion of each particle is tracked as it interacts with a succession of turbulent eddies. The result of this stochastic approach to turbulent particle tracking is that no two particles, released from the same location and time, will travel the same path. As such, turbulent particle simulations require repetitive simulations to identify a mean and variance of particle concentration. Studies looking at releasing blocks of particles should continue until convergence is assured, knowing that this may take up to tens of thousands of particle releases to assure convergence. Graham and Moyeed provide an excellent reference for determination of the number of particles needed for Lagrangian simulations.[8]

Figure 15.4 illustrates the differences in simulating particles laminarly or turbulently. Figure 15.4a represents particles traveling into an aspirating human head over a 1 cm^2 surface: particle trajectories are laminar. Figure 15.4b represents 20 particles traveling into the same form, but this time turbulent particle motion was simulated from a single release point.

15.3 Examples of Studies

Numerous CFD studies are published in the scientific literature. Table 15.2 provides a brief list of papers that use CFD techniques to investigate some aspect of contaminant transport. The top of the table provides a snapshot of how the models have become more geometrically complex, beginning with simple geometrical shapes to represent a human form to models with complex facial structure to examine particle inhalability. The bottom half of the table lists various studies that used CFD to investigate transport phenomenon with different physical geometries. This table is not a comprehensive review of CFD studies useful to the examination of estimating worker exposures, but is merely provided to give readers an idea of the applicability of CFD tools to evaluating contaminant transport and estimating exposures.

15.4 Usefulness of CFD

Although this chapter has not provided readers with specific tools to perform a CFD analysis for a specific physical problem of concern, it hopefully leaves the reader with an appreciation for the field of CFD modeling and an understanding of the requirements needed to conduct these simulations. The time required to set-up these simulations, solve the equations, and verify and validate the results may take weeks to months, depending on the complexity of the geometry needed in a CFD simulation. This often precludes the use of CFD in work where a quick answer is

Table 15.2 — CFD Studies to Review

Reference	*Description*
Evolution of Human Form Used in CFD modeling	
Tanaka and Murata, 1999[9]	Steady-state flow behind circular cylinders, used to represent worker in high velocity
Okamoto and Sunabashiri, 1992[10]	Vortex shedding behind circular cylinders
Flynn and Eisner, 2004[6]	Validation of wake flow for elliptical cylinder with experimental data
Richmond-Bryant, et al., 2006[3]	Evaluation of near wake estimates with circular cylinder
Kulmata, et al., 1996[11]	Representation of human as series of rectangular shapes
Brohus and Nielson, 1996[12]	Representation of human as series of rectangular shapes, in indoor air velocity
Murakamia, et al., 1999[13]	Rounded human torso simulations with thermal effects
Hayashi, et al., 2002[14]	Rounded human torso, as above, with inhaling mouth
Hyun and Klenistreuer, 2001[15]	More realistic human form
Anthony and Flynn, 2006[16]	Realistic human features used to model particle inhalability
Contaminant Transport in Rooms	
Svidt, et al., 1996[17]	Natural ventilation due to thermal heating in glass enclosure
Park and Holland, 2001[18]	2-dimensional model examining the effect of the vertical location of a convective heat source
Bennet, et al., 2000 and 2003[19,20]	Performance comparisons of CFD simulations to single-zone, uniform diffusivity, and other models
Flynn and Sills, 2000[21]	Investigation of particle transport during spray painting
Rota, et al., 2001[22]	Evaluation of efficiency of push-pull ventilation systems

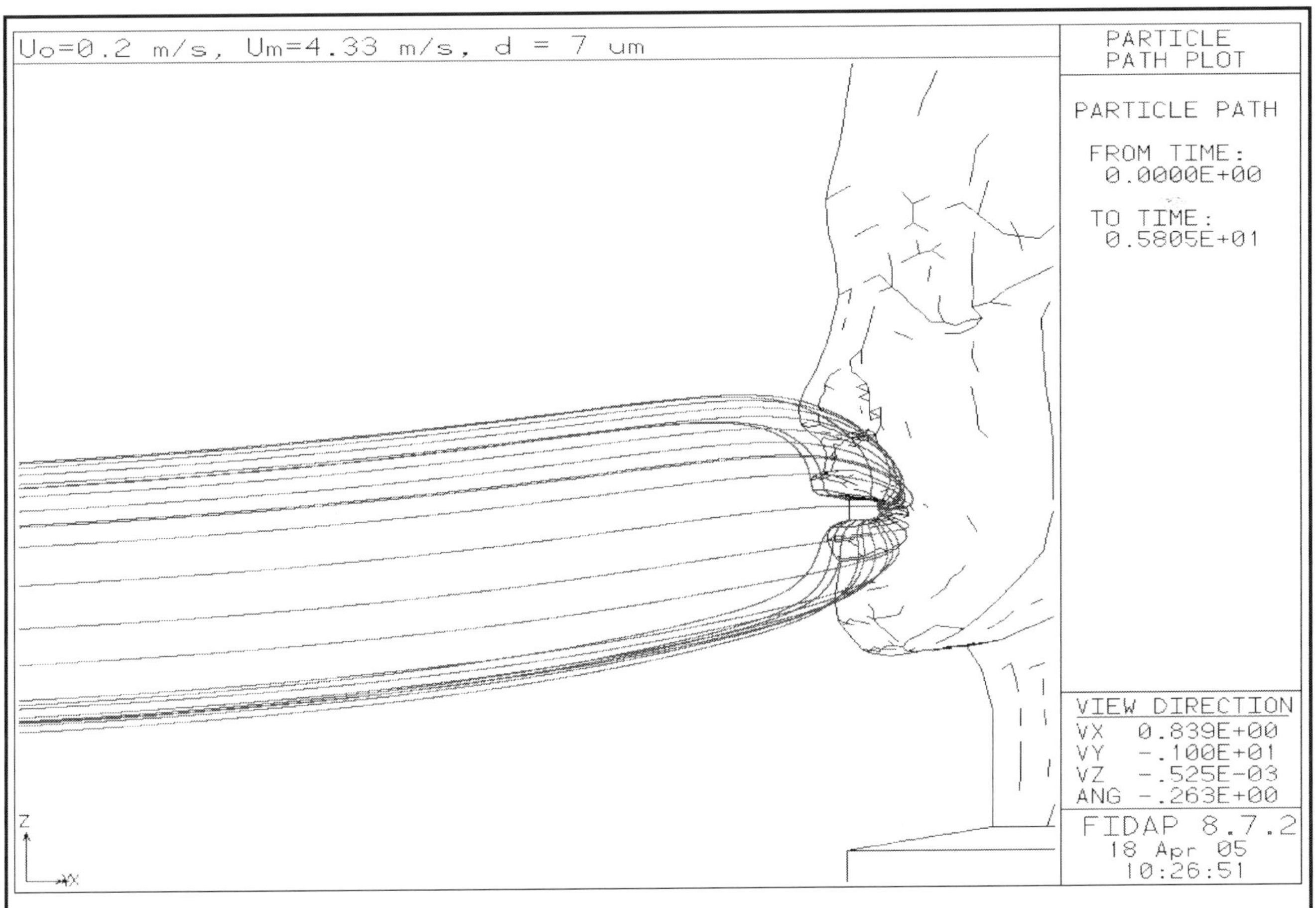

Figure 15.4a — Laminar particle trajectories.

needed. However, these tools may be useful when designing an expensive ventilation systems or new work spaces, determining where to position air supply or exhaust equipment in rooms with complex air motion, or determining how contaminant concentration may vary throughout an existing workplace when other models provide uncertainty intervals that are unacceptably large.

The time required to generate the computer representation of the physical space may take a long time, depending on the complexity of what is being modeled. Solving the flow field and ensuring the results are both verified and validated may also take a long time (e.g., 2-D problems may take up to a week to assess non-linear convergence, whereas 3-D problems may take weeks). However, once the behavior of the fluid is defined, the model can be used to investigate contaminant transport from sources anywhere within the computational "room," and it is here that the time invested in the model becomes quite efficient in investigating exposures.

Packaged CFD software is costly, and shareware is unavailable. Computers typically require over 2 GB RAM to solve complex 3-D problems and computer speed in the 3GHz range becomes important for solving problems quickly. Most problems can now be solved on desk-top machines rather than requiring supercomputers, as the cost of high-memory, high-speed desktops are reasonably priced.

These models take a long time to set up, can take weeks to run with sufficient information to ensure the solutions are valid, and require knowledge of fluid dynamics to ensure the selection of the simulation parameters will yield useful a converged solution and useful results. As such, these models are more suited for use in high-capital projects or when investigation of specific transport phenomenon is critical. CFD will not provide a quick and easy solution, but it can provide a detailed analysis of air movement and contaminant transport within a work environment.

15.5 References

1. **Schey, H.M.:** *Div Grad Curl and All That: An Informal Text on Vector Calculus,* 3rd edition. New York: W.W. Norton and Company, 1997.
2. **Fox, R.W. and A.T. McDonald:** *Introduction to Fluid Mechanics*, 2nd edition. Hoboken, NJ: John Wiley & Sons, Inc., 1985. p 35.

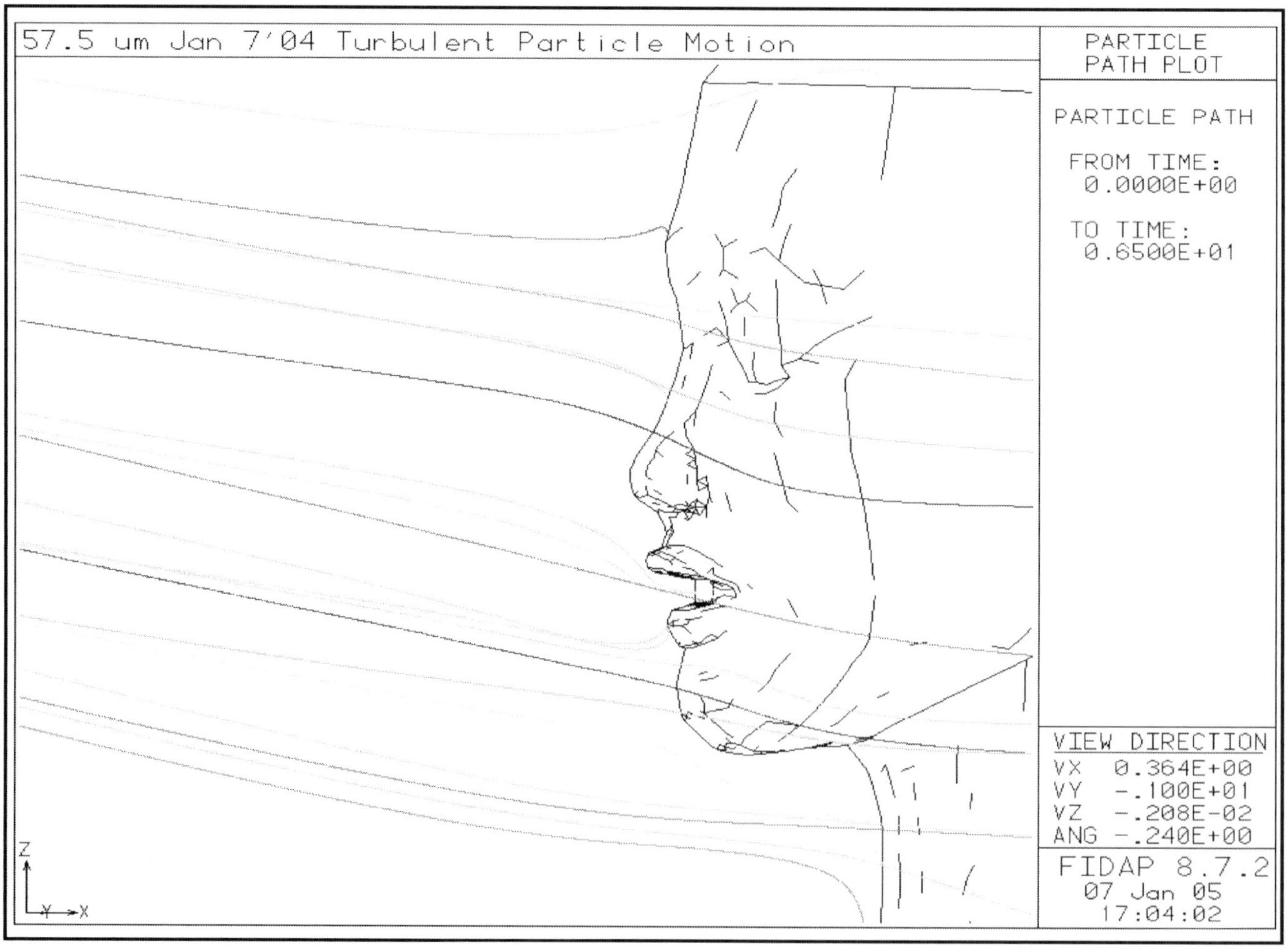

Figure 15.4b — Turbulent particle trajectories.

3. **Richmond-Bryant, J., A.D. Eisner, and M.R. Flynn:** Considerations for modeling particle entrainment into the wake of a circular cylinder. *Aerosol Sci. Tech. 40(1)*:42–51(2006).
4. **Stern, F., et al.:** Comprehensive approach to verification and validation of CFD simulations — part 1: methodology and procedures. *J. Fluids Engin. 123*:793–802 (2001).
5. **Roache, P.J.:** *Validation in Computational Science and Engineering.* Hermosa Publishers, Albuquerque, NM: Hermosa Publishers, 1998.
6. **Flynn, M.R. and A.D. Eisner:** (2004) Verification and validation studies of the time-averaged velocity field in the very near-wake of a finite elliptical cylinder. *Fluid Dyn. Res. 34*:273–288 (2004).
7. **Gosman, A.D. and E. Ioannides:** Aspects of Computer Simulation of Liquid-Fuelled Combustors. AIAA 19th Aerospace Science Mtg., Paper No. 81-0323, St. Louis, MO, 1981.
8. **Graham, D.I. and R.A. Moyeed:** How many particles for my Lagrangian simulations? *Powder Tech. 125 (2–3)*:179–186 (2002).
9. **Tanaka, S. and S. Murata:** An Investigation of the wake structure and aerodynamic characteristics of a finite circular cylinders. *JSME Internatl. J. 42(2)*: 178–187 (1999).
10. **Okamoto, S. and Y. Sunabashiri:** Vortex shedding from a circular cylinder of finite length placed on a ground plane. *J. Fluids Eng. 114*:512–521 (1992).
11. **Kulmala, I., A. Saamanen, and S. Enbom:** The effect of contaminant source location on worker exposure in the near-wake region. *Ann. Occup. Hyg. 40(5)*: 511–523 (1996).
12. **Brohus, H. and P.V. Nielsen:** CFD models of persons evaluated by full-scale wind channel experiments. Proceedings of Roomvent '96. 5th International Conference on Air Distribution in Rooms, Yokohama, Japan, July 17–19, 1996, Vol. 2: 137-144.
13. **Murakamie, S., J. Zeng, and T. Hayashi:** CFD analysis of wind environment around a human body. *J. Wind Eng. Ind. Aerodyn. 83*:393–408 (1999).
14. **Hayasih, T., Y. Ishizu, K. Shinsuke, and S. Murakami:** CFD analysis on characteristics of contaminated indoor air ventilation and its application in the evaluation of the effects of contaminant inhalation by human occupant. *Building Env. 37*: 219–230 (2002).
15. **Hyun, S. and C. Kleinstreuer:** Numerical simulation of mixed convection heat and mass transfer in a human inhalation test chamber. *Internatl. J. Heat Mass Transfer 44*:2247–2260 (2001).
16. **Anthony, T.R. and M.R. Flynn:** Computational fluid dynamics investigation of particle inhalability. *J. Aerosol Sci. 37(6)*:750–765 (2006).
17. **Svidt, K., P. Heiselberg, and O.J. Jendriksen:** Natural ventilation in atria — a case study. Proceedings of Roomvent '96. 5th International Conference on Air Distribution in Rooms, Yokohama, Japan. July 17–19, 1006.
18. **Park, H.J. and D. Holland:** The effect of local ventilation on a convective heat source on displacement ventilation: a CFD study. *Building Env. 36*:883–889 (2001).
19. **Bennett, J.S., C.E. Feigley, J. Kahn, and M.H. Hosni:** Comparison of mathematical models for exposure assessment with computational fluid dynamic simulation. *Appl. Occup. Env.. Hyg. 15(1)*:131–144 (2000).
20. **Bennett, J.S., C.E. Feigley, J. Kahn, and M.H. Hosni:** Comparison of emission models with computational fluid dynamic simulation and a proposed improved mode. *Am. Ind. Hyg. Assoc. J. 64*:739–754 (2003).
21. **Flynn, M.R. and E.D. Sills:** On the use of computational fluid dynamics in the prediction and control of exposure to airborne contaminants — an illustration using spray painting. *Ann. Occup. Hyg. 44(3)*:191–202 (2000).
22. **Rota, R., G. Nano, and L. Canossa:** Design guidelines for push-pull ventilation systems through computational fluid dynamics modeling. *Am. Ind. Hyg. Assoc. J. 62*:141–148 (2001).

Air Dispersion Plume Models

16

By
Thomas W. Armstrong, PhD, CIH

16.0 Overview

The following provides a short introduction to atmospheric plume dispersion modeling. Covering all the theory and applications discussed in such sources is beyond the scope of an introductory chapter, and many texts[1,2] on the topic are available. Air pollution modeling is a discipline unto itself. Knowledge of thermodynamics, kinetics, atmospheric chemistry, and meteorology are a few of the skills fundamental to the field. The field also has significant distinctions between local, regional and global models. We will provide a very basic introduction to the "local" models, point to further information sources, and then focus on a few aspects and examples of plume models with particular relevance to industrial hygiene exposure assessments. The near field and mid field plume models presented below have had limited validation, and their results are an estimate that may or may not compare well to either more recent and more parameter intensive models[3] or measurements. On the other hand, the models presented in this chapter were developed for outdoor conditions, and the models presented in other chapters of this book were developed for and are generally used to (but are not all necessarily all limited to) estimate exposure in indoor conditions. References are illustrative and this chapter is not a comprehensive review of publications on this diverse topic. Additional information, plume model descriptions and guidance to parameter selection for plume modeling is available elsewhere.[1,2]

16.1 Plume Model Development

One can trace the evolution of plume models for air pollution modeling back well over 60 years. The plume model function was to estimate the downwind concentrations of contaminants in the "plumes" from sources such as industrial smokestacks. The Gaussian function (i.e., normal distribution) has had central utility in describing the dispersion of contaminants in a plume emitted from a contaminant source, with first descriptions of this circa 1947.[4,5] The Gaussian function is used to describe the concentration gradient in the plume resulting from turbulent diffusion and mixing. Further development of the Gaussian plume model is largely attributed to Briggs[6], Pasquill, and Gifford.[7,8]

An example[9] of the classic equation commonly referred to as the Pasquill-Gifford equation follows as Eq. 16-1. Many subtly different variants, mainly in the mathematical conventions, exist in earlier and more recent publications.

$$X(x,y,z,H) = \frac{Q}{2\pi\sigma_y\sigma_z U} \exp -\left[\frac{1}{2}\left[\frac{y}{\sigma_y}\right]^2\right] \left[\exp\left[-\frac{1}{2}\left[\frac{z-H}{\sigma_z}\right]^2\right] + \exp\left[-\frac{1}{2}\left[\frac{z+H}{\sigma_z}\right]^2\right]\right] \quad (16\text{-}1)$$

where

- X = concentration, mass/volume
- Q = source strength, mass/time
- U = average wind velocity, distance/time
- σ_y, σ_z = diffusion coefficients in the y and z directions, respectively, distance
- H = effective height of emission source, distance

Note that in this equation, there is no *explicit* parameter for distance downwind from the source to the location of interest. However, the values of σ_i vary with distance, atmospheric turbulence, solar insolation and other factors, so one must enter the values for the diffusion coefficients at the distance of interest and for other relevant conditions such as atmospheric stability. Traditionally, these coefficient values are found via lookup in tables, nomographs, or charts[10] that account for the distance and atmospheric stability aspects. Generally, currently available commercial plume dispersion models have this lookup function built in. There are also estimation algorithms for these parameters.[10–12] Additionally, there is no term for time in this equation or in subsequent models below in this chapter. These models then predict *steady state* concentrations.

The units system must be consistent, and traditional environmental plume modeling uses mass in grams, distance in meters, time in seconds, and volume in meters cubed. The diffusion coefficients are usually based on work by Pasquill[7] and Gifford[8] as presented by Turner[10] or others. Algorithms have been also been developed to estimate these atmospheric diffusion coefficients.

Fully developed Gaussian plume models can include consideration of plume rise or plume fall, ground reflection, ground absorption, reactions, surface roughness, effects of building wakes, averaging times, atmospheric turbulence, inversions, particulate settling, and many other factors. For our purposes in this chapter, we will restrict ourselves to the more simple neutral buoyancy and non-reactive systems.

16.2 Plume Dispersion Models with Particular Industrial Hygiene Relevance

Simplification[9] of the above model that arguably has the greatest industrial hygiene utility is:

$$X(x,0,0,0) = \frac{Q}{\pi\sigma_y\sigma_z U} \tag{16-2}$$

where

- X = concentration at downwind distance x on the plume centerline, with x extending in the main direction of flow and where the vertical distance within the plume is zero, the side-wind distance zero, and at the same height as the emission source (H = zero).
- σ_i = diffusion coefficient in the x, y, or z direction
- G = emission rate, mg/min
- U = time-averaged wind velocity, meters/second

An appropriate value for U can be difficult, as it typically varies with time, height from the ground (z) and surface characteristics. Suggestions include using the mean through the plume[13] or using an algorithm to estimate it.[9]

The above simplification assumes that diffusion is similar in x, y, and z directions, there is no plume rise, and the concentration of interest is at ground level along the plume center line. This equation is perhaps more applicable to estimating concentrations over scales larger than in typical occupational settings, such as in local community exposure studies. Even then, more complex models that include additional parameters (e.g., plume buoyancy, surface roughness, surface reflection, temperature, humidity, reactivity, deposition, time variations) and thus fewer simplifications are more applicable. See for example the compilation of models at the U.S. Environmental Protection Agency Support Center for Regulatory Atmospheric Modeling web site14 for possible models for local scale, community assessments.

16.3 Near Field Plume Model

This model was developed to estimate worker exposure at up to 3 meters distance from leaking process equipment, but not from high pressure, highly directional jets. One purpose of the model was to aid in selecting the level of containment and level of control on equipment leak rates in engineering design of new facilities. The available documentation for this model[15] provides more on its background and derivation than is available for the mid-field equation discussed below. The simplifications in this plume model include average values of diffusion constants for Pasquill-Gifford "C" stability, with the worker downwind and on the plume centerline. This particular choice of C stability is a compromise between the stability class ranges and represents somewhat typical atmospheric conditions with some solar insolation and slight instability. For night time and highly stable conditions, or for other extremes, this model may not perform satisfactorily and more sophisticated approaches may be more appropriate. Nevertheless, this model is presented as an informative estimator for "typical" conditions. It is unlikely that any simple model formulation can meet all use conditions.

$$C = \frac{0.1225 \bullet G}{(r^{1.84}) \bullet U} \tag{16-3}$$

where

- G = emission rate, mg/min
- U = wind velocity, meters/second
- r = distance from source to breathing zone, meters

In this version of the equation, the constant has been modified so the parameters G is in mg/min, U in meters/second and result in mg/m^3. Further details on this model are available[15] such as the equation derivation and discussion of its comparison to exposure measurements.

Example 16.1 Application of the near-field plume model in engineering control selection

This is a hypothetical scenario, but one with parameters chosen to represent a plausible situation. An engineering design in review specifies a single mechanical seal pump for handling an intermittently used process additive composed entirely (100%) of a chemical with a STEL exposure limit of 0.1 mg/m³. The literature from the seal manufacturer suggests the particular seal type has a loss rate of 1 to 10 mg/minute. Presuming a potential worker distance of 1.5 meters, a typical wind velocity of 1 meter/second, and using the upper range of the suggested seal loss rate as the generation rate, what is the predicted concentration? Does this suggest this is a wisely selected pump seal technology for handling this material? Substituting the suggested parameters in the equation:

$$C = \frac{0.1225 \bullet 10}{((1.5)^{1.84}) \bullet 1}$$

Which results in C= 0.58 mg/m³.

This model, although it calculates a "worst case" centerline plume gives an estimated concentration above the STEL OEL. This suggests that better containment technology might be prudent.

16.4 Mid-field Plume Model

This approximation plume model was developed to estimate concentrations in the distance approximately 3 to 100 meters from the emission source. Further information on this model is limited.[16]

$$C = \frac{0.0502 \bullet G}{(r^{1.81}) \bullet U} \qquad (16\text{-}4)$$

where

C = concentration, mg/m³
G = emission rate, mg/min
r = downwind distance of interest, meters
U = wind velocity, meters/second

In this version of the equation, the initial constant, 0.0502, considers conversion units so that the parameter G is in mg/min, and the resulting estimated concentration is in mg/m³. This model incorporates values for the diffusion constant, most likely as an average of the diffusion constants applicable for 3 to 100 meters, and also based on Pasquill-Gifford "C" stability. The diffusion constant values are then incorporated in the exponent on the downwind distance, r, in the equation above. This equation is for plume centerline concentration and does not consider plume meander, the rapid drop of concentration from the centerline of a "small" plume and many other variables.

Example 16.2 Application of the Mid-field plume model in engineering control selection

This is also a hypothetical scenario with parameters chosen to represent a plausible situation. Another engineering project specifies the same single mechanical seal pump for handling the same intermittently used process additive as in Example 16.1 above. The question this time, however, involves estimating the concentration possible at the fenceline that is 50 meters from the proposed installation. Presuming the same wind velocity and generation rate as above, what is the resulting estimate for the "mid-field" concentration? Does the estimate suggest that the selected pump seal technology is appropriate for handling this material for this location? Substituting the suggested parameters in the equation:

$$C = \frac{0.0502 \bullet 10}{((50)^{1.81}) \bullet 1}$$

results in C= 0.00042 mg/m³ or 0.4 micrograms/m³

In this case, the best information available is an estimated community exposure standard based on the full shift TLV® of 0.01 mg/m³ divided by 100, or 0.1 micrograms/m³. This suggests that better containment technology might be prudent for this situation. Additional modeling using more sophisticated models more appropriate for community and fenceline impact studies[14] should be considered to better evaluate this scenario.

16.5 References

1. **Arya, S.P.:** *Air Pollution Metorology and Dispersion.* New York: Oxford University Press, 1999.
2. **Zannetti, P. (ed.):** *Air Quality Modeling: Theories, Computational Methods and Available Databases and Software.* Vol. I. Pittsburgh, PA: The EnviroComp Institute and Air & Waste Management Association, 2005.
3. **Drivas, P.J., P.A. Valberg, B.L. Murphy and R. Wilson:** Modeling Indoor Air Exposure from Short-Term Point Source Releases. *Indoor Air 6(4)*:271–277. (1996).
4. **Sutton, O.G.:** The problem of diffusion in the lower atmosphere. *QJRMS 73*:257 (1947).
5. **Sutton, O.G.:** The theoretical distribution of airborne pollution from factory chimneys. *QJRMS 73*:426 (1947).
6. **Briggs, G.A.:** A plume rise model compared with observations. *JAPCA 15*:433–438. (1965).

7. **Pasquill, F.:** The estimation of the dispersion of windborne material. *Meteor. Mag. 90*:33–49 (1961).
8. **Gifford, F.A., Jr.:** Use of routine observations for estimating atmospheric dispersion. *Nucl. Saf. 2*: 47–57 (1961).
9. **Perkins, H.C.:** *Air Pollution*, 1st edition. New York: McGraw-Hill Book Company, 1974.
10. **Turner, D.B.:** *Workbook of Atmospheric Dispersion Estimates — An Introduction to Dispersion Modeling,* 2nd edition. Boca Raton, FL: Lewis Publishers, 1994.
11. **Degrazia, G.A., D. Anfossi, J.C. Carvalho, C. Mangia, T. Tirabassi and H.F. Campos Velho:** Turbulence parameterisation for PBL dispersion models in all stability conditions. *Atmos. Env. 34(21)*:3575–3583 (2000).
12. **Hemond, H.F. and E.J. Fechner-Levy:** *Chemical Fate and Transport in the Environment*, 2nd edition. San Diego, CA: Academic Press, 2000.
13. **Turner, D.B.:** *Workbook of Atmospheric Dispersion Estimates.* PHS Publ. 999 AP-26. U.S. EPA, 1967. p. 84.
14. **U.S. Environmental Protection Agency (EPA):** *Preferred/Recommended Models*. [Web Page with links]; Available at: http://www.epa.gov/scram001/dispersion_prefrec.htm. [Accessed 2 March, 2009].
15. **Powell, R.:** Estimating Worker Exposure to Gases and Vapors Leaking from Pumps and Valves. *Am. Ind. Hyg. Assoc. J. 45(11)*:7–15 (1984).
16. **Lipton, S. and J. Lynch:** *Handbook of Health Hazard Control in the Chemical Process Industry.* New York: John Wiley & Sons, Inc., 1994.

Exposure Reconstruction

17

By
Thomas W. Armstrong, PhD, CIH
Fred W. Boelter, CIH, PE, BCEE
James O. Rasmuson, PhD, CIH, DABT

17.0 Background on Exposure Reconstruction

The process of evaluating people's exposures, individually or in groups, employs models. In a contemporaneous exposure assessment, we utilize sample collection models, statistical models, and allowable limit models. Models can be useful in describing and estimating exposure ranges and probability distributions given the impossibility of evaluating everyone's exposures associated with every exposure scenario, at all times. This chapter provides models for reconstructing and validating exposures individually or in groups, but emphasizes the exposure assessment process for individuals. The first several sections of this chapter deal with principles and guidelines associated with the exposure reconstruction process followed by five case studies illustrating both application of the exposure assessment process and options for validating exposure reconstructions.

The process of evaluating exposures can be applied retrospectively in estimating exposures. Modeling tools for estimating exposures for similar exposure groups (SEGs) are used in epidemiologic studies, standards setting, toxicological studies and for evaluating source, pathways, and receptors in the general population. Modeling tools for estimating exposures and assessing risk for individuals are used in evaluating control technologies, case report analysis, considerations of the maximum risk individual, and forensic analysis of injury claims. Individuals work in an SEG. An individual's work history and in turn exposure history is the composite of all the SEGs in which they worked. A reasonable objective during the exposure reconstruction process of an individual is to define and collect exposure data representative of the SEGs to which the individual belonged during their working lifetime and to review such data during the validation process. Through the exposure reconstruction process, we can place the individual in an appropriate SEG or a reasonable exposure distribution range.

An estimation of exposure through an exposure reconstruction implies an analysis by assembly of historical activities and exposures. Exposure estimations can also be performed to reconstruct more current conditions such as those which occurred yesterday or the week before as well as those of a prospective nature such as those which may occur in the future. For the most part, the exposure assessment process is the same whether the process applies to the past, present, or future. In any event, the fundamental process of assessing exposures retrospectively is an analysis of time and concentration which may or may not be influenced by other variables or modifying factors. Such variables or modifying factors may include considerations of control technologies as well as physiologic parameters of the individual.

In 1983, the National Academy of Sciences (NAS) published a Four Step Paradigm for Risk Assessment which also serves us well in the process of an exposure reconstruction.[1] The Four Steps are Hazard Assessment, Exposure Assessment, Toxicological Assessment, and Risk Assessment. The NAS Four Step Paradigm closely follows the traditional industrial hygiene paradigm which includes Anticipation, Recognition, Evaluation, and Control. In 2008, the NAS published a report entitled *Science and Decisions — Advancing Risk Assessment*. The 1983 NAS Red Book established a conceptual framework for risk assessments that has been adopted by numerous expert committees, regulatory agencies, and public health institutions. *Science and Decisions — Advancing Risk Assessment* embeds these concepts within a broader framework for risk-based decision-making.[2] The EPA has also published guidelines on exposure and dose assessment as part of the Four Step risk assessment process that does not differentiate whether the exposure assessment pertains to the past, present, or future. Nevertheless, this chapter focuses on some of the special challenges associated with reconstructing past exposures, provides specific tools and guidelines, and introduces validation methods, some of which have only been partially described or evaluated elsewhere.

Exposure reconstruction methodologies will necessarily be somewhat different for a large group epidemiological study than for an individual. Methodologies may also vary by type of study (e.g., cohort versus case control). The level of effort in reconstructing an exposure history or profile for an individual is often unavailable for a large group study. When studying groups, resources are directed to abbreviated methods which nevertheless have statistical power supporting the decision. When assessing individual exposures, more intensive methods may be necessary (such as interviews or exposure simulations). One limitation in assessing the individual is that we do not have statistical methods on self reporting and gathering anecdotal information. When confronted with conflicting or contradictory information involving the individual, the industrial hygienist must reconcile the information with observation, empirical assessment and judgment. However, even with the individual, we have available the scientific method (reproducible, reliable, and predictive) and statistical tools for reducing uncertainty in decision-making.

Exposure reconstructions can be qualitative, semi-quantitative, or quantitative but these terms require definition. A qualitative assessment is based on undefined ranges of no, low, medium or high exposure. A semi-quantitative assessment is based on defined ranges such as "no" meaning <1% of the OEL, "low" meaning 1% to 10% of the OEL, "medium" meaning 10% to 100% of the OEL, and "high" meaning >100% of the OEL. A quantitative assessment is based on exposure measurement data and modifying factors. For an individual, whether the reconstruction is conducted as qualitative, semi-quantitative, or quantitative will depend on resources, available information, and purpose. The level of detail should be commensurate to answer the question being asked.

Exposure reconstructions have utility as well as limitations largely based on the model selected and input data available and utilized. Exposure reconstructions are useful in understanding risk and may be necessary in determining causality. This is true also of epidemiology which studies the strength of associations. For example, exposure reconstructions can be part of the process to address risk as a function of peak, lifetime cumulative average, average annual average, highest annual average, etc. Exposure reconstructions can also be useful in allocating cumulative exposure as a function of time as well as source. Consequently, exposure reconstructions can be useful in allocation of attributable risk.

17.1 Developing the Model for an Exposure Reconstruction

Considerable literature is available regarding exposure reconstructions, modeling, statistical methods and data sources and quality. In the appendix to this chapter, a bibliography has been provided on the topic of exposure reconstruction.

In industrial hygiene, we sometimes use the total exposure and dose as interchangeable terms. The terms exposure and dose, however, are not the same but it is easy to understand their interchangeability as most sampling is performed outside of boundaries of the human body, namely the nose (breathing zone), mouth (ingestion), and skin (dermal contact). Most allowable limit guidelines and standards as well as the dose-response relationships studied in epidemiology are based on external body measurements. When we measure exposure, we are measuring a dose potential. There are other dose terms such as applied dose and internal dose which are relevant in a physiologically based pharmacokinetic (PBPK) model. Unless the PBPK model is understood for a particular compound in an exposure reconstruction, the dose potential is being estimated and is generally proportional to exposure.

17.2 Basic Dose Models for Routes of Entry

Strictly speaking, exposure can be concentration or quantity of an agent that contacts external body parts. Dose is defined as the mass of the agent uptake in the body per unit time. The following exposure and uptake diagrams and principles have been adapted from EPA guideline publications.[3]

The dose models for the dermal, inhalation and oral routes of entry can be expressed as follows:

Dermal

The dose model for dermal absorption (see Figure 17.1) shows Exposure, Potential Dose, and Applied Dose are equivalent at the outside boundary of the skin. The Internal Dose is based on characteristics of absorption through intact or broken skin.

Inhalation

The dose model for inhalation (see Figure 17.2) shows that Exposure is related to the concentration in the breathing zone outside the nose or mouth. Potential Dose is the amount which passes through the upper respiratory system while Applied Dose is the amount at the point of transfer or activity typically in the lower respiratory system. The Internal Dose is related to characteristics of transfer in the alveolar region, the Applied Dose per unit time, and the duration of the dose.

Oral

The dose model for ingestion (see Figure 17.3) shows that Exposure is related to the concentration outside the mouth. Potential Dose is the amount which passes into the mouth and is typically swallowed while Applied Dose is the amount at the point of transfer or activity typically in the gastrointestinal system. The Internal Dose is based on characteristics of transfer in the G.I. tract region.

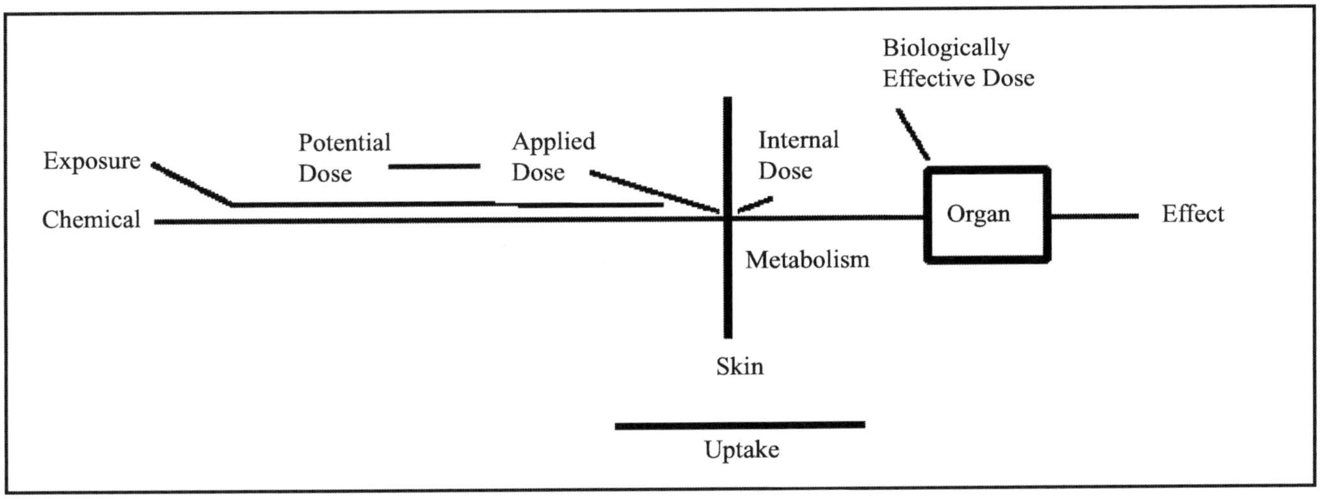

Figure 17.1 — Dermal route of entry.

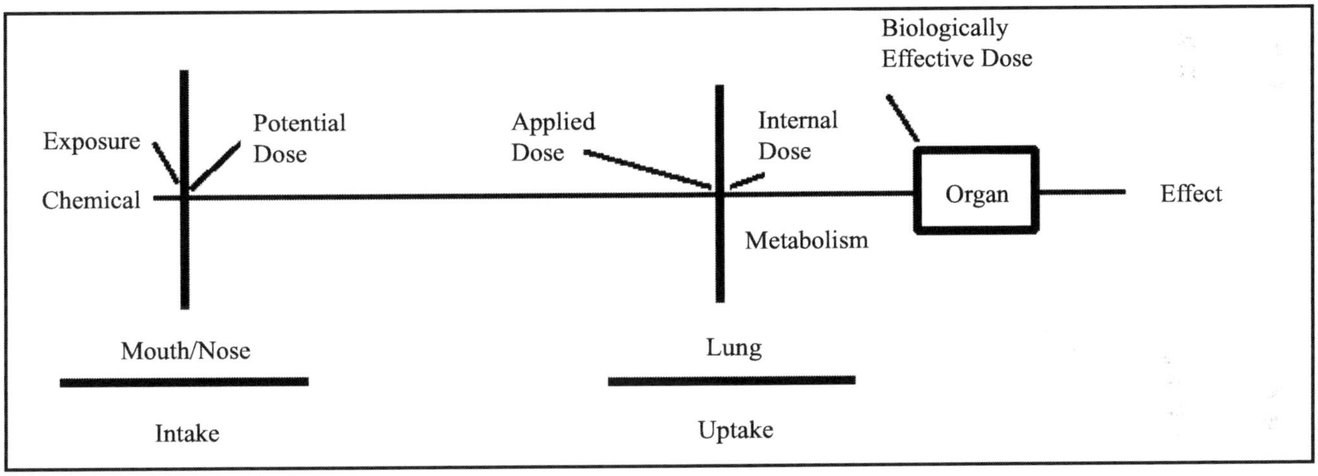

Figure 17.2 — Inhalation route of entry.

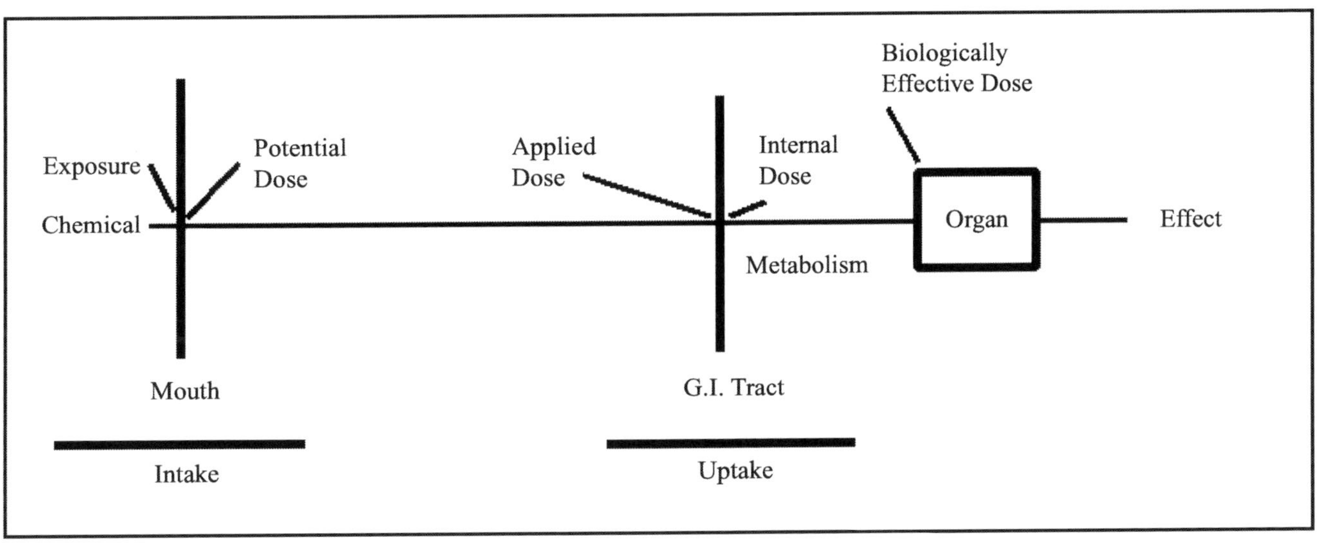

Figure 17.3 — Oral route of entry.

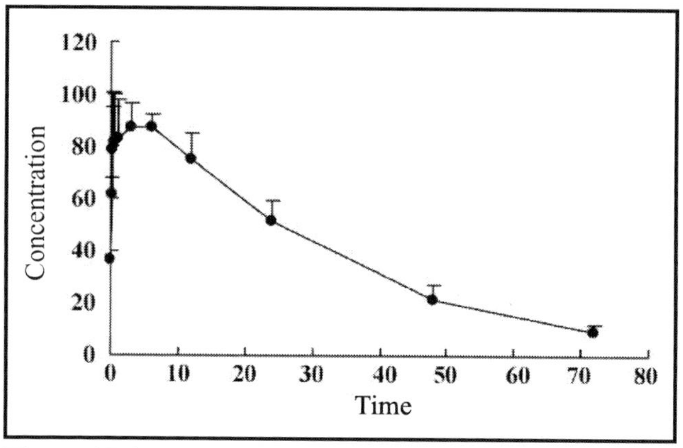

Figure 17.4 — Concentration/Time curve. The area under the curve is the dose or integrated exposure.

17.3 Integrated Exposure Reconstructions

An exposure reconstruction for a single individual, a specific chemical, and a particular pathway or exposure route over a given time period can be displayed in a curve. (See Figure 17.4.)

Integrated or total exposures are useful when a total exposure for a particular route is needed and supported by epidemiological literature and toxicological literature. The integrated exposure or total or cumulative exposure is represented by the area under the curve in Figure 17.4. Units of integrated exposure are concentration times time. For example, smoking dose is pack per day-years (aka pack-years), asbestos dose is fiber/cc-years (aka fiber-years), benzene dose is ppm-years and silica dose is mg/m^3-years.

When a series of reconstructions are performed considering different variables that may represent different individuals within a population, the probability distribution curve is a presentation of an exposure distribution for those individuals or population segment. Such a curve may represent a number of different SEGs.

17.4 Concentration and Time Relationship

A time-weighted average exposure concentration is the integrated exposure divided by the period where exposure occurs. The time-weighted average is useful in the dose estimating equations.

Often, total exposure (proportional to potential dose) is a concentration over an interval of time. The general equation for intake processes (e.g. inhalation and ingestion) is the integration of the chemical intake rate (concentration of the chemical in the medium times the intake rate of the medium, e.g. C times IR) over time. It is given as the formula:

$$D_{pot} = \int_{t_1}^{t_2} C(t)\ IR(t)\ dt \qquad (17\text{-}1)$$

The above formula can thus be transformed to the following general formula:

$$D_{pot} = \bar{C} \cdot \overline{IR} \cdot ED \qquad (17\text{-}2)$$

Where

D_{pot} = Dose Potential outside the body
C = Average concentration (8-hr TWA every day over the course of a year)
IR = Average intake rate (breathing rate, ingestion rate, transfer rate)
ED = Exposure duration (years)

As mentioned earlier, Dose Potential is an amount (mass, number, etc.) as opposed to a concentration. It is easier for an industrial hygienist to determine an amount or Dose Potential with the dermal and oral route of entries but it is more difficult to determine an amount with the inhalation route of entry. For the inhalation route of entry, we typically do not have the individual's intake rate (e.g. breathing rate) and thus do not use intake rate when assessing the significance of inhalation exposures of individuals. For the same reasons, epidemiologic studies do not typically consider intake rate for the inhalation route of entry when assessing populations. Nevertheless, for the inhalation route of entry, epidemiological and industrial hygiene literature expresses total exposure (often interchangeably referred to as dose) in the units of concentration times time as opposed to an amount (e.g. mass or number). Thus for the inhalation route of entry, the above equation reduces to:

$$E_{Total} = C \times ED\ \alpha\ D_{pot} \qquad (17\text{-}3)$$

Where

E_{Total} = Total Exposure outside the body
C = Average concentration (8-hr TWA every day over the course of a year)
ED = Exposure duration (years)
D_{pot} = Dose Potential
α = Proportional to

This formula serves as the basic model for conducting an exposure reconstruction assessment for the inhalation route of entry.

17.5 Exposure Reconstruction — A 12 Step Methodology

Twelve logical steps can be built off a specific exposure assessment strategy for conducting an exposure reconstruction. These steps are shown in Figure 17.5.

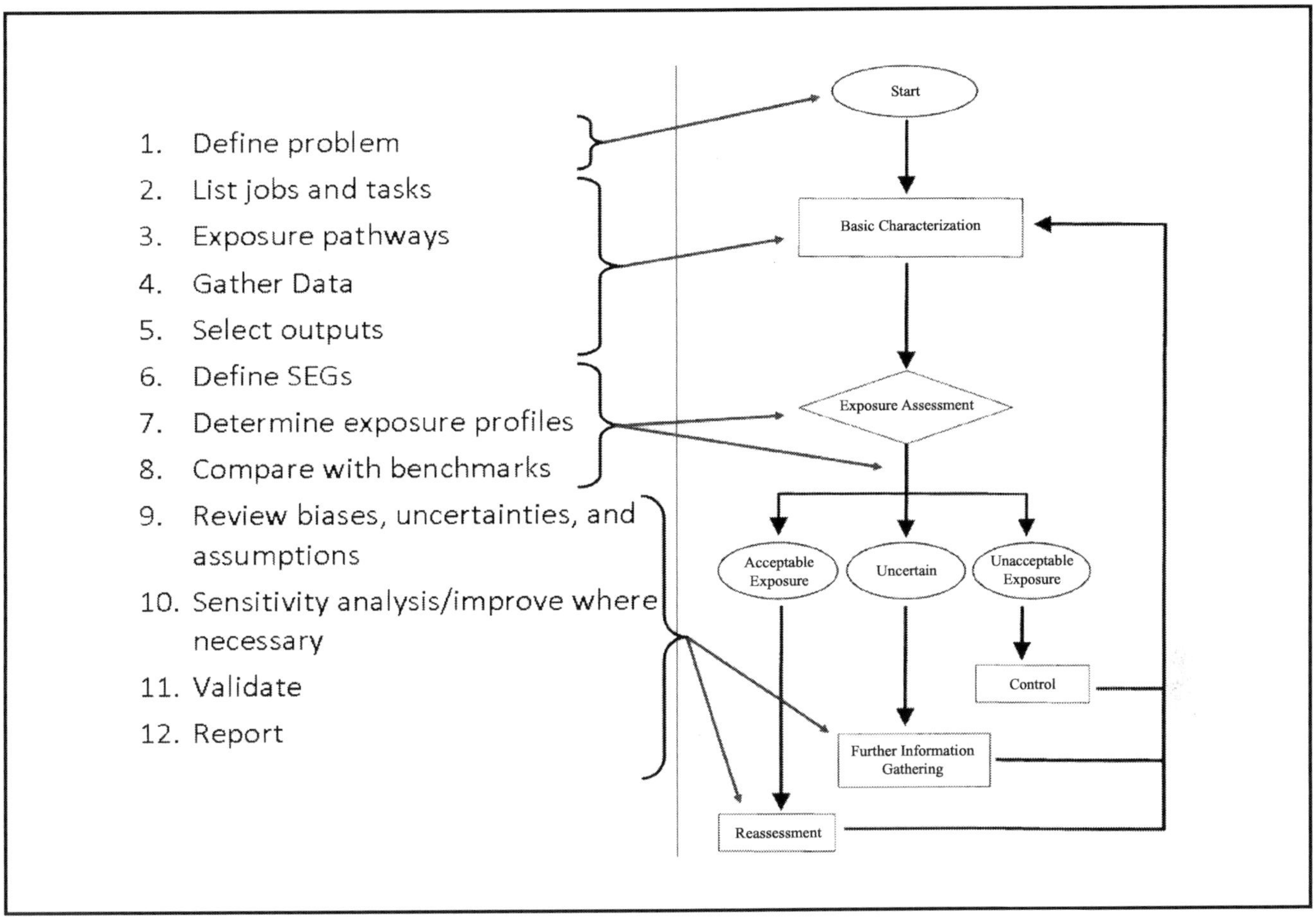

Figure 17.5. Twelve-Step methodology.*

Step 1: Define problem — Determine questions to be answered and establish goals. What is the purpose, what is the hypothesis, what information is necessary to test the hypothesis, identify subjects, chemicals and/or agents to be evaluated?

Step 2: List jobs and tasks — Determine locations, processes, products, and time periods as part of a basic characterization. The probability of exposure should be included when considering whether a specific job or task needs to be included.

Step 3: Exposure pathways — Air, water, soil, and direct contact are all potential exposure pathways. Routes of entry include dermal, inhalation and oral. One pathway may dominate the model inputs and outputs.

Step 4: Gather data — This foundational data includes work histories, monitoring data, process flow charts, SOPs, process standards, production data, personnel, engineering data, environmental reports, management reports, site visits, interviews, literature, etc.

Step 5: Select outputs — If monitoring is to be done, determine appropriate, validated methods. Is the assessment numerical, single point, or stochastic? What is the unit or dose metric? Does the literature report an exposure or dose metric?

Step 6: Define SEGs — SEGs are groups having the same general exposure profile because of the similarity and frequency of the tasks, the materials and processes, and similar way they are performed. SEGs are defined by environmental agent, process, job class, and task.

Step 7: Determine exposure profiles — Set up equations representing exposures or doses of interest that include the relevant exposure parameters; express equations in the basic form is $C \times t$ for each event for estimating total exposures; include modifying factors; use monitoring or surrogate data; use average 8-hr TWA; and verify data quality.

Step 8: Compare with benchmarks — Use the occupational exposure limit (OEL) expressed as a concentration (e.g. ppm, f/cc, mg/m^3, etc.) or expressed as total exposures (e.g. 45yr × OEL). Compare exposure profiles with disease thresholds or typical, total, lifetime ambient cumulative exposures. Describe exposure profiles in terms of risk.

Step 9: Review biases, uncertainties, assumptions — Are the exposure assessment outputs (exposure profiles) reasonable? Is there a need for revisions? Consider performing an uncertainty analysis.

* The Twelve Step Methodology has been derived from the authors of this chapter, publications of the AIHA Exposure Strategies Assessment Committee, and earlier work by Larry Birkner and Charles Redinger.

Step 10: Sensitivity analysis/Improve the analysis where appropriate and necessary — Which non-sensitive exposure parameters can be discounted after screening level analysis? Possible revisions may be needed for the most sensitive exposure parameters. Sensitivity analysis should be included as part of a stochastic analysis.

Step 11: Validate — Do the results make sense? Compare the subject biomarkers/pathology or medical evaluation with results of the exposure reconstruction. Possibly perform exposure re-creation studies. Compare with alternative assessments performed by other individuals or groups; and subject analysis to per review.

Step 12: Report results — Summarize results and their significance. Craft a risk statement. Document and reference all source data.

17.6 Sources of Data

What data we need and what data we can obtain determine what type of exposure reconstruction we can perform. Both the availability of data and the reconstruction objectives are considerations when deciding on the investigation and modeling approach. For example, if the reconstruction is for an epidemiology study seeking possible association between a chemical and a disease, then an "ever exposed/never exposed" analysis may be all that the epidemiologist needs for exposure in order to complete the study objectives. If so, there may be little reason to do an extensive investigation for data to support quantitative exposure modeling. On the other hand, for a study intended to support quantitative health standards setting, then quantitative exposure estimates may need to draw upon a more detailed mathematical model. If the goal is reconstruction of the possible exposure of an individual under specific circumstances, appropriate modeling can be informative. The modeling approach is nevertheless always determined by the availability of information and data for the model parameters.

Often, a quantitative exposure assessment may be the goal, and if so, a broad range of information of reliable quality is desired, not all of which may be available, or the information may not cover all relevant time periods. Generally, the more limited the information, the more difficult it is to deliver well-supported quantitative exposure assessments. The use of surrogate data or the use of determinants of exposure may be valuable in conducting quantitative exposure assessments. For example, maybe CO or NO_x measurements could used as a surrogate for elemental carbon measurements in a diesel study or horsepower of diesel equipment may correlate to elemental carbon. If certain key data cannot be found or cannot be derived from other sources, the exposure assessment strategy and study goals may need realignment. This caveat applies to all the possible assessment approaches.

In this section, we discuss the ramifications of data availability and data quality, and the impact on the ability to move from qualitative to semi-quantitative to quantitative assessments models.

17.6.1 Work History Details

An understanding of what work an individual did is a starting point for an exposure assessment. This history should be both chronological as well as durational. Several levels of detail about the work history might be encountered. An industry worked in is less informative than is a job title, which is less informative than substantiated descriptions of work schedules, tasks performed, their frequency and their duration. The most complete profile on work history would include empirically-based logs of all jobs, all employers, and all tasks/activities.

Some of the details desired to estimate exposure from tasks and changes in these over time include:

- Workplace production process, workflow and a time history of changes in these
- Materials and quantities used giving rise to exposures and time history of changes
- Task assignments or frequency, duration and time history of changes
- Environmental conditions and exposure controls such as ventilation systems and a time history of changes

Only some of these details may be in work history records. A broader investigation may be needed to gather the additional information to meet the exposure assessment plan and level of desired detail.

17.6.2 Exposure Assessment Principles and Guidelines

For operations, jobs and tasks that have been evaluated by exposure monitoring, such data may provide a valuable basis for estimating exposures. It may also serve as a basis for further modeling to predict exposures under differing conditions. The details of the survey strategies and the workplace conditions (see Tables 17.1 and 17.2) associated with the data may determine how informative, reliable and relevant the data are for use in estimating exposures back through time. It is important to understand the reasons for and strategy behind the collection of the available monitoring data. Was the survey to identify "worst case" exposure profiles? If so, the data may over-estimate typical conditions. Particularly for short-term (not full-shift) measurements, was the survey intended to evaluate primary tasks or sources causing exposure to help identify where exposure controls were needed? These aspects of the survey strategy must be considered if the goal is reconstructing typical long-term average exposures. For that purpose, the ideal is to have data that were taken in a statistically random manner, yielding a set of data that are sufficiently robust to accurately estimate the long term average as well as upper and lower limits of exposures. Given limited measurements and limited information, mathematical modeling of exposures may be of utility in putting the measurements in better perspective.

Details on exposure assessment strategies and construction/validation of SEGs have been discussed elsewhere and in the *Guideline on Occupational Exposure Reconstruction*.[4] Exposure monitoring records for SEGs may be particularly informative if the underlying determinants are documented or can be reconstructed. If so, changes in the determinants back through time can be used to model the changes in exposure over time. The modifying factor approach and the mathematical modeling discussed in other chapters of this book work well together. The models can improve the projections of the impacts of changes over time. Understanding the model input parameters data needs can effectively guide the information assembly efforts.

When limited exposure measurement data and other information are available for use in estimating the exposure potential for an operation's personnel, a number of alternative approaches may be used. This presumes the industrial hygienist does not have the opportunity to collect sufficient exposure data or cannot design a modern testing protocol for an historical activity on which to base a quantitative exposure assessment. Industrial hygiene professional judgment is used widely in deciding where, when and on which workers to conduct exposure measurements. This may involve a semi-quantitative exposure assessment, using a comparative mental model. The observed exposure situation is compared to other operations that the industrial hygienist has experienced and for which measured exposure data were available. Modifications are made in the comparative mental model to account for both exposure increasing and exposure decreasing factors. Since a different mix of factors, and different branches of logic may be used in different situations, a general process description and general set of modification considerations is difficult to describe. A conceptual framework can help organize the many considerations. One structure (Table 17.1) considers the source, pathway and receptor aspects of transfer to a route of entry.

The certainty of knowledge about the factors is also considered in the comparative model. If any of the key aspects are highly uncertain, either some exposure measurements or other information may be used to reduce the uncertainty. The variability of the various aspects is also considered, with the first pass assessment commonly following "reasonable worst case" assumptions. Additional refinement in the considerations may be applied until the exposure assessment is sufficiently certain for the intended use of the assessment information.

Another categorizing approach for the many modifying factors (and a way that complements the source-pathway-receptor concept) follows in Table 17.2.

A preliminary, review of these can help with decisions on how to further progress the assessment. For example, generally categorizing the direction and magnitude of change and the confidence in the assessment might be summarized as follows in Table 17.3.

This can support a plan for information gathering to support quantitative modeling using the approaches described in other chapters of this book. This type of factor and change analysis can also direct efforts on simulation of the conditions in order to conduct current exposure measurements that may mirror past conditions and exposures.

Table 17.1 — Source, Pathway, Receptor

SOURCE →	PATHWAY →	RECEPTOR
Source or generation includes the physical and chemical aspects of the contaminants (e.g. vapor pressure, reactivity). Process related variables are also included (e.g. system temperature, pressure) as are controls which are at worker discretion (e.g., work practices)	Pathway or transport includes dilution in the transport media (e.g., air), distance and direction to the receptor, other mitigating factors (e.g., exhaust ventilation) or enhancing or confounding aspects (e.g., other sources, other similar agents, temperature, pressure)	Receptor factors include the worker time, distance and activity in the contamination zone.

Table 17.2 — Modifying Factors Categorization

Workplace	Materials	Work Methods	Environment
ventilation,	physical state,	worker rotation,	especially if outside ...
temperature,	vapor pressure,	task frequency and duration,	temperature,
pressure,	% composition,	distance from source,	humidity,
process enclosure,	quantity used,	direction from source,	wind velocity,
continuous vs. batch,	warning properties,	PPE use, condition	wind direction,
housekeeping	other agents used		enclosure

Table 17.3 — Categorizing Assessment Change and Confidence

Work period 1 Year begin __________ Year end __________

AVERAGE exposure	Up	Down	%	Same	DNK*	Confidence**
Workplace design and production	❑	❑	_____	❑	❑	❑ L ❑ M ❑ H
Materials used quantities	❑	❑	_____	❑	❑	❑ L ❑ M ❑ H
Work methods / tasks	❑	❑	_____	❑	❑	❑ L ❑ M ❑ H
Environment, ventilation / other controls	❑	❑	_____	❑	❑	❑ L ❑ M ❑ H

*Do Not Know, or cannot be determined.
**Confidence is on a qualitative basis of Low, Moderate, High.

Several "rules of thumb" may be used in this comparative process. Some of the rules are quantitative, and many are qualitative. The remainder of this section focuses on further describing the modifying factors, "rules of thumb", use of mathematical modeling, and other tools for qualitative exposure assessment.

Several approaches to qualitative assessment are listed below, somewhat in an order of suggested preference, but the preference depends on the circumstances at hand.

1. Estimation based on screening level measurements of the operation to be estimated.

   1. Screening measurements

2. Estimation based on another agent present which is similarly used and for which data are available.

   2. Other agent in same operation

3. Estimation based on the same agent in another but related unit or operation and used in a similar way to the operation to estimate.

   3. Same agent in similar operation

4. Estimation based on another agent used in a similar way in another but related unit or operation

   4. Similar agent in similar operation

5. Predictive modeling based on physical and chemical properties, methods of use and controls in effect.

   5. Predictive modeling, physical and chemical principles

6. Predictive modeling based on operating system description and method of use, with limited consideration of the physical/chemical properties.

   6. Predictive modeling process and use based

7. Simulations (e.g. re-creation) of specific tasks and operations with measurements of resulting exposure

   7. Simulations and measurement

8. Estimation based on observation and subjective evaluation (e.g., taste, odor, irritation).

   8. Subjective evaluation

9. Application of a default exposure rating and subsequent modification process.

   9. Default rating

Further details on each approach follow.

1. Estimation based on screening level measurements

If suitable instantaneous measurement methods exist for the agents of interest, or suitable surrogates for the agents of interest, screening level measurements may be used to guide the estimation process. Techniques such as detector tubes, photoionization detectors, respirable aerosol monitors, air velocity measurements, etc., can provide quick results helping to calibrate the industrial hygienist's comparative mental model. These screening tools may be used to check "worst case" tasks, or typical exposure levels. While these direct reading and short-term sampling results may not directly give a time weighted average exposure measurement, they may be a good basis for a qualitative or initial quantitative assessment. Nevertheless, short-term samples are likely not good surrogates for long-term or 8-hr TWA samples.

2. Estimation based on another agent present similarly used and data are available

One can estimate exposures for an agent based on measured data for another agent used similarly in an operation. The estimation is based on the relative concentrations of the agents in parent bulk materials, and on the relative differences in vapor pressure. An example is given below. Algorithms exist for predicting vapor pressures at various temperatures, or for predicting vapor pressure from boiling points.

For example, let us assume we have exposure data for substance X, (X_{ppm}) but need to estimate the exposure for substance Y (Y_{ppm}). Both are components in the process, at temperature T and system pressure P. We know their % composition in the stream ($\%_x$ and $\%_y$), their molecular weights, and their vapor pressures VP_X and VP_Y. We can then calculate the mole fraction of each. There are a few other simplifying assumptions in this example, but they are likely often not significant given other assumptions and uncertainties in the estimation process.

$$X_{ppm}\left[\frac{\text{molefraction}_y}{\text{molefraction}_x}\right]\left[\frac{VP_y}{VP_x}\right] = Y_{ppm} \qquad (17\text{-}4)$$

Note: Modifications could also be made for differences in temperature and pressure but in the normal ambient temperature and atmospheric pressure range, these have trivial significance.[5]

3. Estimation based on the same agent in another but related unit or operation

One may need to consider differences in the operating environment or work practices. Rules of thumb for several main considerations follow in the section on rules of thumb. These may better be applied to the main tasks with exposure potential, and then those task assessments combined to yield the exposure assessment for the job.

4. Estimation based on another agent used in another but related unit or operation

This is basically a combination of approaches 2 and 3. Note: This approach introduces a greater number of uncertainties and extrapolations than for the prior approaches. These uncertainties, assumptions, etc. increase the potential for an erroneous estimate.

5. Predictive modeling based on physical and chemical properties, contaminant generation rate, methods of use and controls in effect.

The simpler models and simplifying assumptions may be applied first. For example, assume exposure to saturated vapor. If the resulting concentration and the task duration and frequency lead to nil exposure, then further modeling is not likely to be needed. Otherwise, more detailed modeling and fewer simplifying assumptions may be required. There are various predictive modeling tools ranging from spreadsheets to commercial software. A few of the models discussed in other chapters of this book include:

- Saturated vapor concentration
- Simple box model
- Gaussian plume dispersion
- Eddy diffusion model
- Near field/far field (two zone) model

6. Predictive and/or comparative modeling based on operating system and method of use, with semi-quantitative consideration of the physical/chemical properties

This approach relies on a comparative process and considers, qualitatively, the volume of material, proximity to people, degree of containment, system temperature and pressure, etc. to subjectively rank the potential for exposure. If containment is complete and exposure opportunity during sampling and other open systems operation is nil, then the overall exposure is nil. If the systems are open, use relatively high vapor pressure materials, and controls are absent, the exposure may be rated "> OEL". This may be an adequate approach for the first assessment, but may require later refinement via more detailed modeling or via measurement.

7. Simulation of Relevant Conditions and Measurement

If sufficient information is available to understand the workplace equipment, conditions, materials, and task methods, physical recreation of the scenario might be informative. This hinges on having accurate reproduction of the factors involved; this may be difficult to verify. In some situations, obtaining the equipment and materials is either not possible or is not feasible to time or costs. On occasion, industrial technology museums and older technology journals may provide information on the processes and conditions to simulate. If done on a reduced scale, the validity of extrapolation to full scale conditions may be difficult to verify.

8. Estimation based on observation and subjective evaluation (e.g., taste, odor, irritation)

Some agents provide warning properties, or other subjective input that can be processed, with professional judgment, to rank exposure. This presumes the industrial hygienist has professional experience in the operations evaluated, or in sufficiently similar operations so that the judgment is formed in comparison to past data and current observation of the determinants of exposure. This approach may require later refinement via more detailed modeling or via measurement.

9. Application of a "default" exposure scheme for a qualitative only rating

By using a "high exposure" default, the priority for the agent-job being evaluated should fall into an upper priority zone and lead to further investigation. The uncertainties may then be resolved and the exposure rating established on a more rigorous basis. This may involve qualitative modifications by various process, agent, control, and work method considerations. For example, starting with a default rating of "HIGH", one reviews the various aspects listed below, and keeps a tally of the modifications on the plus side, negative side, or no effect side. The final rating is then modified down from the "HIGH" rating, or may be left unchanged (see Table 17.4).

Some of the mathematical exposure prediction models discussed earlier may be used to make more refined modifications (even to derive comparative modifying factors rather than predict concentrations) than these rules of thumb for other modifiers such as differences in ventilation, temperature, pressure, distance from source, etc.

Table 17.4 — Qualitative Modifying Factors

Intensity Modifier Category	Modifying Factor	Intensity Modifier Category	Modifying Factor
Containment		**Aerosol formation**	
• full	-	• nil/none	-
• partial	0	• low potential	0
• none	+	• present	+
Ventilation Control		**Dust suppression**	
• effective local	-	• effective	-
• dilution/outdoors	0	• moderate	0
• none/ineffective	+	• none/ineffective	+
Work Procedures		**Quantity of Agent**	
• good	-	• small	-
• minimal	0	• moderate	0
• none/ineffective	+	• large	+
Volatility		**Dermal Absorption**	
• low	-	• none	-
• moderate	0	• slight/little	0
• high	+	• extensive	+

These rules of thumb generally assume that all other factors are held constant. Other rules may be applied sequentially, each time holding all other factors constant.

A. *Ventilation.* As a general rule, the concentration will vary directly with the volume of diluting air. Two times the velocity (or two times the volume) of air past a generation point is assumed to divide the downwind concentration by a factor of two (assuming the generation rate holds constant).

B. *Volume of material released or handled.* The concentration will vary directly with the volume of material released (time, ventilation, distance held constant). Two times the volume of material released is assumed to double the concentration.

C. *Number and duration of exposure opportunities.* The exposure will vary directly with the number and duration of exposure opportunities (e.g. generation, distance, ventilation, duration, etc. held constant). Two times the frequency and two times the duration is assumed to increase exposure by a factor of four.

D. *Process conditions.* Vapor pressure varies with temperature. The difference may be calculated and then used to modify the exposure estimates. For example, Process X is at X C° for Compound A. Process Y, with exposure to A at Y ppm, is at Y C°. The predicted exposure for Process X (other factors held constant) is then the ratio of the calculated vapor pressures times the measured exposure for Process Y:

$$X_{ppm} = \left[\frac{VP_x}{VP_y} \right] Y_{ppm} \qquad (17\text{-}5)$$

E. *Differing composition.* Differences in the composition of process streams or other parent materials (which give rise to air contamination) will lead to different exposure potential. Raoult's law relates a substance's mole fraction in a mixture to its equilibrium partial pressure above the bulk mixture. Not all substances and mixtures will follow ideal behavior sufficiently for this to be a completely satisfactory prediction. However, in a comparative manner, Raoult's law can guide exposure assessment considerations. That is, in current operations the material used has compound i present at x%. In the past, the composition was 5x% of compound i. Thus, the exposure, other factors the same, should be roughly 5 times higher in the past. Raoult's law is:

$$P_A = \gamma \bullet X_A \bullet P_A^0 \qquad (17\text{-}6)$$

where

P_A = the partial pressure of the solvent in the mixture
γ = the activity coefficient
X_A = the mole fraction of solvent in the mixture
P_A^0 = vapor pressure of pure solvent

Particularly for very dilute solutions, polar molecules, other substances with hydrogen bonding, or where other interaction between solvent and solutes occur, deviations from ideal behavior can be substantial. If so, activity coefficients (which equals 1 in an ideal solution) may be used, and calculation routines for these are available in the public domain, including a

UNIFAC program.[6] This aspect of compositional differences and partial pressure changes is discussed further elsewhere and software for calculations is also available.[5]

F. *Distance from source.* The concentration with distance from a source may be predicted using exposure estimating equations. For a first approach, the near field variant of the Gaussian plume dispersion model might be considered, with the eddy diffusion model as an alternate. Note: The former predicts equilibrium (steady state) concentration, but the latter can provide a means of following the time course of concentration changes in a non-steady-state situation. If the operation has numerous sources and a "well mixed" atmosphere, then a well-mixed room (box) model may be appropriate. The two-zone (near-field, far-field) model can give two estimates based on a close to source, more distant from source dichotomy for the distances.

Estimates of the likely ranges for the model input variables of the models can draw on many sources. Data mining can focus on job specific and industry specific documents. The early industrial hygiene literature has a number of reports on the air velocity distributions and ventilation volumes in various industries. Unfortunately, much of this is not compiled and is time-consuming to find. Nevertheless, a literature review may provide data on historic exposures and the underlying determinants (volume of material, composition, ventilation, etc) at the time.

Interviews of long-service workers, or other experts such as process engineers and industrial hygienists familiar with the operations can give insights into the range of certain factors. It is unusual for persons to know the composition of materials. Unless records are found, this may be constrained to likely functional ranges, e.g. the product must have at least x% but no more than y% in order to fulfill its physical-chemical function. Workers and others in structured interviews can give information on the frequency and duration of tasks that involved exposure to materials of interest.

17.6.3 Missing Data

Missing data are almost a certainty in most exposure assessment studies for epidemiology, and likely for other exposure reconstructions as well. In many but not all circumstances, careful analogies can help fill those voids. Too large a void may give reason to revisit the exposure assessment design, or the study feasibility.

Interpolation between events prior to and following a data gap can be accomplished in several ways, with the choice depending on the circumstances. Often, changes in operations are "step changes" and then a linear rate of change to bridge the gap is not the best choice. The data of the operational change, if it can be established, should then be the point for switching between the prior and post data for the assessment. In another circumstance, an average value of the prior and post conditions may be the best choice.

Source generation rates can be estimated based on material use or general loss rates. Emission factor approaches (rates of loss from pump seals, valves, flanges, etc.) may be used for estimating the rate of a materials loss from process equipment.

Protective equipment use (e.g. respiratory protection) can add considerable uncertainty about actual received dose from estimated air concentrations. The quality, fit, removal capability, care of and use of respiratory protection has varied over the history of the devices. Some operations have excellent compliance with use, but other operations may have a less successful experience.

Mapping a work history in a particular operation can help define the possible time spent in and patterns of movement through contaminated zones. This may be a necessary part of potentially calculating TWA exposures from estimated area concentrations. Ranges of values can be handled in calculations via stochastic (probabilistic) simulations.

17.7 Exposure Reconstruction and the Industrial Hygiene Model

The traditional industrial hygiene model includes anticipation, recognition, evaluation, and control. In an exposure reconstruction evaluation, we utilize the industrial hygiene model by combination information on concentration with time with modifying factors.* (See Figure 17.6).

17.8 Exposure Reconstruction Case Studies

17.8.1 Performing Product Testing

Exposure reconstructions can be conducted for a number of purposes ranging from assessing exposures for epidemiological studies to assessing risk for the most likely highly exposed to apportioning liability. As discussed

* Figure 17.6 regarding the Exposure Assessment Process is attributed to Larry Birkner and was presented in May, 2003 in New Orleans, LA at the first AIHA Professional Development Course on the topic of Exposure and Dose Reconstruction. The course, "Reconstructing Exposure and Dose: Utility for the Practicing Industrial Hygienist", was developed, expanded, refined and taught to other industrial hygienists by combinations of CIHs including Craig Armstrong, Larry R. Birkner, Fred W. Boelter, Joseph B. Coble, James O. Rasmuson, and Charles F. Redinger.

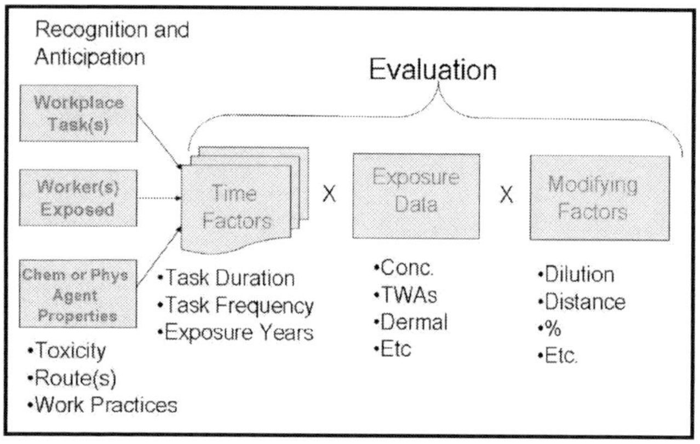

Figure 17.6 — Exposure Reconstruction and the Exposure Assessment Process.

earlier, the exposure reconstruction model (qualitative, semi-quantitative, quantitative) selected will depend on a number of factors. Statistical power may be achievable in a large-population epidemiological study using an ever/never classification. A robust model with quality data and validation becomes an important empirical process in testing and reducing uncertainty associated with anecdotal information. For example, an empirical process could include the use of estimating manuals, Gantt charts, exposure data validation review, setting aside inconsistent data, averages from data distributions, site visits, production records, and specific or simulation (e.g. re-creation) testing. If testing is to be performed, there needs to be a basis to its design which is based on a verifiable empirical process including field observations of tasks as they are performed.

*Case Study Regarding Carbon Monoxide**

Four campers, two in each of two tents, used one portable infrared heater in their respective tents one cold night. In the morning, both of the campers in Tent A were found dead by the two campers from Tent B. Authorities were contacted and the following were reported:

Campers from Tent B had COHb <5%
Campers from Tent A had COHb >75%
Campers from Tent A had been dead about 12 hours
Both portable heaters were off and cold

- Propane tanks from the heater in Tent A were empty
- The two portable infrared heaters looked identical
- None of the campers were smokers

There are a number of ways to conduct an exposure reconstruction and model dose. Using a PBPK Model based on the Coburn-Foster-Kane (CFK) equations would be one way to possibly estimate carbon monoxide inhalation, metabolism, exhalation and carboxyhemoglobin based on physiologic characteristics of the individuals. The carbon monoxide estimations could then be combined with an air change rate model for the tents to correlate with the time line information to back-calculate a generation rate for the heaters. Or was outcome even related to air change rate differences between the tents (e.g. if one tent was tight, was the other tent transparent)? Or was the outcome even related to physiologic differences between the two sets of people in the two tents?

An initial look at carbon monoxide information from Winter and Miller[7], Ellenhorn and Barceloux[8] shows a generalized relationship between COppm and COHb% at steady-state:

CO in atm. (ppm)	COHb in blood (%)	Signs and symptoms (steady state)
10	2	Asymptomatic
70	10	No appreciable effect, except shortness of breath on vigorous exertion; possible tightness across the forehead; dilation of cutaneous blood vessels.
120	20	Shortness of breath on moderate exertion; occasional headache with throbbing in temples
220	30	Decided headache; irritable; easily fatigued; judgment disturbed; possible dizziness; dimness of vision.
350–520	40–50	Headache, confusion; collapse; fainting on exertion
800–1220	60–70	Unconsciousness; intermittent convulsion; respiratory failure, death if exposure is continued
1950	80	Rapidly fatal

Apparently, the concentrations of CO in the two tents were orders of magnitude different from one another. Tent B where the COHb of the campers was less than 5% would suggest CO levels of less than 70ppm. Tent A where the COHb of the campers was greater than 75% would suggest CO levels of greater than 1200 ppm. Could physiologic differences or air change rates explain such differences? Could refinements utilizing the C-F-K equations help analyze this exposure reconstruction?

The real issue here is how could two seemingly identical heaters apparently be so different? Could a portable propane infrared heater generate carbon monoxide at a rate sufficient to achieve lethal concentrations in a tent in such a short time that the campers would have been dead for 12 hours by morning? Why were the other two campers apparently not affected?

Testing of the two heaters was proposed and conducted. A 128 cubic foot test chamber was built approximating the volume of the respective tents. The chamber was sealed and unventilated. Tests were run closely approximating an ASME method used by UL in 1964 which monitored rate of rise until oxygen levels were below 15%. Direct reading multipoint instruments

* The authors acknowledge the help and assistance of Gary N. Crawford and Glenn J. Filip.

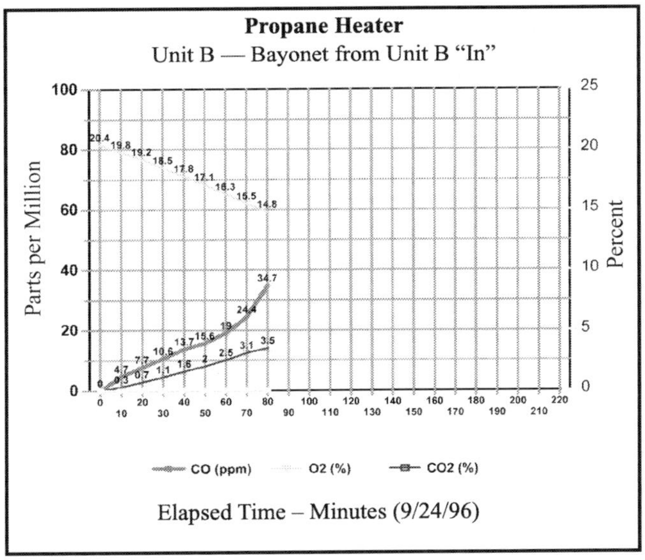

Figure 17.7 — Unit B, Bayonet "In."

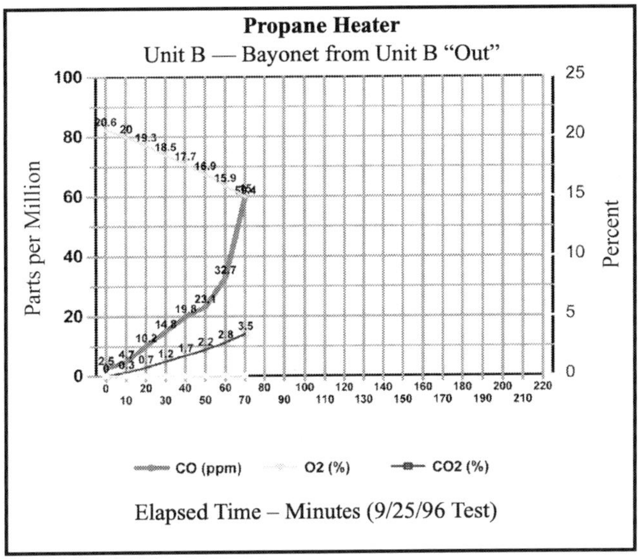

Figure 17.8 — Unit B, Bayonet "Out."

measuring CO, O_2, and CO_2 were placed and calibrated. A CO alarm was also established outside of the test chamber (this proved fortuitous when testing Unit A).

There are two settings for the heater bayonet: in to light and out to run. When the bayonet is in, the combustion is fuel-rich. When the bayonet is out, combustion is fuel/oxygen balanced.*

Testing Unit B

Testing commenced with heater Unit B from Tent B where the campers were fine. Results were found as follows with the bayonet respectively "in" (Figure 17.7) or "out" (Figure 17.8):

For Unit B whether bayonet "In" or bayonet "out", the carbon monoxide concentration rate of rise seems unremarkable reaching 30ppm to 60ppm before oxygen levels were below 15%.

Testing Unit A

Testing then commenced with heater Unit A from Tent A where the campers died. There were a few physical differences with the bayonet and regulator in comparison to Unit B. However, testing of Unit A was also performed with the bayonet "in" and "out". Results were found as follows with the bayonet respectively "in" (Figure 17.9) or "out" (Figure 17.10):

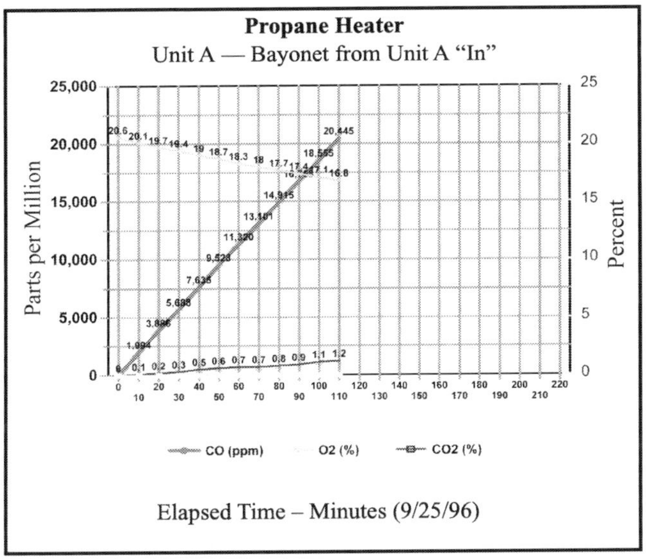

Figure 17.9 — Unit A, Bayonet "In."

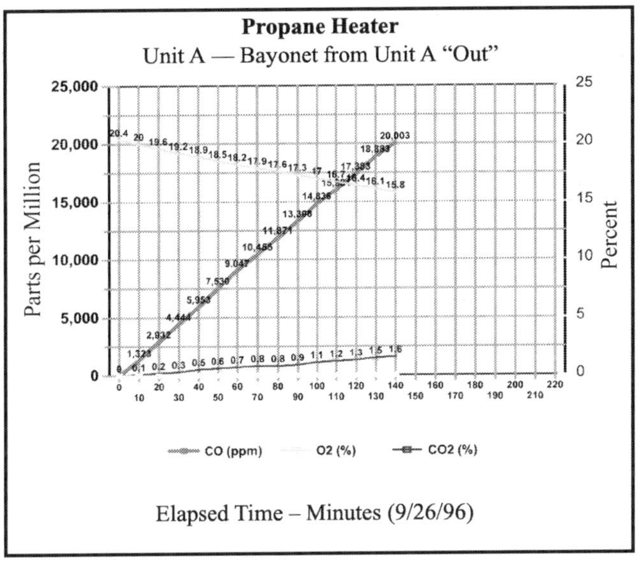

Figure 17.10 — Unit A, Bayonet "Out."

* Carbon monoxide was intended to be fully oxidized to carbon dioxide by an integral catalytic screen surrounding the combustion gases. The glowing catalytic screen is the source of infrared radiation.

For Unit A, the carbon monoxide concentration rate of rise was dramatic reaching 20,000 ppm and the oxygen levels never reached 15%.

Clearly there were differences between Unit A and Unit B. Unit A was generating carbon monoxide at a rate capable of achieving lethal concentrations in a tent after 10 to 20 minutes. Unit B was generating carbon monoxide at non-lethal rate orders of magnitude less than Unit A.

Conclusions

From an exposure reconstruction standpoint, testing of these unit heaters was invaluable in evaluating the potential for achieving lethal concentrations of carbon monoxide in a short length of time. A relationship between CO and COHb has been modeled in the literature and served as a reasonable basis for the testing hypothesis. Physiologic evaluation using the C-F-K model would not likely have assisted in the analysis of why there were two outcomes between the occupants using "identical" heaters in the two tents.

There were a number of subsequent evaluations done with the two portable propane unit heaters to understand why they were not identical. While those evaluations, outcomes, and lessons are interesting from an industrial hygiene, engineering and consumer perspective and important in ultimately allocating responsibility, we will not explore those findings here as they do not substantively address the exposure reconstruction objectives of this chapter.

17.8.2 Performing Exposure Reconstructions for Individuals

Exposure reconstructions conducted by a qualified industrial hygienist can be a useful application whether helping an epidemiologist study a cohort or helping to understand the significance or insignificance involving concerns or claims of disease related to potential past occupational and non-occupational exposures. In the case of asbestos, disease tends to involve chronic exposures occurring over long periods of time. From the epidemiological and toxicological literature, we know the metrics for industrial hygiene considerations include total exposure (concentration x time), fiber type, and latency. We also know even with a very significant dose, a potent fiber type, and a long latency, many humans do not develop a disease related to asbestos. Therefore, some individuals are susceptible and some are not susceptible.

The unit of dose metric regarding asbestos is the f/cc-year (or f/mL-year) often abbreviated "fiber-year." This metric is the average daily concentration (8 hr-TWA in f/cc) in the breathing zone every working day over the course of a year times the number of years (years) at that average daily concentration.

The process of conducting a reconstruction for an individual and calculating dose in fiber-years begins with a reconciled timeline or history. Sources of information for developing a timeline range from recollection by the individual, medical records, work records, military records, union records, and Social Security records. Individual recall is anecdotal information subject to bias whereas records may yield testable empirical information. If primarily anecdotal information is available to the industrial hygienist, consider performing a reconstruction on the basis of "even if the anecdotal information is correct…" would the conclusions drawn be significant or insignificant regarding the questions being asked.

From this reconciled history, the experienced industrial hygienist begins to identify exposure events defined as activities involving potential airborne asbestos fiber exposures as well as the event duration. The industrial hygienist will consider a wide variety of factors and variables before estimating a breathing zone concentration for the exposure event. These factors include those traditionally considered by the industrial hygienist when establishing an SEG. Factors range from proximity to the source, engineering controls such as ventilation, material friability, and activities being performed. These factors serve as modifiers to the estimated breathing zone concentration.

Estimated concentrations can be a single point value such as an average of a data set range and distribution. An estimated concentration using the average from a set of 8-hr TWA results can yield a meaningful and useful value in a retrospective exposure reconstruction. Should a more refined, quantitative analysis be desired, more sophisticated analyses can be performed using the probability distributions themselves (stochastic analyses). Nevertheless, all estimations of breathing zone concentrations rest on data quality, data that is testable scientifically, the result of good industrial hygiene practice, and are not outliers to a distribution.

It is valuable to remember that most or at least a majority of the exposure scenarios are assessed with a worst case monitoring strategy.* The distribution of industrial hygiene data for an SEG is typically log normal with a GSD not far from two. It is improbable that data representing an individual's breathing zone concentrations would day after day fall above the 95th percentile of exposures. Consequently, the use of worst case data in a retrospective exposure reconstruction would probably lead to erroneous and misleading estimations of total exposure. Another important consideration regarding data quality involves use of information based on recollections of occasions a number of decades earlier. Unless this information can be empirically corroborated, analyzed and averaged, or the recall bias can be adequately considered, self-reported information is potentially low quality data.

* The use of a representative monitoring strategy, as opposed to a worst case monitoring strategy, will generate a huge number of samples. This means that although the total number of exposure scenarios assessed in a particular company may be small compared to the total number of possible exposure scenarios in given industry, the total number of measurements collected using a representative monitoring strategy may be as large or larger than the number of samples collected using a worst case strategy.

Site visits, physical materials, historical photographs, operating records, sampling results even if not for the agent in question, construction documents, and military specifications are all examples of sources for improving data quality and addressing missing or censored data. Specific testing can also be performed; however, great care must be taken in the design and fulfillment of testing. Testing must be based on empirical information regarding time and motion, reformulations based on recipe cards or chemical analyses, and equipment of the time period. Such a systematic design to the testing will increase its relevance, reliability and reproducibility.

*Case Study Regarding Single-Point Estimation of Asbestos Total Exposure**

Using the following hypothetical example based on a composite of actual statements, let us examine the process for calculating, using single points, an estimated cumulative or total exposure to asbestos.

Background Information

Diagnosis: Mesothelioma; May 31, 2006

Mr. John Doe was born on April 3, 1937 in Minot, North Dakota. John graduated from high school in 1955.

John started working on Friday, July 1, 1955 as an Apprentice Insulator with AAA Asbestos Insulation, a unionized national insulation contractor. He was assigned work at the Big Hills Generating Facility and attained Journeyman status after two years. Over a period of five years, he insulated "everything" related to the two new units at this coal fired facility (Units 3 & 4). These were commercial units, probably 10 stories tall. He recalls mostly working with Kaylo and Unibestos while insulating. "About 50:50, I'd say." He worked an 8-hour shift and recalls it was like "Christmas on Independence Day" due to the falling discarded insulation pieces. He does not recall there being any difference between his activities as an Apprentice and those as a Journeyman. "They did what they had to do to get the units insulated."

John left AAA and was hired by the utility where he did maintenance as an in-house insulator and mechanic from July 1, 1960 until June 30, 1965. It wasn't a union shop so he could pretty much do "whatever." Mostly he replaced insulation that was removed by others on Units 1 & 2. These units looked a lot like Units 3 & 4 but they were older. He spent about 4 days a week replacing insulation here and there. Most of it was the same kind of insulation that "I put on Units 3 & 4. Ya know, cut the block, wire it on, and trowel it smooth with cement."

John wanted a career change so on July 1, 1965 he went to work for The Northern Construction Company out of Minot. He had always been good at welding and metal working as a hobby so he was hired on as a pipefitter doing new construction. About half his day was spent laying out piping systems, welding seams and welding flanges. About an hour a day was spent handling valves and/or bolting flanges together. The valves were always pre-packed but he had to drop in a gasket before he could bolt up the flange. The gaskets were mostly of the precut variety. About half were spiral-wound and about half were greyish and flexible. The insulators were always right on his heels and he recalled them working no more than five or ten feet away. They worked with the same kind of insulation he remembers from his days at AAA. "Mostly the half round rigid kind. Mostly it was indoors work."

On July 1, 1970, he shifted from new construction to maintenance. Northern Construction was doing a lot of overhaul work in various places. They had us knocking off the insulation and leaving it on the ground. "We would have to walk on it and kick it out of our way." It seemed like one or two days a week were spent working on piping systems out of doors, the rest was indoors. Sometimes it rained or snowed when he was working out of doors but mostly not. Some of the insulation was "that jacketed corrugated insulation" but "most of it was the block kind." Every day "I would have to remove at least one gasket before I could reroute the piping system." John said he would scrape off the old gasket and use a wire brush on the residual. He found the gasket would stick "sometimes... about as often as normal." He would use preformed gaskets whenever he could but sometimes he would have to make one using a ball peen hammer by carefully tapping out the shape. Probably about one in ten would have to be made that way.

On July 1, 1976, John started his own auto paint shop, Doe's Paint Barn. He used body putty which he sanded by hand. He had to fill holes and sand putty at least twice a day. He didn't have a spray booth, "just a big open barn." He used all kinds of paints and solvents. He did the painting on Fridays to allow for curing over the weekend. He never liked the water soluble types, "not a hard enough finish." He "never sanded on Fridays though. Dust would ruin the paint jobs."

He decided to sell the business in 1985 and felt he got lucky with being hired to manage a paint shop for Dakota Joe's All Wheel dealership starting on July 1, 1985 where he worked until retirement on June 30, 2000.

Process and Analysis

Let us develop a first cut exposure reconstruction by developing a timeline or work chronology upon which to build an exposure event history. As appropriate, a basic form of the spreadsheet could be built as a location, time and concentration relationship such as in Table A.

In the example case study, the basic form of the above spreadsheet can be formatted into a spreadsheet to capture specific activity, time and concentration information. A spreadsheet can be built with numerous columns for information and modifying factors as shown in Table B.

The potential exposure event and work history chronology can be compiled in columns A though F.

* The authors acknowledge the help and assistance of Catherine E. Simmons and Daniel M. Podraza.

Table A: BASIC FORM SPREADSHEET CAPTURING LOCATION, TIME, AND CONCENTRATION

Facility Code	Department Code	Mean (conc.) Year 1	Mean (conc.) Year 2	Mean (conc.) Year 3	Mean (conc.) Year 4	Mean (conc.) Year 5	Mean (ppm) Year 6
Plant 1	Dept 1						
Plant 1	Dept 2						
Plant 2	Dept 1						
Plant 2	Dept 2						
Plant 3	Dept 1						
Plant 3	Dept 2						

Table B: SIMPLIFIED DOSE CALCULATION SPREADSHEET (Example)

Pot.Exp. Event	A Empl.	B Loc	C Job	D Start	E Stop	F Equiv. Years	G Prox	H Freq	I Note	J 8-Hr TWA	K Dose (FxJxMF)
1											
2											
3											
4											
5											
6											
7											
8											
9											
10											
Total											

(Note: An equivalent year is based on 250 working days equaling one work year.) Columns G through I are a few examples of modifying factors (MF) that could modify the breathing zone concentration. Column J is the average 8-hr TWA for an individual performing the activity based on quality data. Column K is the event dose calculated by multiplying column F (equivalent years) × J (8-hr TWA) × the modifying factors. The product calculated in column K is expressed as fiber-years for the potential exposure event. A more sophisticated analysis could take into account modifying factors such as fiber type, latency, overtime, ventilation, distribution probability, etc. and thus refine the analysis.

Employer History

A first cut employer history and chronology could be completed in Columns A through F as shown in Table C:

Modifying Factors

When considering modifying factors, it may be clearer to compile the modifying factors by potential exposure event. For a given employer during a given time period, there may be multiple potential exposure events based on job, activity or materials. Consequently, a total of 9 potential exposure events were identified and listed. A first cut on the potential exposure events and modifying factors could be completed for Columns G through I as shown in Table D.

8-hr TWA

Estimating the 8-hr TWA for an exposure reconstruction is one of the most challenging tasks confronting the experienced industrial hygienist. While it may be perceived as preferable to have actual sampling data associated with a specific individual performing specific tasks, industrial hygienists know that the basis of most industrial hygiene decision-making is the utilization of limited data collected or analogized to exposures associated with an SEG. We can use descriptive and inferential statistics to estimate exposure ranges and distributions based on technical knowledge and field experience as well as quality data which is transparent. The less transparent or more compromised the data, the greater the potential such data and resulting exposure reconstruction will not make sense. Such data should be potentially set aside as the reliability, reproducibility, and relevance would be in doubt. In a retrospective exposure assessment or dose reconstruction, we should only use data that can stand up to scientific scrutiny and ultimately validate the

TABLE C: EMPLOYER HISTORY AND CHRONOLOGY

	A	B	C	D	E	F
Potential Exp event	Employer	Location	Job	Start Date	Stop Date	Equiv. years
1	AAA Asbestos Insulation	Big Hills Generating, Minot, ND	Apprentice Insulator on Units 3&4	July 1, 1955	June 30, 1957	2
2	AAA Asbestos Insulation	Big Hills Generating, Minot, ND	Journeyman Insulator Units 3&4	July 1, 1957	June 30, 1960	3
3	Big Hills Generating	Minot, ND	In-house maint,; Insulator & Mechanic; Units 1&2	July 1, 1960	June 30, 1965	5
4	Northern Construction	Minot, ND	Pipefitter; New Construction	July 1, 1965	June 30, 1970	5
5	Northern Construction	Minot, ND	Pipefitter; Maintenance	July 1, 1970	June 30, 1976	6
6	Doe's Paint Barn	Minot, ND	Owner: Paint Shop	July 1, 1976	June 30, 1985	9
7	Dakota Joe's Allwheel	Minot, ND	Manager: Paint Shop	July 1, 1985	June 30, 2000	15
	Total					45

reasonableness of the calculations performed. Ultimately, the calculations should make sense with respect to an SEG.

The average 8-hr TWA for a distribution of data should be used in column J for purposes of this calculation. As mentioned earlier in this chapter, Appendix A has a bibliography of literature related to exposure reconstructions and retrospective exposure assessments as well as a number of sources of industrial hygiene data. Be sure to document the process and have a supportable scientific basis for the data utilized.

Average 8-hr TWA values have been added into column J. Column K has been calculated by multiplying column F × Column J × modifying factors (column G, H, I). Remember, just because the data appears in the peer reviewed literature doesn't mean that the data stands up to scientific reliability, reproducibility, and relevance. The completed first cut dose reconstruction thus is as in Table E.

Conclusions

This process of a systematic exposure reconstruction utilizing a timeline, modifying factors, and quality data are based on the science of industrial hygiene and the art of its application. A systematic exposure (dose) reconstruction provides a helpful assessment to resolve disputes. Further refinement of this analysis can be performed depending on the questions for which answers are desired.

It would be expected that different experienced industrial hygienist performing a calculation with the same information might report different numerical totals. The different values would be related to actual field experience, familiarity with the literature and their limitations, availability of empirical data against which to compare anecdotal information, and incorporation of modifying factors. Nevertheless, experienced industrial hygienists would be expected to reasonability rank potential exposure events similarly and would be expected to produce results that could be correlated with the results of other experienced industrial hygienists.

This process can be utilized for other chemicals, compounds, or contaminants of interest provided there is an accepted dose related metric (ppm-years, mg/m^3-years, pack/day-years, etc.) that is useful from a toxicological perspective and is consistent with the scientific literature.

17.9 Performing Validations of Exposure and Dose Reconstruction Data

Consider the following before accepting an exposure reconstruction:

- Do the results make sense?
- Were applicable average 8-hr TWA data used from reliable, relevant, and reproducible distributions?
- If available, were bio-markers and/or tissue and blood analysis consistent with the results of the exposure assessment?
- Are exposure re-creation (e.g. laboratory or other simulation) studies necessary or available, are results consistent, and were they well-designed following good industrial hygiene principles?

TABLE D: POTENTIAL EXPOSURE EVENT AND MODIFYING FACORS

Pot.Exp. Event	A Empl.	B Loc	C Job	D Start	E Stop	F Equiv. Years	G Prox	H Freq	I Note
1	AAA Asbestos Insulation	Big Hills Generating, Minot, ND	Apprentice Insulator on Units 3&4	July 1, 1955	June 30, 1957	2	Worked w/ insulation	1	Kaylo: Unibestos 50:50
2	AAA Asbestos Insulation	Big Hills Generating, Minot, ND	Journeyman Insulator Units 3&4	July 1, 1957	June 30, 1960	3	Worked w/ insulation	1	Kaylo: Unibestos 50:50
3	Big Hills Generating	Minot, ND	In-house maint,; Insulator & Mech.; Units 1&2	July 1, 1960	June 30, 1965	5	Worked w/ insulation	0.8	Mostly replaced insulation 4 day/week
4	Northern Constr.	Minot, ND	Pipefitter; New Construction	July 1, 1965	June 30, 1970	5	Worked w/ gaskets	0.0625	One half hr; half spiral wound
5	Northern Constr.	Minot, ND	Pipefitter; New Construction	July 1, 1965	June 30, 1970	5[a]	Worked near insulators	0.5	Half the day near insulators
6	Northern Constr.	Minot, ND	Pipefitter; Maintenance	July 1, 1970	June 30, 1976	6	Removed insulation	1	Removed block and aircell
7	Northern Constr.	Minot, ND	Pipefitter; Maintenance	July 1, 1970	June 30, 1976	6[a]	Worked w/ gaskets	0.125	One gasket = one hour
8	Doe's Paint Barn	Minot, ND	Owner: Paint Shop	July 1, 1976	June 30, 1985	9	Worked w/ body putty	0.1	Less than 1hr/day 4day/wk
9	Dakota Joe's Allwheel	Minot, ND	Manager: Paint Shop	July 1, 1985	June 30, 2000	15			
	Total					45[a]			

[a] When inventorying discrete Potential Exposure Event within a common Start and Stop time interval, the Equivalent Years should only be added once to the Total)

- Were other industrial hygiene assessments consistent with respect to exposure and conclusions concerning the significance of the exposure?
- What was the purpose of the exposure reconstruction?
- How does the potential variability in the results potentially influence the conclusion?

This section will consider some of these bulleted items in more detail. The applicability of these considerations may vary depending on whether an exposure reconstruction is being performed for a group or for that of an individual. This section primarily focuses on validation of individual exposure reconstructions or validation of group reconstruction using data associated with individuals. Use of case studies will also explore the precision and accuracy that can be typically obtained for estimating the exposures of individuals compared with group mean values. Comparison with tissue and blood analysis, alternative analysis by other industrial hygienists and what constitutes reasonable agreement for such comparisons will also be evaluated. To address these topics, the SEG concept, statistical methods, and rules of thumb for evaluating results will be illustrated.

The first question to be asked when validating an exposure reconstruction is whether the results make sense. For example, if one is assessing the asbestos total exposure of an individual pipe fitter, and the estimate of total exposure is in the SEG-range of an asbestos insulator, it would be wise to redo the assessment or not accept it. If the estimated total exposure for a worker performing gasket removal is close to or in the range of the asbestos total exposure for an asbestos insulator, the results of the exposure reconstruction should be doubted. If limited use of organic solvents in an open area is estimated to result in exposures close to vapor saturation, similarly the results should not be accepted.* Or if one is asked to assess the exposures from mopping out an unventilated railroad tank car with an organic solvent, if the estimated exposure is not within a reasonable percentage of the vapor pressure or if the estimated exposure is higher than what the saturation vapor pressure would predict, it would appear that something is wrong with the exposure reconstruction.

* Many variables come into play in any scenario but it is unlikely that mopping activities will have exposures over 10% of saturation. Exposures in a poorly ventilated confined space are usually on the order of 1 to 5% of saturation. It is difficult to saturate air.

TABLE E: FIRST CUT DOSE RECONSTRUCTION

Pot. Exp. Event	A Empl.	B Loc	C Job	D Start	E Stop	F Equiv. Years	G Prox	H Freq	I Note	J 8-Hr TWA	K Dose (FxJxMF)
1	AAA Asbestos Insul	Big Hills Gen, Minot, ND	Apprentice Insulator on Units 3&4	July 1, 1955	June 30, 1957	2	Worked w/ insulation	1	Kaylo: Unibestos 50:50	9	18
2	AAA Asbestos Insul	Big Hills Gen, Minot, ND	Journey-man Insulator Units 3&4	July 1, 1957	June 30, 1960	3	Worked w/ insulation	1	Kaylo: Unibestos 50:50	9	27
3	Big Hills Gen	Minot, ND	In-house maint,; Insul & Mech.; Units 1&2	July 1, 1960	June 30, 1965	5	Worked w/ insulation	0.8	Mostly replaced insul 4 day/week	9	36
4	Northern Constr.	Minot, ND	Pipefitter; New Construct-ion	July 1, 1965	June 30, 1970	5	Worked w/ gaskets	0.0625	One half hr; half spiral wound	0.02	0.006
5	Northern Constr.	Minot, ND	Pipefitter; New Construct-ion	July 1, 1965	June 30, 1970	5[a]	Worked near insulators	0.5	Half the day near insulators	4.5	11.2
6	Northern Constr.	Minot, ND	Pipefitter; Main-tenance	July 1, 1970	June 30, 1976	6	Removed insulation	0.5	Removed block and aircell	9	22.5
7	Northern Constr.	Minot, ND	Pipefitter; Main-tenance	July 1, 1970	June 30, 1976	6[a]	Worked w/ gaskets	0.125	One gasket = one hour	0.025	0.019
8	Doe's Paint Barn	Minot, ND	Owner: Paint Shop	July 1, 1976	June 30, 19853	9	Worked w/ body putty	0.125	Less than 1 hour/day	0.04	0.045
9	Dakota Joe's Allwheel	Minot, ND	Manager: Paint Shop	July 1, 1985	June 30, 2000	15					
	Total					45[a]					115 f-yr

[a] When inventorying discrete Potential Exposure Event within a common Start and Stop time interval, the Equivalent Years should only be added once to the Total)

The models in this book can be used to provide an alternative assessment of an estimated exposure to determine whether reasonable agreement is present. However, the question remains, "What is reasonable agreement"? To consider this point, it is useful to explore the use of the SEG concept further in exposure reconstructions.

17.9.1 Case Study — Lead Uptake*

For our first case study, consider the SEG associated with lead uptake in children attending a day-care center about one kilometer from a lead smelter in a Central Asian country. Routes of entry include oral ingestion and inhalation. The children's homes are varying distances from the smelter. Play habits, personal hygiene, hand-to-mouth contact, diets, lead on surfaces of home utensils, the presence and condition of lead-based paint in the home, percent indoor and outdoor times, home construction, the extent of open windows, and inhalation rates all vary. Therefore, for this SEG, based on potential child-to-child exposure variability, estimating a range of average daily lead dose for the SEG is more appropriate than estimating a single value.

On the other hand, if we input single-valued average lead concentrations for each exposure pathway of soil, indoor dust, air, and water measured at the school into a

* The authors acknowledge the help and assistance of Dr. Andrey Korchevsky, a noted environmental scientist living in Almaty, Kazakhstan and Daniel Hall, Dr. Bal Kishan Kaul, Dr. Dalmon Larson, Dr. Roger Olsen, Eric Rasmuson, and Robert Strode.

pharmacokinetic model, we will obtain a single-valued estimate of lead-blood concentration unless we assume some underlying variability in the children's exposure. Lead-blood concentration is a surrogate representative for lead uptake or dose and is an indicator to some extent of both fairly recent and past lead uptake. Based on experience, the U.S. EPA has estimated that lead-blood concentrations within groups of lead-exposed children are typically log-normally distributed with a geometric standard deviation (GSD) of 1.6. Therefore although the U.S. EPA Integrated Exposure Uptake Biokinetic Model for Lead in Children (IEUBK)[9] accepts single-valued environmental concentration data, the model yields a range of predicted lead-bloods consistent with a GSD of 1.6 rather than a single-value.

Plotting uptake data in the form of a histogram is useful to examine exposure variability and the frequency distribution of the data. Histograms can be produced from standard statistical packages and Excel, in this case SAS. Figure 17.11 illustrates the histogram of measured-lead blood concentrations at the day-care center in question. The histogram is representative of the varying average lead doses of children within the SEG. Also shown in Figure 17.11 is the log-normal probability distribution of children's lead-blood concentrations predicted by the U.S. EPA IEUBK model described above.

Figure 17.11 illustrates that within the SEG it is possible, utilizing average environmental data, to predict an individual child's lead-blood concentration within a factor of two or three with a reasonable degree of confidence. The variability of the prediction of an SEG individual member's lead blood concentration is a function of the GSD of the various average exposures of members of the SEG, in this case assumed to be 1.6. Often, it is difficult to retrospectively estimate individual exposures closer than that indicated in Figure 17.11. Somewhat higher variability is possible based on how narrowly or how broadly an SEG is defined, taking into account all forms of variability present in the assessment.

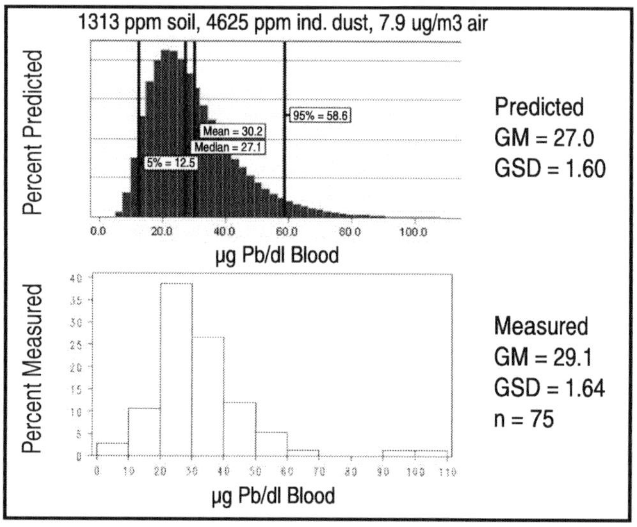

Figure 17.11 — Comparison of IEUBK-Modeled Lead Levels with Histogram of Measured Lead-Blood Concentrations at a Central Asian Day Care Center. The X-Axis (Lead Blood Concentration) is On the Same Scale for Both the IEUBK Probability Density Function and the Histogram. Note: Blood-lead levels more than 100 ug/dl were measured for this SEG.

Combined day-to-day and individual-to-individual variability in industrial hygiene exposure measurements for an SEG often have associated GSDs of about two or slightly lower, but typically not more than four. Long-term exposure averages, however, have less variability than the variability associated with day-to-day air monitoring results. Therefore, it is conceptually possible for the retrospective exposure assessment of individuals to have less variability than the variability of an SEG based upon individual air measurements. The reason for less variability in long-term exposure averages is because the standard deviation of the mean (standard error) is smaller than the standard deviation of individual data points by a factor of the square root of the number of data points. Therefore, the confidence interval based on mean exposure variability is always much smaller than the prediction interval based upon the higher individual air sample variability. Total exposure variability of an individual is related to the confidence interval while day-to-day exposure variability of the individual is more closely associated with the prediction interval of daily exposures related to that individual. However, when performing an exposure reconstruction for an individual, it is important to remember that each individual member of an SEG has associated exposure variability somewhere between the variability of the entire SEG based on the variability of individual air samples and the variability of the mean of the SEG upon which the exposure assessment is based. Nevertheless as Figure 17.11 illustrates, the mean exposure of the group can be determined more precisely than that for a specific individual.

The data shown in Figure 17.11 can also be utilized to illustrate another principle of exposure reconstruction validation: whether bio-markers and/or tissue and blood analysis are consistent with the results of the exposure assessment. Figure 17.11 demonstrates reasonable agreement between the modeled and the measured lead uptake.

Uncertainty in estimating modifying (adjustment) factors also contributes to variability in an exposure estimate. Validation attempts need to be cognizant of such uncertainty. In our experience, exposure data generally remain log-normally distributed (best fit) regardless of how narrowly or how broadly the SEG is defined, whether in fact several different SEGs are combined to form one heterogeneous exposure group composed of several SEGs. Probit plots and goodness-of-fit tests of the combined exposure data can be utilized to test whether the combined data are still log-normal. Individual data points from a collection of studies at different locations for the same process or type of exposure can be combined to form a broader exposure group that can be utilized as an SEG, but the data are still generally log-normally distributed. Underlying variability is important to consider when an exposure reconstruction for an individual is employed for a specific purpose such as determining whether an individual has likely been exposed above or below an occupational exposure limit or a specific toxicological total exposure or dose benchmark. It is essential to consider underlying exposure group variability when judging reasonableness of an exposure assessment and the reliability of associated conclusions. However, especially when estimated total exposure is at a low level relative to significant toxicological consequences, variability associated with the estimated exposure can often be of no practical consequence.

The comparison shown in Figure 17.11 does not necessarily validate the model for differing conditions at other locations such as for areas of differing ratios of lead concentration in the various environmental media. Figure 17.12 compares predicted with modeled geometric mean lead blood concentrations for six separate SEGs with widely varying environmental lead concentration and widely varying ratios of outdoor soil to indoor dust lead concentrations. A regression analysis was performed to determine the correlation of the geometric mean modeled dose for each SEG as a function of measured geometric mean lead blood concentration for each SEG. Although the regression was performed in SAS(10), other statistical software packages can be utilized as well such as Excel.(11) The data in Figure 17.12 represent measurements involving 248 children. It was empirically determined that the closest reasonable agreement between the regression line and the "line of perfect agreement" was with the assumption that the ratio of ingested outdoor soil and indoor dust was seven to three. Therefore, that assumption ratio was used in the pharmacokinetic model when modeling lead blood concentrations near the lead smelter in question. Reasonable agreement between modeled and measured lead-blood concentrations was generally found. Thus, comparison of histograms with predicted probability density functions, and the type of comparisons shown in Figure 17.12 can be not only used to validate the exposure assessment model, but can be utilized to calibrate the model, as well, in this case determining the "best" assumed ratio of ingested outdoor soil and indoor dust in the model. Bioavailability of the lead was determined with a laboratory simulation procedure involving soil samples collected near the smelter.

17.9.2 Case Study — TCDD Contamination during Herbicide Production*

Another example of the usefulness of blood analysis in validating exposure reconstruction results is found in the dose reconstruction of TCDD (2,3,7,8-tetrachlorodibenzodioxin) for former chemical workers who produced 2,4,5-T and 2,4-D herbicides prior to and during the Vietnam War including Agent Orange. The dose reconstruction was completed in the early 1990s, more than 10 years after production stopped in 1979. No historical air monitoring data was available to the industrial hygiene team. Limited process stream TCDD concentrations were available along with process flow sheets and other relevant plant documents. Mass balance calculations were used to expand TCDD concentrations for various media and points in the process for which there were no available TCDD analyses. Forty three previous production workers were interviewed in depth concerning their memory of production details, qualitative aerosol concentration descriptions, quantities of chemicals that contacted skin, and other relevant information. It was determined that dermal absorption was the primary exposure route. SEGs were established for various operator positions.

Daily TCDD uptake was estimated for each of the SEGs based on associated median exposure estimates which in turn were based on compositing the worker interviews. Median rather than average estimates were utilized to reduce the potential for isolated exaggerated claims of exposure. Pre-designed interview forms with non-leading questions were utilized in the interview process. Based on worker recall, the estimated time spent at each of the positions for each of the production workers was estimated, and TCDD uptake was assessed for each of the former workers. Maintenance workers were assumed to have doses equal to the average uptake for all worker positions in the plant. Doses for general supervisors, area supervisors, and area lead men were reduced with modifying factors based on the interviews. Estimated lifetime cumulative TCDD doses for the 43 former workers ranged from 0 to 11,096 micrograms. Taking into account individual estimated TCDD uptake and body weight, risk

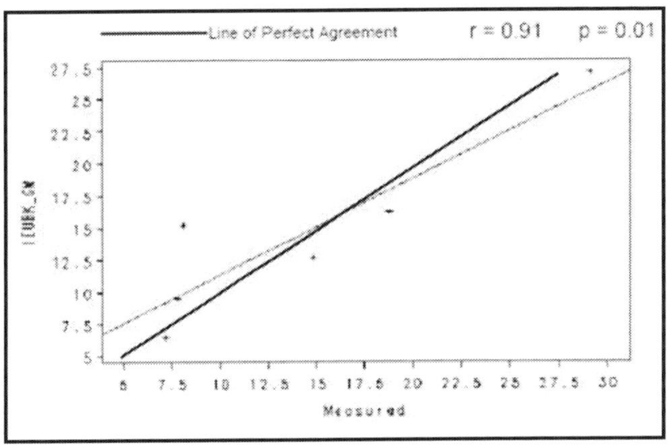

Figure 17.12 — Correlation of IEUBK Predicted Geometric Mean as a Function of Measured Blood-Lead Geometric Mean (50% availability, 70% soil, and 30% indoor dust).

* The authors acknowledge the help and assistance of Dr. Steve Levin who was at Mount Sinai Medical Center at the time of this work. Dr. Levin provided the TCDD blood analyses. Dr. Dalmon Larson, Robert Strode, and Caomhin Connell also participated in the exposure reconstruction.

of cancer was estimated with the then-available USEPA slope factor. Blood sampling and analysis for TCDD in blood serum lipids for a limited number of workers was performed by two different laboratories.

The risk of cancer determined from the dose reconstruction was plotted as a function of the average laboratory result as shown in Figure 17.13. The point that is farthest from the regression line represents a maintenance worker for which the dose reconstruction was more difficult because of some extent of undefined times and tasks at each of the operator position SEGs. The dose reconstruction for a second maintenance worker lies along the fitted line. Figure 17.13 indicates generally good precision for the dose reconstruction. If the seven year half-life determined for TCDD in former Vietnam veterans by CDC is applicable to this population of workers, accuracy within about an order of magnitude was achieved. Utilization of the CDC half-life data thus allowed further calibration of the model. It is clear from the plotted data that not only dose reconstruction for this group of workers is possible, but dose reconstruction for individual workers is a valid process although dose reconstruction was a more difficult process for maintenance workers whose job tasks varied widely from day-to-day.

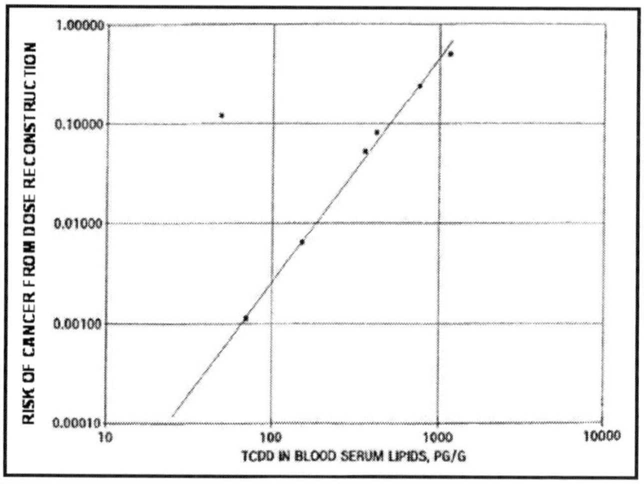

Figure 17.13 — Reconstructed Dose-Predicted Cancer Risk as a Function of Background-Corrected TCDD Serum Levels.

17.9.3 Case Study — Inter-Rater Reliability Study for Assessing Past Asbestos Exposures with Limited Information*

Another example of assessing reconstructed total exposure results involves comparison between assessments of industrial hygienists and comparison of the total exposure results with lung tissue asbestos fiber and asbestos body analysis for individual cases. Lung tissue asbestos fiber and asbestos body burden analysis is performed by pathologists to provide an indication of previous asbestos exposure. Asbestos bodies are ferruginous (iron-rich material)-coated asbestos fibers, produced in the lungs that are typically longer than 10 micrometers. A majority of retained asbestos fibers in the lung are typically not coated. The percent of asbestos fibers that the lung converts to asbestos bodies can vary considerably from individual to individual and is a function of fiber mineral type and fiber dimension. The human lung's ability to clear asbestos fibers varies by orders of magnitude based on whether a fiber is serpentine or amphibole in mineral type. The half-life of asbestos fibers in the lung correspondingly varies from weeks or days for chrysotile (serpentine) to decades for amosite, crocidolite, anthophyllite, actinolite, or tremolite (amphiboles). Another source of variability when attempting to relate lung asbestos burden to past exposure is that the time since asbestos exposure occurred varied with each of the subjects causing variable amounts of decay of retained asbestos bodies and asbestos fibers.

Four industrial hygienists (raters)** were each independently provided a data base from Dr. Victor Roggli† for 363 subjects containing a one to six word description of each subject's potential asbestos exposure history. Age at time of the pathological report, the date of the report, asbestos exposure duration, and the sex of each subject was also provided, but each industrial hygienist estimated asbestos total exposure blind with respect to the other raters and information on asbestos fiber and asbestos body lung tissue analysis. A statistical analysis of the data was performed after each industrial hygienist submitted their results. Log-normal appeared to best describe the distribution of the exposure estimates and tissue analysis concentration data. Therefore, the estimated exposure data from the raters and the lung tissue analysis data were log-transformed prior to statistical analysis.

It is important to assess the reliability and reproducibility of the raters prior to comparing their exposure assessment results with lung tissue analysis. One of the key methods of validating exposure reconstruction data consists of inter-rater comparisons. Out of the 363 case reports, all four experienced Certified Industrial

* The authors wish to thank Dr. Matthew Strand, a statistician with National Jewish Hospital and Instructor in the Graduate Biostatistics Program of The University of Colorado Health Sciences Center, Charles Redinger, and Eric Rasmuson for their assistance with the analysis of the data.

** The CIHs, consisting of Larry Birkner, William Dyson, Fred Boelter, and Jim Rasmuson had over 100 years of combined practical industrial hygiene experience and over 60 years of combined experience performing exposure reconstructions for asbestos-exposed individuals. The industrial hygienists were blind as to each other, worked independently of each other, and did not compare their results before submitting them independently for analysis to another CIH, Charles Redinger.

† Dr. Roggli, MD is a pathologist affiliated with Duke University who has clinical interests in asbestos-related diseases, pneumonoconiosis, pulmonary pathology, and analytical electron microscopy.

Hygienists (CIHs) reported estimated asbestos total exposures for 329 of the subjects. The data was unambiguous for 328 of the subjects, and therefore, the total exposure estimates for the 328 subjects were employed to compare the raters. A statistic, the Intraclass Correlation (ICC)*, has been employed to estimate reliability and reproducibility between independent estimates, in this case, between the different CIH raters. An ICC value less than 0.4 indicates poor reproducibility, An ICC value between 0.4 and 0.75 indicates fair to good reproducibility, and an ICC value greater than 0.75 indicates excellent reproducibility.[11] An ICC value of 0.759, in the good to excellent range was found for the independent exposure estimates based on the Roggli data base. Therefore, comparison of exposure reconstruction results for different industrial hygienists appeared to validate the approach of the raters.

Additional validation of the data can be performed by comparison with tissue analysis data. Comparison of the average log of the estimated total exposure data for 327 subjects with lung-tissue asbestos body burden data are shown in Figure 17.14. Three hundred twenty seven cases (327) were utilized in this analysis because a subsequent check of the exposure data base revealed uncertainty in the laboratory data for one case.

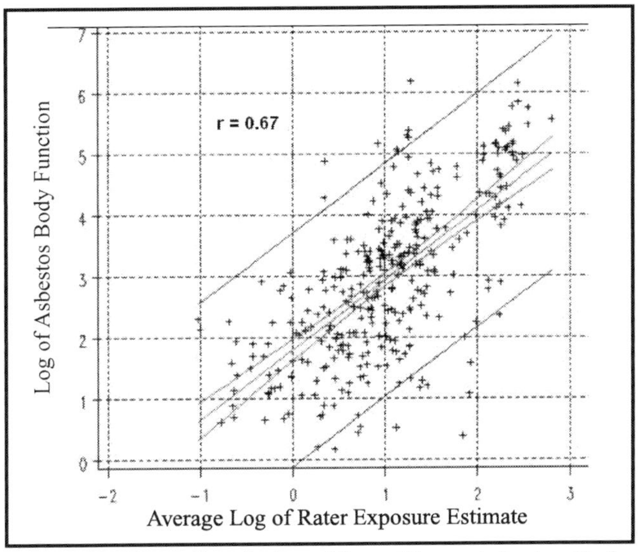

Figure 17.14 — Correlation of Lung Tissue Asbestos Body Burden Analysis with Logarithmically-Averaged Estimated Asbestos Total Exposure Determined by Four CIHs.

Figure 17.14 demonstrates a few key concepts when evaluating exposure data correlation with body burden analysis. First, on a log-log plot, the slope of the line should not be significantly different from 1 if the raters' exposure data are proportional to the total exposure for each individual regardless of exposure level and if the human lung produces asbestos bodies in a linear fashion with respect to total exposure. In this case, a slope of 1.14 with a standard error of 0.07 was found, which is borderline consistent with one. In addition, a fair amount of scatter in the data are found, which is not surprising given all of the variability of factors described above. Nevertheless, the slope of the line is very significantly different from zero ($p < 0.0001$), which is another test of the correlation. Also shown in Figure 17.14 are the confidence interval lines shown to be close around the regression line as well as the prediction interval lines. The prediction interval lines indicate variability associated with individual data points as opposed to variability of the central regression line, which is indicated by the inner confidence interval lines. A key question is whether the observed variability is in large part due to the industrial hygiene estimates or whether there are other factors at play that can be controlled to reduce variability and further assess precision and accuracy of the assessment. We will examine this question further.

The y-axis of Figure 17.14 represents the "logarithm of asbestos body function." This depiction refers to the fact that the numerical values for this axis were actually determined as a function of asbestos bodies by light microscopy (LM) per gram of wet lung tissue, asbestos bodies by scanning electron microscopy (SEM) per gram of wet lung tissue, and total asbestos fibers per gram of wet lung tissue by SEM. The latter two analyses were converted into equivalent asbestos bodies per gram of wet lung tissue through regression of the LM data as a function of both of the SEM values (asbestos bodies and asbestos fibers) to convert all data to the same LM units. Then, logarithms of the values from the three methods (in LM units) were averaged to determine the asbestos body function referenced in Figure 17.14. Determination of an ICC value between the three laboratory methods resulted in reproducibility in the same general range as the reproducibility between the four raters. Therefore, scatter in the data in Figure 17.14 due to CIH rater variability is roughly the same as that derived from the laboratory variability between the three methods. Besides those two sources of variability, there are other factors causing the variability as discussed above. But is there another source for the variability and the difference from the slope being slightly greater than one? Besides individual differences in the human lung's ability to retain and coat asbestos fibers, there remains a question about the asbestos fiber type to which each subject was exposed.

The raters estimated asbestos total exposure for each of the cases; however, some of the subjects were probably primarily exposed to chrysotile asbestos while others were exposed to significant levels or exclusively to amosite and other amphiboles. Since the human lung does not efficiently coat but rapidly eliminates chrysotile fibers, there should be little if any increase in retained asbestos bodies and asbestos fibers with increasing chrysotile total exposure. Namely, in

* Intraclass correlation (ICC) is used to measure inter-rater reliability for two or more raters when data involve continuous variables. It may also be used to assess test-retest reliability. ICC may be conceptualized as the ratio of between-groups variance to total variance. (http://faculty.chass.ncsu.edu/garson/PA765/reliab.htm)

terms of the type of plot shown in Figure 17.14, one would expect a slope equal to zero for exclusive or close to exclusive chrysotile-exposed subjects as opposed to those who were exposed to significant amphibole concentrations. The chrysotile effect tends to focus data points for such subjects downward on the plot. To test the fiber type hypothesis, quantitatively, each case was classified into four groups as shown in Figure 17.15 based on SEM analysis of fibers type for 321 of the cases. In statistical terms, a class variable was added to describe fiber type.

Shown in Figure 17.15 are the results of an ANCOVA correlation of the data taking into account fiber type as the class variable.* The ANCOVA analysis depicted in Figure 17.15 can be thought of as performing four simultaneous regression analyses for each of the fiber classifications to determine an overall correlation. Several results are noteworthy. First of all for Class Variable 1 data, representing the vast majority of the cases, a slope equal to 1.106 with a standard error of 0.085 was determined for this amphibole-exposed class. Thus, the slope for the majority of cases was not significantly different from one. Slopes for regression lines 1 ($p < 0.0001$) and 2 ($p = 0.0007$) were significantly different from zero, but as expected, Slopes 3 ($p = 0.18$) and 4 ($p = 0.20$) were not significantly different from zero. Therefore, the hypothesis that commercial chrysotile-exposed subjects would not have significantly increased asbestos bodies and asbestos fibers with increasing exposure appears to be true. In addition, the overall correlation coefficient for the analysis increased from 0.67 to 0.77, which is equivalent to stating that r^2, the fraction of explained variance to total variance in the correlation, increased from about 45% to 60%. The result of a test revealed that the increase in correlation was not due to increasing the number of variables. Including an additional class variable for occupation/means of exposure further increased the ANCOVA correlation, probably related to differing degrees of respiratory protection for the different occupational exposure groups and possibly due to other factors.

Overall, it appears the industrial hygiene raters generally were able to estimate asbestos total exposure of the subjects to within better than an order of magnitude with the very limited data regardless of total exposure level based on an error histogram associated with the ANCOVA analysis and considering both rater and laboratory data. The presence of other sources of variability would seem to

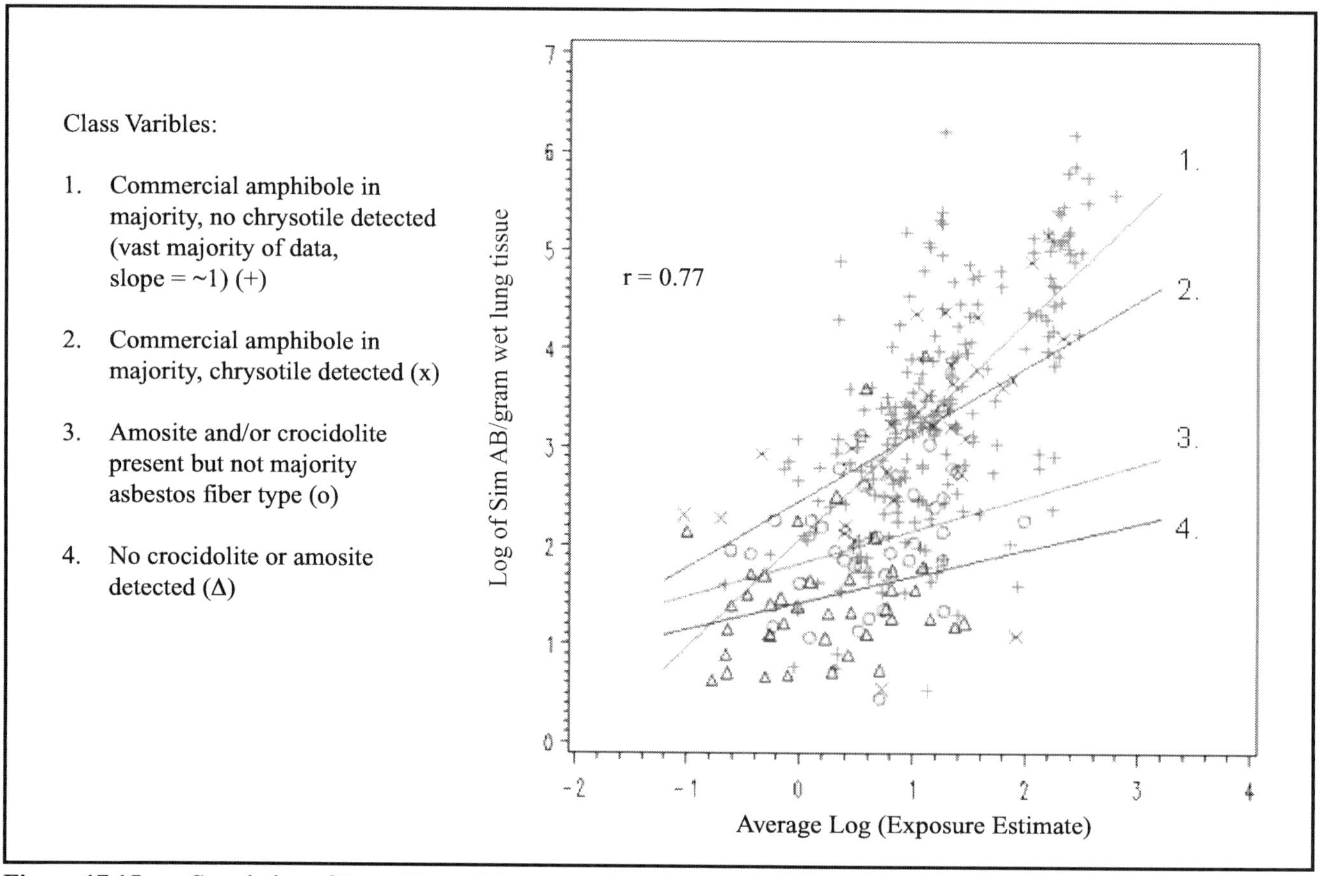

Figure 17.15 — Correlation of Lung Tissue Asbestos Body Burden Analysis with Logarithmically-Averaged Estimated Asbestos Total Exposure Determined by Four CIHs by ANCOVA, with Fiber Type Class Variable.

* Analysis of covariance (ANCOVA) is a more sophisticated method of analysis of variance. The ANCOVA was performed through SAS. (http://www.statistics.com/resources/glossary/a/ancova.php).

indicate even better precision. Overall accuracy of the raters was checked by extrapolating the data to laboratory background levels and verifying the corresponding f/cc-year range was typical of lifetime background exposures in the United States.

It would be expected that there would have been somewhat less scatter in the data if more detail had been available to the raters for each of the cases. But, it is important to remember that there remains underlying variability related to the associated SEG for each of the individual cases. Thus some degree of variability between raters is inherent in the analysis and is unavoidable.

Additional Considerations

It is not always possible to sample blood, other body fluids, or tissue to help validate past levels of exposure. The chemical, metabolite, dust, or fiber must be somewhat persistent in the body for this type of analysis to have meaning unless the exposure occurred somewhat recently. Such is not the case for many volatile organic compounds and many dusts. Therefore, laboratory-assisted studies on the efficacy of the exposure reconstruction process relate to substances such as long durable fibers, certain metals, and semi-volatile and involatile organic compounds.

The presence of asbestos bodies in the case of asbestos can provide useful checks of exposure reconstruction data based on consensus exposure thresholds associated with disease or pleural plaque causation. Clearly, however, alternate disease etiologies besides inhalation exposure of a given substance must be considered.

Measurement of other chemical agents in exhaled air, blood, and urine can be made to verify modeled contemporaneous exposure data. The ACGIH® Biological Exposure Indices (BEIs) are a useful starting point when designing such a study. However, caution must be applied when the analysis of body fluids occurs too long after exposure for chemicals that are rapidly eliminated from the body; when this problem occurs, the results of the analysis are often misinterpreted.

When it is impossible to validate an exposure assessment or an exposure assessment process for a particular individual or group by tissue or body fluid analysis, it is important for industrial hygienists to validate their exposure assessment process with substances for which they can do this. With the knowledge gained through comparison with such data, the industrial hygienist is in a better position to perform exposure reconstructions for a wide variety of substances for which this cannot be done.

The experienced industrial hygienist can provide a critical analysis of past exposure and dose. This analysis is important to understanding risk, establishing a basis for causality, and/or allocation of attributable risk.

17.10 References

1. **National Research Council:** *Risk Assessment in the Federal Government: Managing the Process.* National Academy of Sciences. Commission on Life Sciences, Committee on the Institutional Mean for Assessment of Risks to Public Health. Washington, D.C.: National Academy Press, 1983.
2. **National Academy of Science:** *Science and Decisions — Advancing Risk Assessment.* National Research Council, Committee on Improving Risk Analysis. Washington, D.C.: National Academy Press, 2008.
3. **U.S. Environmental Protection Agency (EPA):** Guidelines for Exposure Assessment. 600Z-92/001. Risk Assessment Forum. Washington, D.C.: U.S. EPA, 1992.
4. **Viet, S.M., M. Stenzel, C.P. Rennix, T.W. Armstrong, and J.R. Couch (eds.):** *Guideline on Occupational Exposure Reconstruction.* Fairfax, VA: American Industrial Hygiene Association, 2008.
5. **Ignacio, J.S. and W.H. Bullock (eds.):** *A Strategy for Assessing and Managing Occupational Exposures*, 3rd edition. Fairfax, VA, 2006.
6. UNIFAC Calculator, V. 3.0, 1996, Choy and Reible (www.hsrc.org/hscr/html/ssw/ssw-downloads.html)
7. **Winter, P.M. and J.N. Miller:** Carbon Monoxide Poisoning. *JAMA 236(13)*:1502 (1976).
8. **Ellenhorn, M.J. and D.G. Barceloux:** *Medical Toxicology. Diagnosis and Treatment of Human Poisoning.* New York, NY: Elsevier Science Publishing Co., Inc., 1988.
9. **U.S. Environmental Protection Agency (EPA):** IEUBK Software and Users' Manuals. (http://epa.gov/superfund/health/contaminants/lead/products.htm). [Accessed on April 13, 2009].
10. **SAS Institute, Inc.:** Statistical Analysis Software — SAS/STAT®. Cary, NC: SAS Institute, Inc., (http://www.sas.com/technologies/analytics/statistics/stat/). [Accessed on April 13, 2009].
11. **Rosner, B.:** *Fundamentals of Biostatistics*, 6th edition. Florence, KY: Cengage Learning, Inc., 2006. pp. 613–617.

17.11 Additional References on Exposure Reconstruction and Retrospective Exposure Assessment

Ahlborg, G.: Validity of Exposure data Obtained by Questionnaire. Two Examples from Occupational Reproductive Studies. *Scand. J. Work Environ. Health 16(4)*:284–288 (1990).

Armstrong, T.W., et al.: Retrospective Benzene and Total Hydrocarbon Exposure Assessment for a Petroleum Marketing and Distribution Worker Epidemiology Study. *Am. Ind. Hyg. Assoc. J. 57(4)*:333–343 (1996).

Axelson, O. and P. Soderkvist: Characteristics of Disease and Some Exposure Constraints. *Appl. Occup. Environ. Hyg. 6(6)*:428–435 (1991).

Ayer, H.E., et al.: A Monumental Study — Reconstruction of a 1920 Granite Shed. *Am. Ind. Hyg. Assoc. J. 34(5)*:206–211 (1973).

Balzer, J.L. and W.C. Cooper: The Work Environment of Insulating Workers. *Am. Ind. Hyg. Assoc. J. 29(3)*: 222–227 (1968).

Barnard, A.E.: Retrospective Beryllium Exposure Assessment at the Rocky Flats Environmental Technology Site. *Am. Ind. Hyg. Assoc. J. 57(9)*: 804–808 (1996).

Benke, G., S. Sim, A. Forbes, and M. Salzberg: Retrospective Assessment of Occupational Exposure to Chemicals in Community-Based Studies: Validity and Repeatability of Industrial Hygiene Panel Ratings. *Int. J. Epidemiol. 26(3)*:635–642 (1997).

Berry, G. and H.C. Lewinsohn: Dose-Response Relationships for Asbestos-Related Disease: Implications for Hygiene Standards: Part I — Morbidity. *Ann. N. Y. Acad. Sci. 330*:185–194 (1979).

Blair, A. and P.A. Stewart: Do Quantitative Exposure Assessments Improve Risk Estimates in Occupational Studies of Cancer? *Amer. J. Ind. Med. 21(1)*:53–63 (1992).

Blake C., G. Dotson, and R.D. Harbison: Assessment of Airborne Asbestos Exposure during the Servicing and Handling of Automobile Asbestos-Containing Gaskets. *Regul. Toxicol. Pharmacol. 45(2)*:214–222 (2006).

Boelter, F.W., G.N. Crawford, and D.M. Podraza: Airborne Fiber Exposure Assessment of Dry Asbestos-Containing Gaskets and Packings Found in Intact Industrial and Maritime Fittings. *Am. Ind. Hyg. Assoc. J. 63(6)*:732–740 (2002).

Boelter, F.W.: Author's reply to "Asbestos Exposure from Gasket Removal" Letter to the Editor. *Am. Ind. Hyg. Assoc. J. 64(5)*:595–597 (2003).

Boelter, F.W., J.W. Spencer, and C.E. Simmons: Heavy Equipment Maintenance Exposure Assessment: Using a Time-Activity Model to Estimate Surrogate Values for Replacement of Missing Data. *J. Occup. Environ. Hyg. 4(7)*:525–537 (2007).

Bond, G.G.: Validation of Work Histories for the Purpose of Epidemiological Studies. *Appl. Occup. Environ. Hyg. 6(6)*:521–527 (1991).

Bourguet, C.C., H. Checkoway, and B.S. Hulka: A Case-Control Study of Skin Cancer in the Tire and Rubber Manufacturing Industry. *Am. J. Ind. Med. 11(4)*:461–473 (1987).

Bouyer, J., J. Dardenne, and D. Hemon: Performance of Odds Ratios Obtained with a Job-Exposure Matrix and Individual Exposure Assessment with Special Reference to Misclassification. *Scand. J. Work Environ. Health 21(4)*:265–271 (1995).

Breslow, N.E. and N.E. Day: *Statistical Methods in Cancer Research, Volume I — The Analysis of Case-Control Studies.* Lyon, France: International Agency for Research on Cancer, 1980.

Burstyn, I., et al.: Statistical Modeling of the Determinants of Historical Exposure to Bitumen and Polycyclic Aromatic Hydrocarbons Among Paving Workers. *Ann. Occup. Hyg. 44(1)*:43–56 (2000).

Camus, M., J. Siemiatycki, and B. Meek: Nonoccupational Exposure to Chrysotile Asbestos and the Risk of Lung Cancer. *New Engl. J. Med. 338(22)*: 1565–1571 (1998).

Camus, M., et al.: Risk of Mesothelioma Among Women Living Near Chrysotile Mines Versus U.S. EPA Asbestos Risk Model: Preliminary Findings. *Ann. Occup. Hyg. 46(Suppl 1)*:95–98 (2002).

Case, B.W., et al.: Preliminary Findings for Pleural Mesothelioma among Women in the Quebec Chrysotile Mining Regions. *Ann. Occup. Hyg. 46(Suppl 1)*:128–131 (2002).

Checkoway, H., D.A. Savitz, and N.J. Heyer: Assessing the Effects of Nondifferential Misclassification of Exposures in Occupational Studies. *Appl. Occup. Environ. Hyg. 6(6)*:528–533 (1991).

Checkoway, H.: Rapporteur's Summary: Epidemiologic Evaluation of Exposure-Effect Relationships. In *Exposure Assessment for Epidemiology and Hazard Control.* Rappaport, S.M. and T.J. Smith (eds.). Chelsea, MI: Lewis Publishers, Inc., 1991. pp. 207–216.

Checkoway H. and C. Rice: Time-Weighted Averages, Peaks, and Other Indices of Exposure in Occupational Epidemiology. *Am. J. Ind. Med. 21(1)*:25–33 (1992).

Cohen, H.J. and D.R. Van Orden: Asbestos Exposure of Mechanics Performing Clutch Service on Motor Vehicles. *J. Occup. Environ. Hyg. 5(3)*:148–156 (2008).

Corn, M. and N.A. Esmen: Workplace Exposure Zones for Classification of Employee Exposures to Physical and Chemical Agents. *Am. Ind. Hyg. Assoc. J. 40(1)*:47–57 (1979).

Corn, M.: Historical Perspective on Approaches to Estimation of Inhalation Risk by Air Sampling. *Am. J. Ind. Med. 21(1)*:113–123 (1992).

Corn, J.K. and M. Corn: Changing Approaches to Assessment of Environmental Inhalation Risk: A Case Study. *Milbank Q 73(1)*:97–119 (1995).

Coughlin, S.S. and L. Chiazze, Jr.: Job-Exposure Matrices in Epidemiologic Research and Medical Surveillance. *Occup. Med. 5(3)*:633–646 (1990).

Crawford-Brown, D.J.: *Theoretical and Mathematical Foundations of Human Health Risk Analysis.* Boston, MA: Kluwer Academic Publishers, 1997.

Dement, J.M., R.L. Harris, Jr., M.J. Symons, and C.M. Shy: Exposure and Mortality Among Chrysotile Asbestos Workers. Part I: Exposure Estimates. *Am. J. Ind. Med. 4(3)*:399–419 (1983).

Dement, J.M., R.L. Harris, Jr., M.J. Symons, and C.M. Shy: Exposure and Mortality Among Chrysotile Asbestos Workers. Part II: Mortality. *Am. J. Ind. Med. 4(3)*:421–433 (1983).

Demers, P.A. et al.: Pooled Reanalysis of Cancer Mortality Among Five Cohorts of Workers in Wood-Related Industries. *Scand. J. Work Environ. Health 21(3)*:179–190 (1995).

Dewar, R., J. Siemiatycki, and M. Gerin: Loss of Statistical Power Associated with the Use of a Job-Exposure Matrix in Occupational Case-Control Studies. *Appl. Occup. Environ. Hyg. 6(6)*:508–515 (1991).

Dosemeci, M., P.A. Stewart, and A. Blair: Evaluating Occupational and Industrial Separately to Assess Exposures in Case-Control Studies. *Appl. Occup. Environ. Hyg. 4(10)*:256–259 (1989).

Dosemeci, M., et al.: Indirect Validation of a Retrospective Method of Exposure Assessment Used in a Nested Case-control Study of Lung Cancer and Silica Exposure. *Occup. Environ. Med. 51(2)*: 136–138 (1994).

Eisen, E.A., et al.: Estimation of Long Term Dust Exposures in the Vermont Granite Sheds. *Am. Ind. Hyg. Assoc. J. 45(2)*:89–94 (1984).

Esmen, N.A.: Retrospective Industrial Hygiene Surveys. *Am. Ind. Hyg. Assoc. J. 40(1)*:58–65 (1979).

Esmen, N.A.: Analysis of Strategies for Reconstructing Exposures. *Appl. Occup. Environ. Hyg. 6(6)*: 488–494 (1991).

Esmen, N.A. and M. Corn: Airborne Fiber Concentrations During Splitting Open Boxes and Boxing Bags of Asbestos. *Toxicol. Ind. Health 14(6)*:843–856 (1998).

Esmen, N.A.: Exposure Estimation in Four Major Epidemiologic Studies in the Acrylonitrile Industry. *Scand. J. Environ. Health 24(Suppl 2)*:63–70 (1998).

Fleischer W., F. Viles, R. Gade, and P. Drinker: A Health Survey of Pipe Covering Operations in Constructing Naval Vessels. *J. Ind. Hyg. Toxicol. 28(1)*:9–16 (1946).

Gamble, J. and R. Spirtas: Job Classification and Utilization of Complete Work Histories in Occupational Epidemiology. *J. Occup. Med. 18(6)*:399–404 (1976).

Gerin, M. and J. Siemiatycki: The Occupational Questionnaire in Retrospective Epidemiologic Studies: Recent Approaches in Community Studies. *Appl. Occup. Environ. Hyg. 6(6)*:495–501 (1991).

Goldberg, M., et al.: Job Exposure Matrices in Industry. *Int. J. Epidemiol. 22(Suppl 2)*:S10–S15 (1993).

Hallock, M.F., T.J. Smith, S.R. Woskie, and S.K. Hammond: Estimation of Historical Exposures to Machining Fluids in the Automotive Industry. *Am. J. Ind. Med. 26(5)*:621–634 (1994).

Hansen, J., N.H. de Klerk, A.W. Musk, and M.S.T. Hobbs: Individual Exposure Levels in People Environmentally Exposed to Crocidolite. *Appl. Occup. Environ. Hyg. 12(7)*:485–490 (1997).

Harries, P.G.: Asbestos Dust Concentrations in Ship Repairing: A Practical Approach to Improving Asbestos Hygiene in Naval Dockyards. *Ann. Occup. Hyg. 14(3)*:241–254 (1971).

Harris, R.L.: Measures to Facilitate Future Epidemiologic Studies. *Appl. Occup. Environ. Hyg. 6(6)*:555–559 (1991).

Hawkes, A.P. and J.R. Wilkins: Assessing Agreement between Two Job-Exposure Matrices. *Scand. J. Work Environ. Health 23*:140–148 (1997).

Heederik, D. and F. Hurley: Occupational Exposure Assessment: Investigating Why Exposure Measurements Vary. *Appl. Occup. Environ. Hyg. 9(1)*:71–73 (1994).

Hemon, D., et al.: Retrospective Evaluation of Occupational Exposures in Cancer Epidemiology: A European Concerted Action of Research. *Appl. Occup. Environ. Hyg. 6(6)*:541–546 (1991).

Henn, S.A., et al.: Task- and Time-Dependent Weighting Factors in a Retrospective Exposure Assessment of Chemical Laboratory Workers. *J. Occup. Environ. Hyg. 4(2)*:71–79 (2007).

Herrick, R.F. and P. Stewart: International Workshop on Retrospective Exposure Assessment for Occupational Epidemiologic Studies — Preface. *Appl. Occup. Environ. Hyg. 6(6)*:417–420 (1991).

Hertzman, C., K. Teschke, H. Dimich-Ward, and A. Ostry: Validity and Reliability of a Method for Retrospective Evaluation of Chlorphenate Exposure in the Lumber Industry. *Amer. J. Ind. Med. 14(6)*:703–713 (1988).

Hewett, P., et al.: Rating Exposure Control Using Bayesian Decision Analysis. *J. Occup. Environ. Hyg. 3(10)*: 568–581 (2006).

Hoppin, J.A., et al.: Use of a Life Events Calendar Approach to Elicit Occupational History From Farmers. *Am. J. Ind. Med. 34(5)*:470–476 (1998).

Hornung, R.W.: Statistical Evaluation of Exposure Assessment Strategies. *Appl. Occup. Environ. Hyg. 6(6)*:516–520 (1991).

Hornung, R.W., et al.: An Experimental Design Approach to Retrospective Exposure Assessment. *Am. Ind. Hyg. Assoc. J. 57*:251–256 (1996).

Iwatsubo, Y., et al.: Pleural Mesothelioma: Dose-Response Relation at Low Levels of Asbestos Exposure in a French Population-Based Case-Control Study. *Am. J. Epidemiol. 148(2)*:133–142 (1998).

Jarvholm, B. and A. Sanden: Estimating Asbestos Exposure: A Comparison of Methods. *J. Occup. Med. 29(4)*:361–363 (1987).

Johnson, J.V.: Measuring Work Organization Exposure Over the Life Course with a Job-Exposure Matrix. *Scand. J. Work Environ. Health 19(1)*:21–28 (1993).

Karjalainen, A., et al.: Pulmonary Asbestos Bodies and Asbestos Fibers as Indicators of Exposure. *Scand. J. Work Environ. Health 22(1)*:34–38 (1996).

Kauppinen, T. and T. Partanen: Use of Plant- and Period-Specific Job-Exposure Matrices in Studies on Occupational Cancer. *Scand. J. Work Environ. Health 14(3)*:161–167 (1988).

Kauppinen, T.P.: Development of a Classification Strategy of Exposure for Industry-Based Studies. *Appl. Occup. Environ. Hyg. 6(6)*:482–487 (1991).

Kauppinen, T.P., B. Pannett, D.A. Marlow, and M. Kogevinas: Retrospective Assessment of Exposure Through Modeling in a Study on Cancer Risks Among Workers Exposed to Phenoxy Herbicides, Chlorophenols and Dioxins. *Scand. J. Work Environ. Health 20(4)*:262–271 (1994).

Kauppinen, T.P.: Exposure Assessment — A Challenge for Occupational Epidemiology. *Scand. J. Environ. Health 22(6)*:401–403 (1996).

Kriebel, D., N.L. Sprince, E.A. Eisen, and I.A. Greaves: Pulmonary Function in Beryllium Workers: Assessment of Exposure. *Br. J. Ind. Med. 45(2)*:83–92 (1988).

Kunzli, N., T. Kelly, J. Balmes, and I.B. Tager: Reproducibility of Retrospective Assessment of Outdoor Time-Activity Patterns as an Individual Determinant of Long-Term Ambient Ozone Exposure. *Int. J. Epidemiol. 26(6)*:1258–1271 (1997).

Lave, C.A. and J.G. March: *An Introduction to Models in the Social Sciences*. Lanham, MD: University Press of America, Inc., 1993.

Leung, H. and D.J. Paustenbach: Techniques for Estimating the Percutaneous Absorption of Chemicals Due to Occupational and Environmental Exposure. *Appl. Environ. Hyg. 9(3)*:187–197 (1994).

Linet, M.S., et al.: Comparison of Methods for Determining Occupational Exposure in a Case-Control Interview Study of Chronic Lymphocytic Leukemia. *J. Occup. Med. 29(2)*:136–141 (1987).

Lippmann, M.: Research Needs: Retrospective Exposure Assessment for Occupational Epidemiology. *Appl. Occup. Environ. Hyg. 6(6)*:550–554 (1991).

Macaluso, M., et al.: Leukemia and Cumulative Exposure to Butadiene, Styrene and Benzene among Workers in the Synthetic Rubber Industry. *Toxicology 113 (1-3)*:190-202 (1996).

Mangold C., K. Clark, A. Madl, and D. Paustenbach: An Exposure Study of Bystanders and Workers During the Installation and Removal of Asbestos Gaskets and Packing. *J. Occup. Environ. Hyg. 3(2)*:87–98 (2006).

McLaughlin, J.K., et al.: A Nested Case-Control Study of Lung Cancer Among Silica Exposed Workers in China. *Br. J. Ind. Med. 49(3)*:167–171 (1992).

McMichael, A.J., R. Spirtas, J.F. Gamble, and P.M. Tousey: Mortality Among Rubber Workers: Relationship to Specific Jobs. *J. Occup. Med. 18(3)*:178–185 (1976).

Miettinen, O.S. and C.E. Rossiter: Man-Made Mineral Fibers and Lung Cancer. Scand. *J. Work Environ. Health 16(4)*:221–231 (1990).

Miligi, L., and G. Masala: Methods of Exposure Assessment for Community-Based Studies: Aspects Inherent to the Validation of Questionnaires. *Appl. Occup. Environ. Hyg. 6(6)*:502–507 (1991).

Morinaga, K., K. Nakamura, N. Kohyama, and T. Kishimoto: A Retrospective Cohort Study of Male Workers Exposed to PVA Fibers. *Ind. Health 37(1)*:18–21 (1999).

Moulin, J.J.: A Meta-Analysis of Epidemiologic Studies of Lung Cancer in Welders. *Scand. J. Work Environ. Health 23(2)*:104–113 (1997).

Nicholson, W.J., D.A. Holaday, and H. Heimann: Direct and Indirect Occupational Exposure to Insulation Dusts in United States Shipyards. In *Safety and Health in Ship Building and Ship Repairing*, Proceedings of a Symposium, Aug-Sept 1971, Helsinki, Finland. Occupational Safety and Health Series No.27. Geneva, Switzerland: International Labor Office, 1971. pp. 37–47.

Nolan R.P., A.M. Langer, and R. Wilson: A Risk Assessment for Exposure to Grunerite Asbestos (Amosite) in an Iron Ore Mine. *Proc. Natl. Acad. Sci. USA 96(7)*:3412–3419 (1999).

Paik, N.W., R.J. Walcott, and P.A. Brogan: Worker Exposure to Asbestos During Removal of Sprayed Material and Renovation Activity in Buildings Containing Sprayed Material. *Am. Ind. Hyg. Assoc. J. 44(6)*:428–432 (1983).

Pastides, H., R. Austin, S. Lemeshow, J. Klar, and K. Mundt: A Retrospective-Cohort Study of Occupational Exposure to Hexavalent Chromium. *Am. J. Ind. Med. 25*:663–675 (1994).

Pastides H.: Managing Measurable and Perceived Risk in the Occupational Setting. *J. Ambulatory Care Manage. 17(2)*:44–52 (1994).

Park, R.M., M.A. Silverstein, and F.E. Mirer: Characteristics of Worker Populations: Exposure Considerations in the Selection of Study Populations and Their Analysis. *Appl. Occup. Environ. Hyg. 6(6)*:436–440 (1991).

Paustenbach, D.J.: Health Risk Assessment and the Practice of Industrial Hygiene. *Am. Ind. Hyg. Assoc. J. 51(7)*:339–351 (1990).

Paustenbach D.J., et al.: Reevaluation of Benzene Exposure for the Pliofilm (Rubberworker) Cohort (1936–1976). *J. Toxicol. Environ. Health 36(3)*: 177–231 (1992).

Paustenbach, D.J.: The Practice of Exposure Assessment: A State-of-the-Art Review. *J. Toxicol. Environ. Health B. Crit. Rev. 3(3)*:179–291 (2000).

Paustenbach D.J., R.O. Richter, B.L. Finley, and P.J. Sheehan: An Evaluation of the Historical Exposures of Mechanics to Asbestos in Brake Dust. *Appl. Occup. Environ. Hyg. 18(10)*:786–804 (2003).

Peto, J.: Dose-Response Relationships for Asbestos Related Disease: Implications for Hygiene Standards Part II. Mortality. *Ann. N.Y. Acad. Sci. 330*:195–203 (1979).

Plato N., S. Krantz, P. Gustavsson, T. Smith, and P. Westerholm: Fiber Exposure Assessment in the Swedish Rock Wool and Slag Wool Production Industry in 1938–1990." *Scand. J. Work Environ. Health 21(5)*:345–352 (1995).

Plato, N., T. Tornling, C. Hogstedt, and S. Krantz: An Index of Past Asbestos Exposure as Applied to Car and Bus Mechanics. *Ann. Occup. Hyg. 39(4)*: 441–454 (1995).

Quinn, M.M., et al.: Historical Cohort Study of US Man-Made Vitreous Fiber Production Workers: VIII. Exposure-Specific Job. Analysis. *J. Occup. Environ. Med. 43(9)*:824–834 (2001).

Ramachandran, G., and J.H. Vincent: A Bayesian Approach to Retrospective Exposure Assessment. *Appl. Occup. Environ. Hyg. 14(8)*:547–557 (1999).

Ramachandran, G.: Retrospective Exposure Assessment using Bayesian Methods. *Ann. Occup. Hyg. 45(8)*:651–667 (2001).

Ramachandran, G.: *Occupational Exposure Assessment for Air Contaminants; Reconstructing Historical Exposures* (Chapter 18). Boca Raton, FL: CRC Press, 2005. pp. 299–321.

Rappaport, S.: Exposure Assessment Strategies. In *Industrial Hygiene Science Series: Exposure Assessment for Epidemiology and Hazard Control.* Rappaport, S.M., and T.J. Smith (eds.). Chelsea, Michigan: Lewis Publishers, 1991. pp. 219–249.

Rappaport, S.M.: Selection of the Measures of Exposure for Epidemiology Studies. *Appl. Occup. Environ. Hyg. 6(6)*:448–457 (1991).

Rice, C.H., R.L. Harris, Jr., J.C. LUmsden, and M.J. Symons: Reconstruction of Silica Exposure in the North Carolina Dusty Trades. *Am. Ind. Hyg. Assoc. J. 45(10)*:689–696 (1984).

Rice, C.H.: (1991); Retrospective Exposure Assessment: A Review of Approaches and Directions for the Future. In *Exposure Assessment for Epidemiology and Hazard Control.* Rappaport, S.M. and T.J. Smith (eds.). Chelsea, MI: Lewis Publishers, Inc., 1991. p. 185–197.

Rinsky, R.A., R.J. Young, and A.B. Smith: Leukemia in Benzene Workers. *Am. J. Ind. Med. 2*:217–245 (1981).

Ripple, S.R.: Looking Back at Nuclear Weapons Facilities: The Use of Retrospective Health Risk Assessments. *Environ. Sci. Technol. 26(7)*:1270–1277 (1992).

Ripple, S.R., T.E. Widner, and T.R. Morgan: Past Radionuclide Releases from Routine Operations at Rocky Flats. *Health Phys. 71(4)*:502–509 (1996).

Rom, W.N., et al.: A Study of Dermatitis in Trona Miners and Millers. *J. Occup. Med. 25(4)*:295–299 (1983).

Romundstad, P.R., A. Ronneberg, H.L. Leira, and T. Bye: Health Survey of Former Workers in a Norwegian Coke Plant: Part 1. Estimation of Historical Exposures. *Occup. Environ. Med. 55(9)*:616–621 (1998).

Rybicki, B.A., et al.: Comparability of Different Methods of Retrospective Exposure Assessment of Metals in Manufacturing Industries. *Am. J. Ind. Med. 31(1)*: 36–43 (1997).

Rybicki, B.A., et al.: Intra-and Inter-Rater Agreement in the Assessment of Occupational Exposure to Metals. *Int. J. Epidemiol. 27*:269–273 (1998).

Sanderson, W.T., M.R. Petersen, and E.M. Ward: Estimating Historical Exposures of Workers in a Beryllium Manufacturing Plant. *Am. J. Ind. Med. 39(2)*:145–157 (2001).

Saracci, R.: Ten years of Epidemiologic Investigations on Man-Made Mineral Fibers and Health. *Scand. J. Work Environ. Health 12(Suppl 1)*:5–11 (1986).

Schneider, S.: Past Approaches to Retrospective Epidemiological Occupational Exposure Assessment: The Use and Misuse of Exposure Assessment. *Appl. Occup. Environ. Hyg. 6(6)*:547–549 (1991).

Schneider, T., I. Olsen, O. Jorgensen, and B. Lauersen: Evaluation of Exposure Information. *Appl. Occup. Environ. Hyg. 6(6)*:475–481 (1991).

Seidman, H., I.J. Selikoff, and S.K. Gelb: Mortality Experience of Amosite Asbestos Factory Workers: Dose-Response Relationships 5 to 40 Years After Onset of Short-Term Work Exposure. *Am. J. Ind. Med. 10(5-6)*:479–514 (1986).

Seixas, N.S., N.J. Heyer, E.A.E. Welp, and H. Checkoway: Quantification of Historical Dust Exposures in the Diatomaceous Earth Industry. *Ann. Occup. Hyg. 41(5)*:591–604 (1997).

Seixas, N.S., T.G. Robbins, and M. Becker: A Novel Approach to the Characterization of Cumulative Exposure for the Study of Chronic Occupational Disease. *Am. J. Epidemiol. 137(4)*:463–471 (1993).

Seixas, N.S. and H. Checkoway: Exposure Assessment in Industry Specific Retrospective Occupational Epidemiology Studies. *Occup. Environ. Med. 52(10)*:625–633 (1995).

Seixas, N.S. and L. Sheppard: Maximizing Accuracy in Precision Using Individual and Grouped Exposure Assessments. *Scand. J. Work Environ. Health 22(2)*:94–101 (1996).

Shen, N., et al.: Epidemiology of Occupational and Environmental Risk Factors Related to Ovarian Cancer. *Scand. J. Work Environ. Health 24(3)*: 175–182 (1998).

Shy, C.M.: Epidemiologic Principles and Methods for Occupational Health Studies. In *Occupational Respiratory Diseases*. Merchant, J.A., (ed.). Washington, D.C.: Government Printing Office, 1986. pp. 103–136.

Siemiatycki, J. et al.: Reliability of an Expert Rating Procedure for Retrospective Assessment of Occupational Exposures in Community-Based Case-Control Studies. *Am. J. Ind. Med. 31(3)*:280–286 (1997).

Siemietycki, J. and P. Boffetta: Invited Commentary: Is it Possible to Investigate the Quantitative Relation between Asbestos and Mesothelioma in a Community-based Study? *Am. J. Epidemiol. 148(2)*:143–147 (1998).

Smith, T.J., S.K. Hammond, F. Laidlaw, and S. Fine: Respiratory Exposures Associated with Silicon Carbide Production: Estimation of Cumulative Exposures for an Epidemiological Study. *Br. J. Ind. Med. 41(1)*:100–108 (1984).

Smith, T.J.: Development and Application of a Model for Estimating Alveolar and Interstitial Dust Levels. *Ann. Occup. Hyg. 29(4)*:495–516 (1985).

Smith, T.J.: Exposure Assessment for Occupational Epidemiology. *Am. J. Ind. Med. 12(3)*:249–268 (1987).

Smith, T.J., S.K. Hammond, M. Hallock, andS.R. Woskie: Exposure Assessment for Epidemiology: Characteristics of Exposure. *Appl. Occup. Environ. Hyg. 6(6)*:441–447 (1991).

Smith, T.J.: Occupational Exposure and Dose Over Time: Limitations of Cumulative Exposure. *Am. J. Ind. Med. 21(1)*:35–51 (1992).

Smith, T.J., S.K. Hammond, and O. Wong: Health Effects of Gasoline Exposure. I. Exposure Assessment for U.S. Distribution Workers. *Environ. Health Perspect. 101(Suppl 6)*:13–21 (1993).

Smith, T.J., et al.: Historical Cohort Study of US Man-Made Vitreous Fiber Production Workers: VII. Overview of the Exposure Assessment. *J. Occup. Environ. Med. 43(9)*:809–823 (2001).

Steenland, K. (ed.): *Case Studies in Occupational Epidemiology.* Oxford, U.K.: Oxford University Press, 1993.

Stewart, P.A., et al.: Estimating Historical Exposure to Formaldehyde in a Retrospective Mortality Study. *Appl. Ind. Hyg. 1(1)*:34–41 (1986).

Stewart, P. and R.F. Herrick: Issues in Performing Retrospective Exposure Assessment. *Appl. Occup. Environ. Hyg. 6(6)*:421–427 (1991).

Stewart, P.A., et al.: Highlights of the 1990 Leesburg, Virginia International Workshop on Retrospective Exposure Assessment for Occupational Epidemiology Studies. *Scand. J. Work Environ. Health 17(4)*:281285 (1991).

Stewart, P.A., et al.: Evaluation of Several Retrospective Exposure Assessment Methods for Estimating Exposure Levels in a Study of Workers Exposed to Acrylonitrile. *Rev. Epidemiol. Sante Publique 40(Suppl 1)*:S83 (1992).

Stewart, P.A. and M. Dosemeci: A Bibliography for Occupational Exposure Assessment for Epidemiologic Studies. *Am. Ind. Hyg. Assoc. J. 55(12)*:1178–1187 (1994).

Stewart, P.A., et al.: Exposure Assessment for a Study of Workers Exposed to Acrylonitrile. II. A Computerized Exposure Assessment Program. *Appl. Occup. Environ. Hyg. 10(8)*:698–706 (1995).

Stewart, P.A., P.S.J. Lees, and M. Francis: Quantification of Historical Exposures in Occupational Cohort Studies. *Scand. J. Work Environ. Health 22(6)*:405–414 (1996).

Stewart, P.A., et al.: Exposure Assessment for a Study of Workers Exposed to Acrylonitrile. *Scand. J. Work Environ. Health 24(Suppl 2)*:42–53 (1998).

Stewart, P.: Challenges to Retrospective Exposure Assessment. *Scand. J. Work Environ. Health 25(6)*:505–510 (1999).

Stewart, P.A., R. Carel, C. Schairer, and A. Blair: Comparison of Industrial Hygienists' Exposure Evaluations for an Epidemiologic Study. *Scand. J. Work Environ. Health 26(1)*:44–51 (2000).

Suarez-Almazor, M.E., C.L. Soskolne, K. Fung, and G.S. Jhangri: Empirical Assessment of the Effect of the Effect of Different Summary Worklife Exposure Measures on the Estimation of Risk in Case-Referent Studies of Occupational Cancer. *Scand. J. Work Environ. Health 18(4)*:233–241 (1992).

Takahashi, B., et al.: Relation between Lung Asbestos Fibre Burden and Exposure Indices Based on Job History. *Occup. Environ. Med. 51(7)*:461–469 (1994).

Teschke, K., et al.: A Comparison of Exposure Estimates by Worker Raters and Industrial Hygienists. *Scand. J. Work Environ. Health 15(6)*:424–429 (1989).

Theriault, G.P., W.A. Burgess, L.J. DiBerardinis, and J.M. Peters: Dust Exposure in the Vermont Granite Sheds. *Arch. Environ. Health 28(1)*:12–17 (1974).

Theriault, G.P., J.M. Peters, and L.J. Fine: Pulmonary Function in Granite Shed Workers of Vermont. *Arch. Environ. Health 28(1)*:18–22 (1974).

Tielemans, E., et al.: Assessment of Occupational Exposures in a General Population: Comparison of Different Methods. *Occup. Environ. Med. 56(3)*:145–151 (1999).

Tuomi, T., et al.: Occupational Exposure to Asbestos as Evaluated from Work Histories and Analysis of Lung Tissues from Patients with Mesothelioma. *Brit. J. Ind. Med. 48(1)*:48–52 (1991).

van Tongeren, M., et al.: Retrospective Exposure Assessment for a Cohort Study into Respiratory Effects of Acid Anhydrides. *Occup. Environ. Med. 55(10)*:692–696 (1998).

Vetter, R., P.A. Stewart, M. Dosemeci, and A. Blair: Validity of Exposure in One Job as a Surrogate for Exposure in a Cohort Study. *Am. J. Ind. Med. 23(4)*:641–651 (1993).

Widner, T.E., S.R. Ripple, and J.E. Buddenbaum: Identification and Screening Evaluation of Key Historical Materials and Emission Sources at the Oak Ridge Reservation. *Health Phys. 71(4)*:457–469 (1996).

Williams, P., D. Paustenbach, J.L. Balzer, and C. Mangold: Retrospective Exposure Assessment of Airborne Asbestos Related to Skilled Craftsmen at a Petroleum Refinery in Beaumont, Texas (1940–2006). *J. Toxicol. Environ. Health A 70(13)*:1076–1107 (2007).

Elementary Calculus Principles Used in Modeling

By
Mark Nicas, PhD, CIH

I. Introduction

Understanding and using mathematical models does not require a working knowledge of calculus. On the other hand, the ability to formulate new models appropriate for specific applications of interest does require calculus skills. The purpose of this section is to: (i) review basic concepts, notation, and rules of calculus operations; (ii) demonstrate the application of calculus principles to exposure assessment computations, including the derivation of a well mixed room model with a simple contaminant emission rate function; and (iii) explain how a steady-state concentration can be found from a mass balance equation without relying on calculus. Note that the review of calculus presented here is mostly limited to techniques used to derive several mathematical models discussed in the monograph.

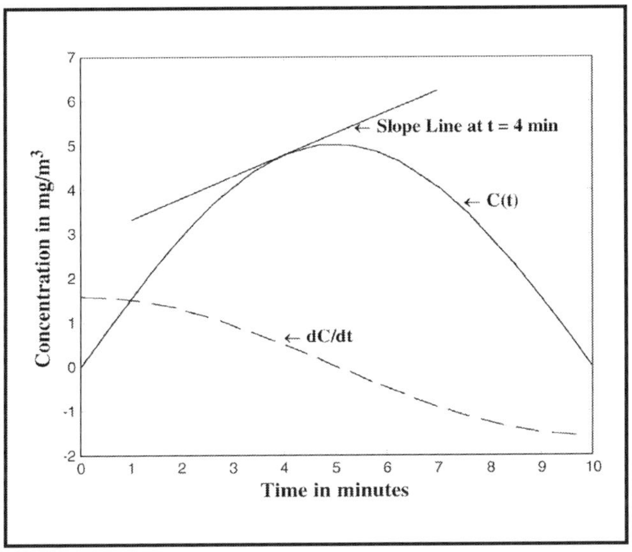

Figure I.1 — The first derivative of C(t) has a simple geometric meaning — it is the slope of the tangent line to the C(t) curve. Figure I.1 depicts a tangent line drawn to C(t) at t = 4.0 min. The valve of C(4.0) is 4.76 mg/m^3. The value of the slope of the tangent line is 0.485 mg/m^3-min, and is equal to the function dC/dt evaluated at t = 4.0 min.

Notation: If the contaminant concentration C varies with time t, then C is a function of t which we denote C(t). In some books, C is termed a dependent variable and t is termed an independent variable. The unit of C used in these notes is mg/m^3 (mass by volume), but C can also be expressed in ppm (volume by volume). The unit of t used in these notes is minutes. The curve labeled "C(t)" in Figure I.1 depicts a hypothetical function that starts at zero, increases to a maximum value at t = 5 min, then decreases to zero by t = 10 min.

II. Differentiation

The first derivative of C(t) is the instantaneous rate of change in the value of C(t). Because C(t) has the unit mg/m^3, the first derivative of C(t) has the unit mg/m^3 · min (or mg/m^3 per minute). If the first derivative has a positive value at some reference point in time t_r, then C is increasing in value at time t_r. If the first derivative has a negative value at t_r, then C is decreasing in value at t_r. If the first derivative is zero at t_r, then C is neither increasing nor decreasing in value at t_r, which indicates that $C(t_r)$ is either a maximum value or a minimum value.

The first derivative of C(t) is variously denoted:

$$\frac{dC}{dt}, \quad \frac{dC(t)}{dt}, \quad C'(t)$$

The curve labeled "dC/dt" in Figure I.1 depicts the first derivative of the function C(t) shown in Figure I.1. From t = 0 to t = 5 min when C(t) is increasing in value, dC/dt is positive. When C(t) reaches a maximum value at t = 5 min, dC/dt is zero. From t = 5 to t = 10 min when C(t) is decreasing, dC/dt is negative. Note that dC/dt is also a function of t, because its value varies with time.

The first derivative of C(t) has a simple geometric meaning — it is the slope of the tangent line to the C(t) curve. Figure I.1 depicts a tangent line drawn to C(t) at t = 4.0 min. The value of C(4) is 4.76 mg/m^3. The value of the slope of the tangent line is 0.485 mg/m^3 · min, and is equal to the function dC/dt evaluated at t = 4.0 min.

Rules of Differentiation:

* Consider the function $f(t) = t^n$, where n is a real number, but $n \neq 0$. The first derivative of f(t) is:

Rule #1:

$$\frac{df}{dt} = \frac{d[t^n]}{dt} = n \cdot t^{n-1}$$

The power of t is decreased by subtracting 1 from n, and the remaining term t^{n-1} is multiplied by the original number n.

Example:

$$\text{Let } f(t) = t^3\text{. Then: } \frac{df}{dt} = \frac{d[t^3]}{dt} = 3 \cdot t^2$$

Example:

$$\text{Let } f(t) = t\text{. Then: } \frac{df}{dt} = \frac{d[t]}{dt} = 1 \cdot t^0 = 1$$

* Consider the function $f(t) = e^{k \cdot t}$, where k is a constant, real number. The quantity e is often denoted exp, and is the base of the natural logarithm. The quantity e is a number approximately equal to 2.718. The first derivative of f(t) is:

Rule #2:

$$\frac{df}{dt} = \frac{d[e^{k \cdot t}]}{dt} = k \cdot e^{k \cdot t}$$

The constant k is brought down from the power term and becomes a multiplier of the original term $e^{k \cdot t}$.

Example:

$$\text{Let } f(t) = e^{3t}\text{. Then: } \frac{df}{dt} = \frac{d[e^{3 \cdot t}]}{dt} = 3 \cdot e^{3 \cdot t}$$

Example:

Let $f(t) = e^{-3t}$. Consider the constant k to equal -3. As before:

$$\text{Let } f(t) = e^{3t}\text{. Then: } \frac{df}{dt} = \frac{d[e^{-3 \cdot t}]}{dt} = -3 \cdot e^{-3 \cdot t}$$

* Consider the function $g(t) = k \cdot f(t)$, where k is a constant and f is a function of t. The first derivative of g(t) is:

Rule #3:

$$\frac{dg}{dt} = \frac{d[k \cdot f(t)]}{dt} = k \cdot \frac{d[f(t)]}{dt}$$

The derivative of a constant times a function is the constant times the derivative of the function.

Example:

$$\text{Let } g(t) = 3 \cdot t^2\text{. Then: } \frac{dg}{dt} = \frac{d[3 \cdot t^2]}{dt} = 3 \cdot \frac{d[t^2]}{dt} = 3 \cdot 2 \cdot t^1 = 6 \cdot t$$

* Consider the function $f(t) = k$, where k is a constant. The first derivative of f(t) is:

Rule #4:

$$\frac{df}{dt} = \frac{d[k]}{dt} = 0$$

The derivative of a constant is zero. By definition, the value of a constant does not change over time. Because the first derivative is the instantaneous rate of change, the first derivative of a constant must be zero.

* Consider the function $h(t) = f(t) + g(t)$. The first derivative of h(t) is:

Rule #5:

$$\frac{dh}{dt} = \frac{d[f(t) + g(t)]}{dt} = \frac{d[f(t)]}{dt} + \frac{d[g(t)]}{dt}$$

The derivative of the sum of two functions is the sum of the derivatives of the individual functions.

Example:

Let $h(t) = e^{3t} + t^3$.

$$\text{Then: } \frac{dh}{dt} = \frac{d[e^{3 \cdot t} + t^3]}{dt} = \frac{d[e^{3 \cdot t}]}{dt} + \frac{d[t^3]}{dt} = 3 \cdot e^{3 \cdot t} + 3 \cdot t^2$$

* Consider the function $h(t) = f(t) \cdot g(t)$. The first derivative of h(t) is:

Rule #6:

$$\frac{dh}{dt} = \frac{d[f(t) \cdot g(t)]}{dt} = \frac{df}{dt} \cdot g(t) + f(t) \cdot \frac{dg}{dt}$$

The derivative of the product of two functions is the sum of the derivative of the first function times the second function, and the first function times the derivative of the second function. Rule #6 is termed the "product rule."

Example:

Let $h(t) = C(t) \cdot e^{\left(\frac{Q}{V}\right) \cdot t}$, where C(t) is an unspecified function, and the function $e^{\left(\frac{Q}{V}\right) \cdot t}$ is of the general form $e^{k \cdot t}$ (in this case, $k = Q/V$). Then:

$$\frac{dh}{dt} = \frac{d[C(t) \cdot e^{\left(\frac{Q}{V}\right) \cdot t}]}{dt} = \frac{dC}{dt} \cdot e^{\left(\frac{Q}{V}\right) \cdot t} + C(t) \cdot \frac{d[e^{\left(\frac{Q}{V}\right) \cdot t}]}{dt} =$$

$$\frac{dC}{dt} \cdot e^{\left(\frac{Q}{V}\right) \cdot t} + C(t) \cdot \frac{Q}{V} e^{\left(\frac{Q}{V}\right) \cdot t}$$

The above answer comes into play in finding the concentration function for a one-box model.

---

* Consider the function $g(t) = \ln[f(t)]$, where ln denotes the natural logarithm. To explain the meaning of g(t), if we wished to evaluate g(t) at some specified value of t denoted t_r, we would first evaluate f(t) at t_r, and then find the natural logarithm of this number. The first derivative of g(t) is:

Rule #7:

$$\frac{dg}{dt} = \frac{d[\ln[f(t)]]}{dt} = \frac{\frac{df}{dt}}{f(t)}$$

The numerator is the first derivative of f(t), while the denominator is f(t).

Example:

Let $g(t) = \ln[t]$, in which case $f(t) = t$. The first derivative of g(t) is:

$$\frac{dg}{dt} = \frac{d[\ln[t]]}{dt} = \frac{1}{t}$$

---

* Consider the functions $f(t) = \cos(t)$, and $g(t) = \sin(t)$. The first derivatives are:

Rule #8a:

$$\frac{df}{dt} = \frac{d[\cos(t)]}{dt} = -\sin(t)$$

Rule #8b:

$$\frac{dg}{dt} = \frac{d[\sin(t)]}{dt} = \cos(t)$$

These two trigonometric functions can be useful in modeling emissions that have an inherent periodicity. For purposes of illustration, the C(t) curve depicted in Figure 1 is the function $C(t) = 5 \cdot \sin(\pi \cdot t/10)$, where the argument $\pi \cdot t/10$ is in radians. For example, at t = 5 min, C(5) = $5 \cdot \sin(\pi/2) = 5$ mg/m^3, because $\sin(\pi/2) = 1$. This particular C(t) function is not realistic for t > 10 min, because C(t) would take on negative values.

III. Integration

Integration can be considered the process of anti-differentiation, at least with regard to indefinite integrals that are not evaluated over specific time points. If f(t) denotes a function of time t, and f(t) is the first derivative of another function of time F(t), then F(t) is the *indefinite integral* of f(t) and is defined by the expression:

$$F(t) = \int f(t) \cdot dt = \int \frac{dF}{dt} \cdot dt$$

where the quantity dt is an infinitesimal interval of time. Because f(t) is the first derivative of F(t), or dF/dt = f(t), the anti-derivative or indefinite integral of f(t) must equal F(t).

The *definite integral* of f(t) over the interval $a \le t \le b$ is represented by the term:

$$F(b) - F(a) = \int_a^b f(t) \cdot dt$$

The meanings of the expressions for the indefinite and definite integrals will be made clear in the following examples.

Rules Of Integration:

* Consider the function $f(t) = t^n$, where n is a real number, but $n \neq -1$. The indefinite integral of f(t) is:

Rule #9a:

$$F(t) = \int t^n \cdot dt = \frac{t^{n+1}}{n+1} + I$$

where I is a constant of integration. The value of I is determined by an initial value condition that must be specified. In the examples that illustrate the rules of integration, we will usually assume that I = 0.

Example:

Find the indefinite integral of $f(t) = t^2$. Assume I = 0.

$$F(t) \int = t^2 \cdot dt = \frac{t^3}{3}$$

---

* Consider the function $f(t) = \frac{1}{t}$. The indefinite integral of f(t) is:

Rule #9b:

$$F(t) \int \frac{1}{t} \cdot dt = = \ln[t] + I$$

where I is a constant of integration. Recall that in the example illustrating Rule #7, we saw that $\frac{d[\ln[t]]}{dt} = \frac{1}{t}$. Therefore, the anti-derivative (i.e., integral) of $\frac{1}{t}$ must be ln[t] (plus a constant).

---

* Consider the function $f(t) = e^{k \cdot t}$, where k is a constant. The indefinite integral of f(t) is:

Rule #10:

$$F(t) = \int e^{k \cdot t} \cdot dt = \frac{e^{k \cdot t}}{k} + I$$

The numerator is the original term $e^{k \cdot t}$ and the denominator is the constant k.

Example:

Find the indefinite integral of $f(t) = e^{-3t}$. Assume that I = 0.

$$F(t) = \int e^{-3 \cdot t} \cdot dt = -\frac{e^{-3 \cdot t}}{3}$$

---

* Consider the function $g(t) = k \cdot f(t)$, where k is a constant. The indefinite integral G(t) is:

Rule #11:

$$G(t) = \int k \cdot f(t) \cdot dt = k \cdot \int f(t) \cdot dt = k \cdot F(t) + I$$

The indefinite integral of a constant times a function is the constant times the indefinite integral of the function.

Example:

Find the indefinite integral of $g(t) = 3 \cdot t^2$. Assume that I = 0.

$$G(t) = \int 3 \cdot t^2 \cdot dt = 3 \cdot \int t^2 \cdot dt = 3 \cdot \frac{t^3}{3} = t^3$$

In the above example, the function $3 \cdot t^2$ is of the form $n \cdot t^{n-1}$. Note that by Rules #8 and #10:

$$\int n \cdot t^{n-1} \cdot dt = n \cdot \int t^{n-1} \cdot dt = n \cdot \frac{t^{n-1+1}}{n-1+1} = t^n$$

Recall that in differentiation Rule #1, the first derivative of t^n was $n \cdot t^{n-1}$. Therefore, the anti-derivative (i.e., integral) of $n \cdot t^{n-1}$ must be t^n (plus a constant which we've assumed here to be zero).

---

* Consider the function f(t). The definite integral of f(t) over the interval $a \le t \le b$ is:

Rule #12:

$$\int_a^b f(t) \cdot dt = F(t)\Big|_{t=a}^{t=b} = F(t=b) - F(t=a)$$

Note that we do not need to consider a constant of integration I in evaluating F(a) and F(b), because the constant would end up canceling from the answer.

Example:

Find the definite integral of $f(t) = t^2$ over the interval $0 \le t \le 2$.

$$\int_0^2 t^2 \cdot dt = \left[\frac{t^3}{3}\right]\Bigg|_{t=0}^{t=2} = \left[\frac{2^3}{3}\right] - \left[\frac{0^3}{3}\right] = \frac{8}{3}$$

Example:

Find the definite integral of $f(t) = e^{-3t}$ over the interval $0 \le t \le 1$.

$$\int_0^1 e^{-3 \cdot t} \cdot dt = \left[-\frac{e^{-3 \cdot t}}{3}\right]\Bigg|_{t=0}^{t=1} = \left[-\frac{e^{-3}}{3}\right] - \left[-\frac{e^{0}}{3}\right] = \frac{1}{3} - \frac{.0498}{3} = 0.317$$

---

* Consider the function f(t) = k, where k is a constant. The indefinite integral F(t) is:

Rule #13:

$$F(t) = \int k \cdot dt = k \cdot t + I$$

The indefinite integral of a constant with respect to the independent variable t is the constant times t. Another way to think about this rule is that $k = k \cdot t^0$. By Rule #10, the indefinite integral of $k \cdot t^0$ is k times the indefinite integral of t^0. By Rule #8, the indefinite integral of t^0 is t. Therefore, the indefinite intergral of $k \cdot t^0$ is $k \cdot t$ (plus a constant).

---

* Consider the function h(t) = f(t) + g(t). The indefinite integral H(t) is:

Rule #14:

$$H(t) = \int h(t) \cdot dt = \int [f(t) + g(t)] \cdot dt = \int f(t) \cdot dt + \int g(t) \cdot dt$$

The indefinite integral of the sum of two functions is the sum of the indefinite integrals of the individual functions.

Example:

Let $h(t) = e^{-3 \cdot t} + t$. The indefinite integral H(t) is:

$$H(t) = \int e^{-3 \cdot t} \cdot dt + \int t \cdot dt = -\frac{e^{-3 \cdot t}}{3} + \frac{t^2}{2}$$

III. Applications of Calculus Principles

(a) Computing a Time-Weighted Average Concentration:

Let C(t) denote a concentration function. The time-weighted average concentration over the interval $a \le t \le b$, denoted $\overline{C}_{a\to b}$, is the following:

Rule #15:

$$\overline{C}_{a\to b} = \frac{1}{b-a} \cdot \int_a^b C(t) \cdot dt$$

The overbar on C signifies that an average quantity is being considered (in this case a TWA).

Example:

Let $C(t) = 100 \cdot e^{-.05\cdot t}$, in mg/m³. This function corresponds to the exponential purging of airborne contaminant from a well mixed room receiving .05 air changes per minute, or 3 air changes per hour. What is the TWA concentration from t = 0 to t = 30 min?

$$\overline{C}_{0\to 30} = \frac{1}{30-0} \cdot \int_0^{30} 100 \cdot e^{-.05\cdot t} \cdot dt$$

$$= \frac{100}{30} \cdot \int_0^{30} e^{-.05\cdot t} \cdot dt \qquad \text{(by Rule \#11)}$$

$$= \frac{100}{30} \cdot \left[-\frac{e^{-.05\cdot t}}{.05} \right] \Bigg|_{t=0}^{t=30} \qquad \text{(by Rule \#12)}$$

$$= \left[\frac{10}{3} \cdot \frac{e^{0}}{.05} \right] - \left[\frac{10}{3} \cdot \frac{e^{-1.5}}{.05} \right]$$

$$= 51.8 \text{ mg/m}^3$$

Example:

Let $C(t) = 100 \cdot [1 - e^{-.05\cdot t}]$, in mg/m³. This function corresponds to the buildup of airborne contaminant in a well mixed room receiving .05 air changes per minute. What is the TWA concentration from t = 0 to t = 30 min?

$$\overline{C}_{0\to 30} = \frac{1}{30-0} \cdot \int_0^{30} 100 \cdot [1 - e^{-.05\cdot t}] \cdot dt$$

$$= \frac{100}{30} \cdot \int_0^{30} [1 - e^{-.05\cdot t}] \cdot dt \qquad \text{(by Rule \#11)}$$

$$= \frac{10}{3} \cdot \left[\int_0^{30} 1 \cdot dt - \int_0^{30} e^{-.05\cdot t} \cdot dt \right] \qquad \text{(by Rules \#11 and \#14)}$$

$$= \frac{10}{3} \cdot \left[t \Big|_{t=0}^{t=30} - \left(-\frac{e^{-.05\cdot t}}{.05} \right) \Bigg|_{t=0}^{t=30} \right] \qquad \text{(by Rules \#12 and \#13)}$$

$$= \frac{10}{3} \cdot (30-0) - \frac{10}{3} \cdot \left[\frac{e^{0}}{.05} - \frac{e^{-.05\cdot t}}{.05} \right]$$

$$= 100 \text{ mg/m}^3 - 51.8 \text{ mg/m}^3$$

$$= 48.2 \text{ mg/m}^3$$

(b) The Relation between the Contaminant Emission Rate Function and the Contaminant Mass Remaining or Emitted:

Consider the situation in which a room contains a finite source of contaminant with an initial mass M_0 in mg. For example, M_0 might equal 10,000 mg of a volatile chemical applied to a room surface. The mass of contaminant remaining on the surface decreases over time as the chemical evaporates into room air. For example, at t = 5 min after the start of emission perhaps only 5,000 mg remains on the surface. Denote the remaining mass function as M(t). The function that describes the instantaneous rate of contaminant emission into room air (in mg/min) is denoted G(t). The functions G(t) and M(t) are related as follows:

Rule #16:

$$G(t) = \left| \frac{dM}{dt} \right|$$

In words, G(t) is the absolute value of the first derivative of the remaining mass function.

This relation makes intuitive sense. Consider a small time interval from t to t + Δ t. The mass lost from the surface in this interval is M(t) - M(t + Δ t). However, mass lost from the surface is emitted into room air, so the mass emitted in this same time interval is M(t) - M(t + Δ t). The average *rate* of emission (in mg/min) is the mass emitted divided by the interval length Δ t:

$$\text{Average Rate of Emission} = \frac{M(t) - M(t + \Delta t)}{\Delta t}$$

As the interval length gets smaller and smaller, or $\Delta t \to 0$, the average rate of emission becomes the instantaneous rate of emission G(t) in mg/min:

$$G(t) = \frac{M(t) - M(t + \Delta t)}{\Delta t}, \text{ as } \Delta t \rightarrow 0$$

The mathematical definition of the first derivative of a function M(t) is:

$$\frac{dM}{dt} = \frac{M(t + \Delta t) - M(t)}{\Delta t}, \text{ as } \Delta t \rightarrow 0$$

Therefore, $G(t) = -\, dM/dt$. Because the contaminant mass remaining on the surface is always decreasing, dM/dt is always negative and G(t) is always positive. In our application, Rule #15 is another way of writing $G(t) = -\, dM/dt$.

Example:

Assume that the remaining mass function is described by $M(t) = 10000 \cdot (1 - t/60)$. This signifies that at t = 0, the initial mass = 10,000 mg, and that by t = 60 min, no mass remains in the source (i.e., all the contaminant mass has been emitted into room air). What is the emission rate function?

$$G(t) = \left| \frac{d\left[10000 \cdot \left[1 - \frac{t}{60}\right]\right]}{dt} \right|$$

$$= \left| \frac{d[10000]}{dt} - \frac{d\left[\frac{10000 \cdot t}{60}\right]}{dt} \right| \qquad \text{(by Rules \#3 and \#5)}$$

$$= \left| 0 - \frac{10000}{60} \right| \qquad \text{(by Rules \#1 and \#4)}$$

$$= 167 \text{ mg/min}$$

The emission rate function is a constant value, because it corresponds to the slope of a linearly decreasing remaining mass function.

The emission rate function G(t) has an analogous relationship to the function EM(t), which describes the total (cumulative) contaminant mass emitted into room air from the start of the emission process Let t_r denote some reference point in time after the start of emission. Then:

Rule #17a:

$$EM(t_r) = \int_0^{t_r} G(t) \cdot dt$$

Example:

Assume that the emission rate function is described by $G(t) = 10000 \cdot e^{-0.25 \cdot t}$. What total mass of contaminant is emitted from t = 0 to t = 5 min?

$$EM(5) = \int_0^5 10000 \cdot e^{-0.25 \cdot t}\, dt$$

$$= 10000 \cdot \int_0^5 e^{-0.25 \cdot t}\, dt \qquad \text{(by Rule \# 11)}$$

$$= 10000 \cdot \left[- \frac{e^{-0.25 \cdot t}}{0.25} \right]\Bigg|_{t=0}^{t=5} \qquad \text{(by Rules \#10 and \#12)}$$

$$= 10000 \cdot \left[\frac{e^{0}}{0.25} - \frac{e^{-1.25}}{0.25} \right]$$

$$= 28{,}540 \text{ mg}$$

Rule #16a can be generalized to consider the cumulative mass emitted in an interval $a \le t \le b$, where $a > 0$.

Rule #17b:

$$EM_{a \rightarrow b} = \int_a^b G(t) \cdot dt$$

Example:

Again consider the emission rate function is described by $G(t) = 10000 \cdot e^{-0.25 \cdot t}$. What is the contaminant mass emitted from t = 1 to t = 2 min?

$$EM_{1 \rightarrow 2} = \int_1^2 10000 \cdot e^{-0.25 \cdot t}\, dt$$

$$= 10000 \cdot \left[- \frac{e^{-0.25t}}{0.25} \right]\Bigg|_{t=1}^{t=2}$$

$$= 10000 \cdot \left[\frac{e^{-0.25}}{0.25} - \frac{e^{-0.5}}{0.25} \right]$$

$$= 6{,}891 \text{ mg}$$

(3) Finding the Concentration Function C(t) for a Well Mixed Room:

Deriving a model that predicts contaminant concentrations usually begins by forming a "mass balance" equation. Consider a very small time interval denoted dt. The change in the contaminant mass in well mixed room air in interval dt equals the mass entering room air minus the mass leaving room air during this interval. If Δ denotes the change in mass in interval dt, then:

Rule #18:

$$\Delta = (\text{Mass Entering Room Air in dt}) - (\text{Mass Leaving Room Air in dt})$$

Let dC denote the change in contaminant concentration in room air (in mg/m^3) over the interval dt. If V denotes the room volume (in m^3), then:

$$\Delta = V \cdot dC$$

The product V · dC yields mass in mg as the quantity, because m^3 (for V) times mg/m^3 (for dC) leaves mg as the unit.

Next, consider the instantaneous contaminant emission rate G(t) due to a source in the room. The mass emitted in the very small interval dt is essentially equal to G(t) · dt. This product yields mass in mg, because mg/min [for G(t)] times min (for dt) leaves mg as the unit. If contaminant does not enter the room in supply or infiltration air, then:

$$\text{Mass Entering Room Air in dt} = G(t) \cdot dt$$

Finally, consider how the contaminant in room air can be lost from room air. First, if the room has an air exhaust rate Q (in m^3/min), the contaminant mass lost due to exhausting room air in the interval dt is Q · C(t) · dt, where C(t) denotes the contaminant concentration in room air. This product yields mass in mg, because m^3/min (for Q) times mg/m^3 (for concentration C) times min (for dt) leaves mg as the unit.

Second, contaminant might be removed from room air by other processes, for example, adsorption onto or absorption into materials in the room, or chemical decomposition or reaction. The rate at which such removal occurs is usually difficult to specify *a priori*. For simplicity, let's denote k_1 as the fraction of contaminant removed from room air per minute due to all these processes. The contaminant mass lost from room air due to these various processes in the interval dt is $k_1 \cdot V \cdot C(t) \cdot dt$. To explain how the units cancel properly, the contaminant mass in room air is V · C(t) in mg, a fraction k_1 in min^{-1} is lost per unit time, and this loss occurs over interval length dt in min. So, $(mg)(min^{-1})(min)$ yields mg as the unit.

Given these two mechanisms of mass loss, we can write:

$$\text{Mass Leaving Room Air in dt} =$$

$$Q \cdot C(t) \cdot dt + k_1 \cdot V \cdot C(t) \cdot dt = [Q + k_1 \cdot V] \cdot C(t) \cdot dt$$

Based on the preceding assumptions regarding mechanisms of mass gain and mass loss, Rule #18 can be mathematically stated as follows:

Rule #19a:

$$V \cdot dC = G(t) \cdot dt - [Q + k_1 \cdot V] \cdot C(t) \cdot dt$$

Note that if the air entering the room via the mechanical supply or infiltration were contami-nated with the substance of interest, we would need an additional term for mass gain in Rule #18. For example, if Q m^3/min of mechanical supply air contained the contaminant at a constant concentration denoted C_S, the additional mass gain term would be $C_S \cdot Q \cdot dt$. For simplicity, this mass gain term will not be considered.

We can rearrange the above expression by: (i) dividing both sides by V and dt, and (ii) moving the term containing C(t) from the right-hand side to the left-hand side. This yields:

Rule #19b:

$$\frac{dC}{dt} + \frac{[Q + k_1 \cdot V]}{V} \cdot C(t) = \frac{G(t)}{V}$$

This expression is an ordinary, linear, first-order differential equation. The "solution" is the equation for C(t). The latter can be found by the following expression:

Rule #20:

$$C(t) = \left[e^{-\left[\frac{Q + k_1 \cdot V}{V}\right] \cdot t}\right] \left[\int \frac{G(t)}{V} \cdot e^{\left[\frac{Q + k_1 \cdot V}{V}\right] \cdot t} \cdot dt\right] + I \cdot e^{-\left[\frac{Q + k_1 \cdot V}{V}\right] \cdot t}$$

where I is a constant of integration which is specified by an initial value condition, most typically $C(0) = C_0$.

Solving the Rule #19 equation for C(t) can be relatively simple, or impossible, depending on the exact expression for the function G(t). If G(t) = G, which signifies a constant contaminant emission rate, the solution is relatively simple. In this case, the integrand in the second bracketed term is handled as follows:

$$\int \frac{G}{V} \cdot e^{\left[\frac{Q + k_1 \cdot V}{V}\right] \cdot t} \cdot dt = \frac{G}{V} \cdot \int e^{\left[\frac{Q + k_1 \cdot V}{V}\right] \cdot t} \cdot dt \qquad \text{(by Rule \#10)}$$

$$= \frac{G}{V} \cdot \frac{V}{(Q + k_1 \cdot V)} \cdot e^{\left[\frac{Q + k_1 \cdot V}{V}\right] \cdot t} \qquad \text{(by Rule \#9)}$$

$$= \frac{G}{(Q + k_1 \cdot V)} \cdot e^{\left[\frac{Q + k_1 \cdot V}{V}\right] \cdot t}$$

When the above term is multiplied by the first bracketed term $e^{-\left[\frac{Q + k_1 \cdot V}{V}\right] \cdot t}$, the exponent terms cancel and become 1. Recall that: $e^x \cdot e^{-x} = e^{x-x} = e^0 = 1$. Therefore, the expression in Rule #19 becomes:

$$C(t) = \frac{G}{(Q + k_1 \cdot V)} + I \cdot e^{-\left[\frac{Q+k_1 \cdot V}{V}\right] \cdot t}$$

To find the constant of integration I, we need to specify the concentration value at some point in time. Because the starting contaminant concentration at t = 0 is usually zero, let C(0) = 0. Further, because the term $e^{-\left[\frac{Q+k_1 \cdot V}{V}\right] \cdot t}$ equals 1 when t = 0, the above expression simpifies to:

$$0 = \frac{G}{(Q + k_1 \cdot V)} + I$$

It's easy to see that:

$$I = -\frac{G}{(Q + k_1 \cdot V)}$$

By using this expression for I, we find that:

$$C(t) = \frac{G}{(Q + k_1 \cdot V)} \cdot \left[1 - e^{-\left[\frac{Q+k_1 \cdot V}{V}\right] \cdot t}\right]$$

This is the equation that predicts the contaminant concentration in well mixed room air. If the mass loss mechanisms quantified by the rate constant k_1 are deemed unimportant, then k_1 is set equal to zero, and the solution becomes the traditional expression for C(t) in a well mixed room starting at a zero concentration value:

$$C(t) = \frac{G}{Q} \cdot \left[1 - e^{-\left[\frac{Q}{V}\right] \cdot t}\right]$$

Note that Rule #20 applies only when G(t) does not depend on the contaminant concentration in room air, which is to say that G(t) does not contain the function C(t). For example, Rule #20 could not be used to derive the backpressure model equation for an organic compound with a low vapor pressure, because the backpressure G(t) function is:

$$G(t) = G_0 \cdot \left[1 - \frac{C(t)}{C_{sat}}\right]$$

where $G_0 = G(0)$, the starting emission rate, and where C_{sat} is the contaminant's saturated vapor concentration in air. In this case, there is a rule that is analogous to Rule #20, but its explanation is beyond the scope of these notes.

(4) The Relation between Rules #6 and #20:

Rule #20 is based on Rule #6, the "product rule" for differentiation. The following discussion explains the relationship. Rule #19b is rewritten here as Equation (a):

$$\text{(a)} \quad \frac{dC}{dt} + \frac{[Q + k_1 \cdot V]}{V} \cdot C(t) = \frac{G(t)}{V}$$

Both sides of Equation (a) are multiplied by the term $e^{\left[\frac{Q+k_1 \cdot V}{V}\right] \cdot t}$ to yield Equation (b):

$$\text{(b)} \quad \frac{dC}{dt} \cdot e^{\left[\frac{Q+k_1 \cdot V}{V}\right] \cdot t} + \frac{[Q + k_1 \cdot V]}{V} \cdot e^{\left[\frac{Q+k_1 \cdot V}{V}\right] \cdot t} \cdot C(t) = \frac{G(t)}{V} \cdot e^{\left[\frac{Q+k_1 \cdot V}{V}\right] \cdot t}$$

Consider the left-hand side of Equation (b). It is equal to the derivative of the product of the functions C(t) and $e^{\left[\frac{Q+k_1 \cdot V}{V}\right] \cdot t}$, or:

$$\frac{d\left[C(t) \cdot e^{\left[\frac{Q+k_1 \cdot V}{V}\right] \cdot t}\right]}{dt} = \frac{dC}{dt} \cdot e^{\left[\frac{Q+k_1 \cdot V}{V}\right] \cdot t} + C(t) \cdot \frac{[Q + k_1 \cdot V]}{V} \cdot e^{\left[\frac{Q+k_1 \cdot V}{V}\right] \cdot t}$$

This is the product rule. In fact, the above expression is similar to the example used to illustrate the product rule on page 4. At this point, Equation (b) is rewritten as:

$$\text{(c)} \quad \frac{d\left[C(t) \cdot e^{\left[\frac{Q+k_1 \cdot V}{V}\right] \cdot t}\right]}{dt} = \frac{G(t)}{V} \cdot e^{\left[\frac{Q+k_1 \cdot V}{V}\right] \cdot t}$$

Now, we find the indefinite integral (with respect to time t) of both sides of Equation (c):

$$\text{(d)} \quad \int \frac{d\left[C(t) \cdot e^{\left[\frac{Q+k_1 \cdot V}{V}\right] \cdot t}\right]}{dt} \cdot dt = \int \frac{G(t)}{V} \cdot e^{\left[\frac{Q+k_1 \cdot V}{V}\right] \cdot t} \cdot dt$$

The term on the left-hand side indicates that one is finding an anti-derivative, or:

$$\int \frac{d\left[C(t) \cdot e^{\left[\frac{Q+k_1 \cdot V}{V}\right] \cdot t}\right]}{dt} \cdot dt = C(t) \cdot e^{\left[\frac{Q+k_1 \cdot V}{V}\right] \cdot t} + I^*$$

where I* denotes a constant of integration. The above identity is substituted into Equation (d) to yield:

$$\text{(e)} \quad C(t) \cdot e^{\left[\frac{Q+k_1 \cdot V}{V}\right] \cdot t} + I^* = \int \frac{G(t)}{V} \cdot e^{\left[\frac{Q+k_1 \cdot V}{V}\right] \cdot t} \cdot dt$$

I* is subtracted from both sides of Equation (e). Next, both sides are multiplied by the term $e^{-\left[\frac{Q+k_1 \cdot V}{V}\right] \cdot t}$ to yield:

$$\text{(f)}\quad C(t) = \left[e^{-\left[\frac{Q+k_1\cdot V}{V}\right]\cdot t}\right]\left[\int \frac{G(t)}{V}\cdot e^{\left[\frac{Q+k_1\cdot V}{V}\right]\cdot t}\cdot dt\right] - I^* \cdot e^{-\left[\frac{Q+k_1\cdot V}{V}\right]\cdot t}$$

If we let I = – I*, we have recovered Rule #20.

Finding the Steady State Concentration without Calculus

C(t) is a dynamic concentration function in that it describes a concentration that changes over time. However, in many exposure assessment applications we are concerned with the steady state concentration value, if it exists. This steady state concentration value, denoted C_{SS}, can be found by evaluating C(t) as $t \rightarrow \infty$, but that approach requires one to first derive C(t). In the alternative, C_{SS} can be derived from the mass balance equation, Rule #19b, without resorting to integration techniques. We simply note that at the steady state the concentration value is not changing by definition, in which case $dC/dt = 0$.

Example:

Consider the emission rate function G(t) = G. In this case, Rule #18b is:

$$\frac{dC}{dt} + \frac{Q + k_1 \cdot V}{V} \cdot C(t) = \frac{G}{V}$$

At the steady state, this expression becomes:

$$\frac{Q + k_1 \cdot V}{V} \cdot C_{SS} = \frac{G}{V}$$

Note that the term dC/dt has disappeared, and that the term C(t) is replaced by C_{SS}. By rearrangement:

$$C_{SS} = \frac{G}{Q + k_1 \cdot V}$$

Example:

Consider finding C_{SS} for the backpressure model. For simplicity, let $k_1 = 0$. The mass balance equation for the backpressure model is:

$$V \cdot dC = G_0 \cdot \left(1 - \frac{C(t)}{C_{sat}}\right) \cdot dt - Q \cdot C(t) \cdot dt$$

After rearrangement:

$$\frac{dC}{dt} + \left[\frac{G_0}{V \cdot C_{sat}} + \frac{Q}{V}\right] \cdot C(t) = \frac{G_0}{V}$$

At the steady state, this expression becomes:

$$\left[\frac{G_0}{V \cdot C_{sat}} + \frac{Q}{V}\right] \cdot C_{SS} = \frac{G_0}{V}$$

By further rearrangement:

$$C_{SS} = \frac{G_0}{Q + \dfrac{G_0}{C_{sat}}}$$

The denominator term is unusual due to the quotient G_0/C_{sat}. However, this quotient yields the correct unit of m^3/min.

Example:

Consider the coupled mass balance equations for the near-field/far-field model with a constant contaminant emission rate G in the near field:

$$\text{(a)}\quad \frac{dC_{NF}}{dt} = \frac{G}{V_{NF}} + \frac{\beta \cdot C_{FF}(t)}{V_{NF}} - \frac{\beta \cdot C_{NF}(t)}{V_{NF}}$$

$$\text{(b)}\quad \frac{dC_{FF}}{dt} = \frac{\beta \cdot C_{NF}(t)}{V_{FF}} - \frac{(Q + \beta) \cdot C_{FF}(t)}{V_{FF}}$$

At the steady state these expressions become:

$$\text{(c)}\quad 0 = \frac{G}{V_{NF}} + \frac{\beta \cdot C_{FF,SS}}{V_{NF}} - \frac{\beta \cdot C_{NF,SS}}{V_{NF}}$$

$$\text{(d)}\quad 0 = \frac{\beta \cdot C_{NF,SS}}{V_{FF}} - \frac{(Q + \beta) \cdot C_{FF,SS}}{V_{FF}}$$

where $C_{NF,SS}$ and $C_{FF,SS}$ denote the respective steady state concentrations. Consider Equation (c). We multiply both sides by V_{NF} (which removes it from the expression), and then rearrange to find $C_{NF,SS}$ in terms of $C_{FF,SS}$:

$$\text{(e)}\quad C_{NF,SS} = \frac{G}{\beta} + C_{FF,SS}$$

We substitute this identity for $C_{NF,SS}$ into Equation (d) to yield:

$$\text{(f)}\quad 0 = \frac{\beta}{V_{FF}} \cdot \left(\frac{G}{\beta} + C_{FF,SS}\right) - \frac{(Q + \beta) \cdot C_{FF,SS}}{V_{FF}}$$

We multiply both sides of Equation (f) by V_{FF} (which removes it from the expression), and then rearrange to find $C_{FF,SS}$:

$$\text{(g)} \quad C_{FF,SS} = \frac{G}{Q}$$

By substituting this identity for $C_{FF,SS}$ into Equation (e), we find that:

$$\text{(h)} \quad C_{NF,SS} = \frac{G}{\beta} + \frac{G}{Q}$$

Equations (g) and (h) are the same steady state concentrations that we found by first deriving $C_{FF}(t)$ and $C_{NF}(t)$, and then evaluating these dynamic concentration functions as $t \rightarrow \infty$. The above approach is simple by comparison.

Appendix II: Sources of Vapor Pressure Data

By
Patricia H. Reinke

1.0 Introduction

Vapor pressures of pure substances are available in most chemical property references, including:

- *NIOSH Pocket Guide to Chemical Hazards*[1],
- *Chemical Engineer's Handbook*, 8th edition[2],
- *Hazardous Chemicals Desk Reference*, 6th edition[3],
- *Handbook of Physical Properties of Liquids and Gases*, 3rd edition[4],
- *Thermodynamic Data for Pure Compounds.*[5]

2.0 Changes in Vapor Pressure with Temperature

As one might anticipate, the number of molecules with sufficient energy to escape from a liquid and evaporate into the gas phase increases with liquid temperature, making vapor pressure a strong function of temperature. It is therefore important to note the temperature when using reference vapor pressure data. An example of temperature dependency is hown in Figure II-1, the vapor pressure/temperature curve for isopropyl alcohol.

2.1 Clausius-Clapeyron Equation

The simplest representations of the vapor pressure-temperature relationships are derived from the Clausius-Clapeyron equation:

$$\frac{d\,(p_{vap})}{dT} = \frac{\Delta H_{vap}}{T(V_g - V_l)} \qquad \text{(II-1)}$$

Where

H_{vap} = difference in enthalpy between the saturated liquid and the saturated vapor, or the heat of vaporization

T = absolute temperature

$(V_g - V_l)$ = the difference between the molar volume of the saturated gas and the saturated liquid

When pressures are low (aroud or below 1 atmosphere), V_g is usually significantly greater than V_l. (At 1 atm and 25°C [77°F], the molar volume of acetone vapor is approximately 421 times the molar volume of the liquid.)

$$V_g >> V_L \qquad \text{(II-2)}$$

At low pressures, the ideal gas law can be used to estimate V_g:

$$V_g = \frac{RT}{p_{vap}} \qquad \text{(II-3)}$$

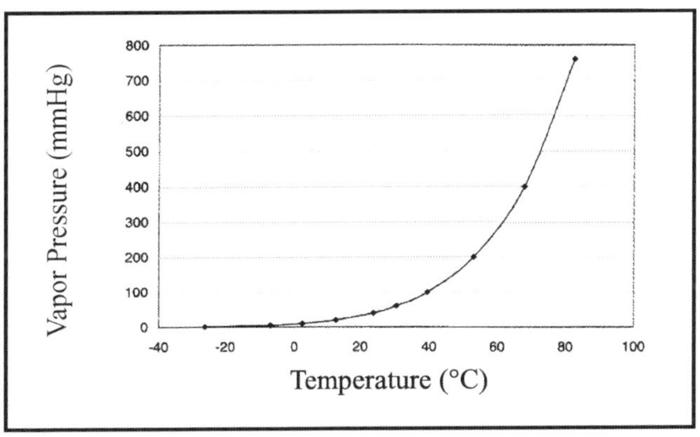

Figure II.1 — Vapor pressure/temperature curve for isopropyl alcohol.

Which, when combined with Equation (II-1), gives:

$$\frac{d\,p_{vap}}{p_{vap}} = \frac{\Delta H_{vap}}{R}\,\frac{dT}{T^2}$$

$$\ln p_{vap} = \frac{-\Delta H_{vap}}{RT}\,1/T$$

$$\log p_{vap} = \frac{-\Delta H_{vap}}{2.303\,R}\,1/T \qquad \text{(II-4)}$$

Equation (II-4) implies that if the heat of vaporization is constant over the temperature interval of interest, a plot of log pvap against 1/T — where T is absolute temperature — should give a straight line with slope ($-\Delta H_{vap}$)/(2.303R). Figure II-2 shows a plot of the isopropyl alcohol data from Figure II-1 in this format. The isopropanol data are very linear when plotted as log pvap against 1/T, illustrating how Equation (II-4) can accurately describe vapor pressure-temperature behavior over small temperature ranges.

Equation (II-4) has been integrated (assuming constant ΔH_{vap}) from the observed temperature to the normal boiling point to give the following:

$$\log\left[\frac{p_{vapnb}}{p_{vapob}}\right] = \frac{\Delta H_{vap}}{2.303\,R}\left[\frac{1}{T_{ob}} - \frac{1}{T_{nb}}\right] \qquad \text{(II-5)}$$

At the normal boiling point, the vapor pressure of the liquid equals standard atmospheric pressure, 760 mmHg (101.3 kPa). Therefore, if the normal boiling point temperature and the heat of vaporization are known, the vapor pressure at the observed temperature can be predicted with Equation (II-5).

The Clausius-Clapeyron equation is an exact thermodynamic relationship. A number of vapor pressure correlation equations are an outgrowth of the Clausius-Clapeyron. These assume constant $\Delta H/_{vap}/R$ and have the form:

$$\mathrm{Ln}(p_{vap}) = A - B/T \qquad \text{(II-6)}$$

Equation (II-6) is fairly good at approximating vapor pressures over small temperature intervals and is therefore good for interpolating between data points (see Figure II.2). For large temperature ranges and strongly associating materials, however, it represents vapor pressure data poorly.[(6)]

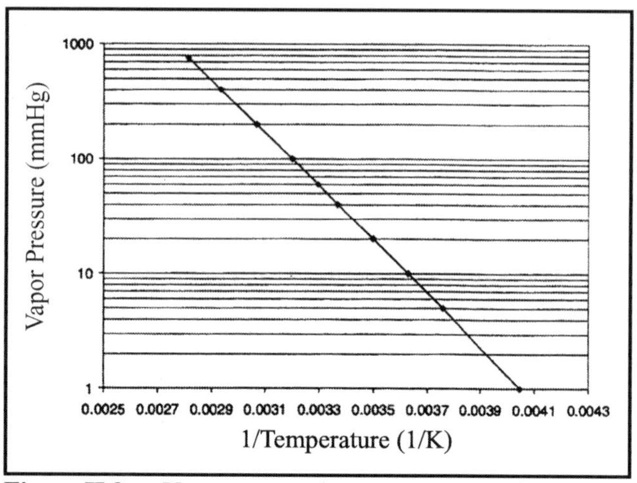

Figure II.2 — Vapor pressure/temperature curve for isopropyl alcohol.

2.2 Antoine Vapor Pressure Correlation

Antoine proposed a modification to empirically describe vapor pressure changes over limited temperature ranges. The Antoine equation has the form:

$$\ln (p_{vap}) = A - \frac{B}{T + C} \qquad \text{(II-7)}$$

Antoine constants (A, B, and C) have been calculated from experimental data on many chemicals. Among the sources of Antoine constants are:

- *The Properties of Gases and Liquids*[(6)],
- *The Vapour Pressures of Pure Substances*[(7)], and
- *Thermodynamic Data for Pure Compounds*.[(5)]

It is important to use the Antoine equation only in the range of temperatures for which the constants were evaluated. Beyond this range, serious errors may result.

2.3 The Wagner Equation

One Successful method for correlating vapor pressure data over wide ranges of temperature is the Wagner equation.[(6)] In its simplest form, the Wagner equation is:

$$\ln \frac{p_{vap}}{p_c} = \frac{a\,(1\ T/T_c) + b\,(1\ T/T_c)^{1.5} + c\,(1\ T/T_c)^3 + d\,(1\ T/T_c)^6}{T/T_c} \qquad \text{(II-8)}$$

Where

p_c = critical pressure of pure substance
T_c = absolute critical temperature of pure substance
T = absolute temperature

(NOTE: The Wagner equation has several forms. The form presented here is the most common, but there are others. Be sure to check which equation constants are to be used before using the Wagner equation).

Constants for the Wagner equation can be found in:

- *The Properties of Gases and Liquids*[(6)],
- *The Vapour Pressures of Pure Substances*[(7)], and
- *Thermodynamic Data for Pure Compounds*.[(5)]

3.0 Predicting Vapor Pressures when Data are Unavailable

When data on vapor pressures or correlations based on actual data are not available, it is best to use a laboratory and obtain accurate experimental data. As can be seen from Figure II.2, a plot of log p_{vap} vs. 1/T can be used successfully to interpolate between a few data points in the temperature range of interest.

If, however, it is not possible to obtain any vapor pressure data, there are a number of options. Equation (II-5) can be used with reasonable accuracy when the normal boiling point and the heat of vaporization are known, and the temperature is near the boiling point. When the normal boiling point, T_b, the molecular weight, M, and the critical temperature, T_c, are known, the Gomez-Thodos method can also be used. This method is recommended for reduced temperatures $0.5 > T/T_c > 1.0$ and involves a number of equations that differ depending on the structure of the chemical of interest. The Gomez-Thodos method is outlined in detail in *The Properties of Gases and Liquids.*[6]

4.0 Multicomponent Vapor — Liquid Equilibria

It has long been known that when a chemical exists in a solution, the equilibrium properties of the chemical differ from those of the pure substance. Estimates of the partial equilibrium properties of mixtures are essential to predicting evaporation rates or worst-case room concentrations using the saturation vapor pressure model. When possible, estimates should be based on experimental data for the mixture at the temperature, pressure, and composition of interest.

Figures II.3 and II.4 are graphs of vapor-liquid equilibrium data. Figure II.3 is a plot of the equilibrium partial pressure of pentane above a pentane-hexane mixture at 25°C (77°F) against the mole fraction of pentane in the mixture. Figure II.4 shows the equilibrium partial pressure of ethanol above an ethanol-benzene mixture at 45°C (113°F).[8]

Vapor-liquid equilibrium diagrams such as those shown in Figures II.3 and II.4 have been developed for a large number of chemical mixtures and can be found in the following references:

- *Vapor-Liquid Equilibria Data Collection*[8],
- *Vapor-Liquid Equilibrium in Mixtures and Solutions*[9], and
- *Vapor-Liquid Equilibrium Bibliographic Database.*[10]

Phase equilibria in multicomponent systems requires that chemical potentials in the two phases are equal. Chemical potential is an abstract quantity that cannot be measured, but an alternative quantity, the fugacity, is equivalent to chemical potential and can be used.

Equilibrium between a liquid and a gas occurs when the fugacities are equal:

$$f_A^v = f_A^l \tag{II-9}$$

The fugacity of a component in a mixture depends on temperature, pressure, and composition of a mixture. Physically, fugacity can be considered a "thermodynamic pressure" that is identical to the partial pressure of a component in an ideal gas. In a real gas, fugacity has no physical representation but can be considered a "corrected partial pressure."[11]

In gases, the fugacity is related to composition and pressure by the fugacity coefficient Φ_A, so:

$$f_A^v = \Phi_A y_A P \tag{II-10}$$

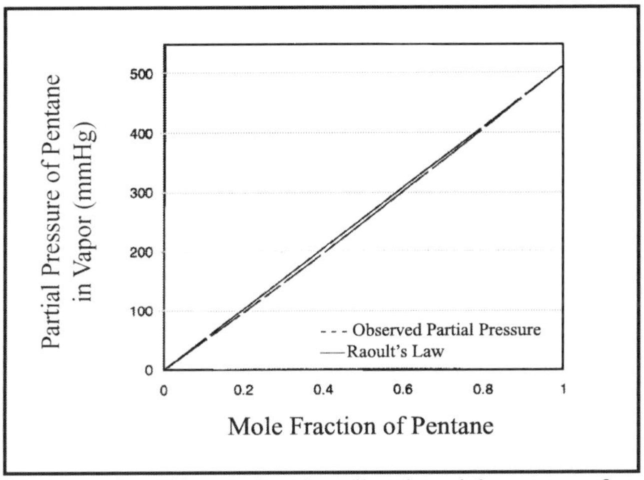

Figure II.3 — Observed and predicted partial pressure of pentane above a pentane-hexane mixture.

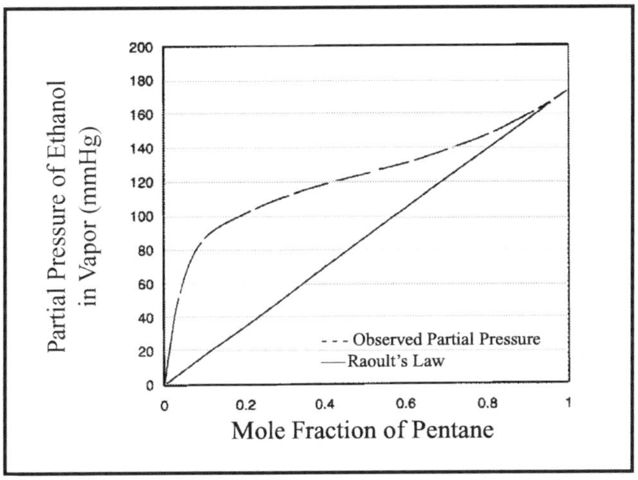

Figure II.4 — Observed and predicted partial pressure of ethanol above an ethanol-benzene mixture.

Fortunately for an industrial hygienist predicting indoor air concentrations, $\Phi_A = 1$ for a mixture of ideal gases. Since most contaminants of concern are at or near atmospheric pressure, where the ideal gas assumption is a very good one, this equation can be reduced to:

$$f_A^v = y_A P = p_A \quad \text{(II-11)}$$

One might recognize p_A as the partial pressure of component A in the vapor phase. Equation (II-11) then becomes:

$$y_A P = p_A \text{ p } f_A^l = \text{fugacity} \in \text{the liquide phase} \quad \text{(II-12)}$$

The fugacity of a liquid phase is related to the composition of that phase through the activity coefficient Φ_A:

$$f_A^l = \gamma_A x_A f_A^o \quad \text{(II-13)}$$

Where

f_A^o = standard state fugacity of component A at the system temperature
x_A = mole fraction of component A in the liquid

The standard state fugacity is defined differently depending on the vapor-liquid equilibrium model used. In this section, the standard state fugacity will be defined as in the Lewis-Randall rule. That is, f_A^o is the fugacity of pure A at the temperature and pressure of the system. In Henry's law, which is discussed in Section 4.2, f_A^o is defined differently.

When components are below their critical temperature, and pressures are low, the f_A^o can be estimated by the vapor pressure of pure A. So, for industrial hygiene applications, the following is usually a good estimate of vapor-liquid equilibria:

$$y_A P = p_A = \gamma_A x_A p_{Avap} \quad \text{(II-14)}$$

The activity coefficient, γ_A, depends on temperature and, to a much lesser extent, pressure. To predict vapor-liquid equilibria of mixtures, we need to estimate γ_A.

4.1 Predicting Activity Coefficients

4.1.1 Raoult's Law

The most obvious selection for the activity coefficient, γ_A, is 1.

$$\gamma_A = 1$$

For solutions in which components in the mixture are very similar in chemical structure, it can be assumed that $\gamma_A = 1$ for all components at all compositions. This simplification is known as Raoult's Law and can be written:

$$y_A P = p_A = x_A p_{Avap} \quad \text{(II-15)}$$

Raoult's Law works reasonably well when components of the mixture are very similar, such as benzene and toluene, or n-butane in n-pentane. It also becomes more valid for concentrated solutions (i.e., when x_A approaches 1). For most mixtures, however, Raoult's Law provides only a rough approximation of vapor-liquid equilibrium behavior.

Raoult's Law predictions are shown in Figures II.3 and II.4. As would be expected, Raoult's Law represents the data very well for the pentane-hexane mixture in Figure II.3. These two substances are very similar, and their behavior in mixture is represented well by Equation (II-15). Raoult's Law poorly represents the ethanol partial pressure in the ethanol-benzene mixture in Figure II.4, except when the mole fraction of ethanol is between 0.9 and 1.0.

4.1.2 Activity Coefficient Equations

Vapor-liquid equilibrium literature usually includes both experimental data on a mixture as well as methods to determine the activity coefficients for that mixture. Parameters for one or more activity coefficient equations are obtained from experimental data. The activity coefficient equations can then be used to estimate activity coefficients for the conditions of interest. Activity coefficient correlations include the van Laar, Wilson, NRTL, and UNIQUAC methods. Information on how to use each method with the constants given is usually provided along with the constants. Sources of constants for the activity coefficient correlations include:

- *Vapor-Liquid Equilibria Data Collection*[8],
- *Computer-Aided Data Book of Vapor-Liquid Equilibria*[12],
- *Vapor-Liquid Equilibrium in Mixtures and Solutions*[9], and
- *Vapor-Liquid Equilibrium Bibliographic Database.*[10]

4.1.3 Activity Coefficients at Infinite Dilution

Another useful piece of data for predicting activity coefficients is the activity coefficient at infinite dilution for a given chemical mixture. The activity coefficient at infinite dilution is reatively easy to obtain and can be used with activity coefficient correlations to predict vapor-liquid equilibria for all compositions.[6]

One method uses the Wilson equation with the infinite dilution g to predict all activity coefficients for the mixture. When the activity coefficients at infinite dilution $g_A^{¥}$ and $g_B^{¥}$ are known, parameters for the Wilson equation, L_{AB} and L_{BA}, can be obtained from simultaneous solution of the following[6]:

$$\ln \gamma_A^\infty = -\ln \Lambda_{AB} - \Lambda_{BA} + 1 \quad \text{(II-16)}$$

$$\ln \gamma_B^\infty = -\ln \Lambda_{BA} - \Lambda_{AB} + 1 \quad \text{(II-17)}$$

Λ_{AB} and Λ_{BA} can then be used in the complete Wilson equation to determine γ_A and γ_B at all compositions.

$$\ln \gamma_A = -\ln(x_A + \Lambda_{AB}x_B) + x_B\left[\frac{\Lambda_{AB}}{x_A+\Lambda_{AB}x_B} - \frac{\Lambda_{BA}}{x_B+\Lambda_{BA}x_A}\right] \quad \text{(II-18)}$$

$$\ln \gamma_B = -\ln(x_B + \Lambda_{BA}x_A) + x_A\left[\frac{\Lambda_{AB}}{x_A+\Lambda_{AB}x_B} - \frac{\Lambda_{BA}}{x_B+\Lambda_{AB}x_A}\right] \quad \text{(II-19)}$$

Activity coefficients at infinite dilution can be found in the sources given in the preceding section (Section 4.1.2) and in the book *Activity Coefficients at Infinite Dilution.*[11]

4.1.4 Predicting Activity Coefficients Using the UNIFAC Method

Although activity coefficients determined from actual data are preferred, the UNIFAC method can be used to obtain reasonable predictions of γ when no experimental data are available. The UNIFAC method predicts activity coefficients based on the interactions between the structural groups on chemicals in mixtures. UNIFAC was introduced by Fredenslund, Jones, and Prausnitz.[13] The method equations and parameters are also given in Reid.[6] UNIFAC calculations are extensive (although not difficult) and require attention to detail to correctly predict activity coefficients. Some of the tedium and potential for error can be reduced by using a UNIFAC-based computer program.[14]

4.2 Henry's Law

Henry's Law is another simplification of the vapor-liquid equilibrium relationship in Equation (12). In Henry's Law, however, the standard state fugacity fi° is selected as the fugacity at the limit when the mole fraction of A, x_A. That is:

$$f_A^0 = k_A\, x_A \quad \text{as } x_A \rightarrow 0 \quad \text{(II-20)}$$

Now this relationship, combined with Equation (12), gives:

$$y_A\, P = p_A = \gamma_A x_A k_A \quad \text{(II-21)}$$

Because the activity coefficient of a pure solution is one, as x_A approaches zero, g_A approaches one. So, for a solution and an ideal gas, Henry's Law is written:

$$y_A\, P = p_A = k_A x_A \quad \text{(II-22)}$$

As would be expected from its development, Henry's Law is most accurate for dilute solutions of component A. In fact, Henry's Law is often used to describe solubility of gases in liquids for gas absorption technologies, and Henry's Law is the preferred method for describing the solubility of gases in water.

In Figures II.3 and II.4, Henry's Law would apply in the low mole fraction region ($x_A < 0.1$) and would be represented by a straight line passing through (0,0) and approximating the data points. One can see in Figure II.4 how a Henry's Law model would differ greatly from the Raoult's Law model shown.

Sources of Henry's Law constants include:

- *Basics of Pump and Treat Ground Water Remediation Technology*[15], and
- *International Critical Tables.*[16]

As with activity coefficients, methods have been presented that predict Henry's Law constants from the structure of the solute and solvent.[17] Computer programs to predict kA based on these methods are also available.

4.3 Nonmiscible Liquid Solutions

Components of nonmiscible liquid solutions are usually considered to behave as pure solutions of each component. In most cases, one can estimate that the components of the solution do not influence each other.

References

1. **National Institute for Occupational Safety and Health (NIOSH):** *NIOSH Pocket Guide to Chemical Hazards.* Cincinnati, OH: NIOSH, 2007.
2. **Perry, R.H. (ed.):** Physical and Chemical Data (Chapter 2). In *The Chemical Engineer's Handbook,* 8th edition. New York: McGraw-Hill & Co., 2007.
3. **Lewis, R.J.:** *Hazardous Chemicals Desk Reference,* 6th edition. New york: John Wiley & Sons, Inc., 2008.
4. **Vargaftik, N.B.:** *Handbook of Physical Properties of Liquids and Gases: Pure Substancnes and Mixtures,* 3rd edition. Redding, CT: Begell House Publishers, 1996.
5. **Smith, B.D., and R. Srivastava:** *Thermodynamic Data for Pure Compounds.* Amsterdam, The Netherlands: Elsevier Science Publishers B.V., 1986.
6. **Reid, R.C., J.M. Prausnitz, and B.E. Poling:** *The Properties of Gases and Liquids,* 4th edition. New York: McGraw-Hill & Co., 1987.
7. **Boublik, T., V. Fried, and E. Hala:** *The Vapour Pressures of Pure Substances,* 2nd edition. Amsterdam, The Netherlands: Elsevier Science Publishers B.V., 1984.
8. **Gmehling, J., U. Onken, and W. Arlt:** *Chemistry Data Series, Volume 1. Vapor-Liquid Equilibrium Data Collection.* Frankfurt, Germany: DECHEMA, 1986.
9. **Wichterle, I.:** *Vapor-Liquid Equilibrium in Mixtures and Solutions.* Montreuil, France: ELDATA, 1994.
10. **Wichterle, I., et al.:** *Vapor-Liquid Equilibrium Bibliographic Database.* Amsterdam, The Netherlands: Elsevier Science Publishers B.V., 1993.

11. **Tiegs, D., et al.:** *Chemistry Data Series*, Volume 9. Activity Coefficients at Infinite Dilution. Frankfurt, Germany: DECHEMA, 1986.
12. **Hirata, M., S. Ohe, and K. Nagahama:** *Computer-Aided Data Book of Vapor Liquid Equilibria.* Amsterdam, The Netherlands: Elsevier Science Publishers B.V., 1975.
13. **Fredenslund, A., R.L. Jones, and J.M. Prausnitz:** Group-Contribution Estimation of Activity Coefficients in Nonideal Liquid Mixtures. *AIChE J. 21*:1086–1099 (1975).
14. UNIFAC Program. Atlanta, GA: BRI Software, Inc., 1993.
15. **Kerr, R.S.:** *Basics of Pump and Treat Ground Water Remediation Technology*. EPA600-8-90-003. Washington, D.C.: U.S. EPA, 1990.
16. *International Critical Tables, Volume III.* New York: McGraw-Hill & Co., 1929.
17. **Meylan, W.M. and P.H. Howard:** Bond Contribution Methods for Estimating Henry's Law Constants. *Environ. Toxicol. Chem. 10*:1283–1293 (1991).
18. **Syracuse Research Corporation:** "Henry's Law Constant Program." Syracuse, NY: Syracuse Research Corp., Chemical Hazard Division, Environmental Chemistry Center, 1991.

Index

E

F

G

H

I

L

M

N

O

P

R

S

T

U

V

W